Genetic Modification and Food Quality

Genetic Modification and Food Quality

A Down to Earth Analysis

Robert Blair

Faculty of Land & Food Systems, University of British Columbia, MacMillan Building, 2357 Main Mall, Vancouver, Canada V6T1Z4

Joe M. Regenstein

Department of Food Science, College of Agriculture and Life Sciences, Cornell University, Stocking Hall, Ithaca, NY 14853-7201, USA

WILEY Blackwell

Registered Office
John Wiley & Sons, Ltd, The Atrium, Southern Gate, Chichester, West Sussex, PO19 8SQ, UK

Editorial Offices
9600 Garsington Road, Oxford, OX4 2DQ, UK
The Atrium, Southern Gate, Chichester, West Sussex, PO19 8SQ, UK
111 River Street, Hoboken, NJ 07030-5774, USA

For details of our global editorial offices, for customer services and for information about how to apply for permission to reuse the copyright material in this book please see our website at www.wiley.com/wiley-blackwell.

Library of Congress Cataloging-in-Publication Data

Blair, Robert, 1933–
Genetic modification and food quality : a down to earth analysis / Robert Blair, Faculty of Land & Food Systems, University of British Columbia, MacMillan Building, 2357 Main Mall, Vancouver, Canada V6TIZ4, Joe M. Regenstein, Department of Food Science, College of Agriculture and Life Science, Cornell University, Stocking Hall, Ithaca, NY 14853-7201 USA.
pages cm
Includes bibliographical references and index.
ISBN 978-1-118-75641-6 (hardback)
1. Genetically modified foods. 2. Food–Quality. I. Regenstein, J. M. (Joe M.) II. Title.
TP248.65.F66B646 2015
664–dc23

2015006385

A catalogue record for this book is available from the British Library.

Wiley also publishes its books in a variety of electronic formats. Some content that appears in print may not be available in electronic books.

Cover image: Corn©filo/iStockphoto

Set in 9.5/13pt Meridien by SPi Global, Pondicherry, India

MIX
Paper from responsible sources
FSC www.fsc.org FSC® C013604

Contents

CHAPTER 1
Introduction

> It is time to reopen the debate about GM crops in the UK but this time based on scientific facts and analysis. We need to consider what the science has to say about risks and benefits, uncoloured by commercial interests and ideological opinion. It is not acceptable if we deny the world's poorest access to ways that could help their food security, if that denial is based on fashion and ill-informed opinion rather than good science.
>
> *Sir Paul Nurse, Royal Society – Richard Dimbleby Lecture, February 2012*

After World War II there was a rapid expansion in food production, supported by advances in agricultural science. This led to an abundance of food in the developed countries, but not in all developing countries. The world population is expected to increase from the current 6.7 billion to 9 billion by 2050. To accommodate the increased demand for food, it has been estimated that world agricultural production needs to increase 50% by 2030 (Royal Society 2009).

As outlined by Ronald (2011) the amount of arable land is limited and is being reduced due to urbanisation, salinisation, desertification and environmental degradation. Another challenge is that water systems are under severe strain in many parts of the world. Thus, increased food production must largely take place on a diminishing land area while using fewer resources. Compounding the challenges are the predicted effects of climate change, limiting crop production and exposing crops to increased damage from pests and disease.

As a result several strategies are being adopted to address these issues, including the improvement of agricultural crops using genetic modification (GM). The first GM (transgenic) crops for food and feed use were introduced in 1996. Later the technique was extended to animals, although the development of superior strains and breeds of animals at this time is still based mainly on traditional selective breeding and cross-breeding techniques.

Genetic Modification and Food Quality: A Down to Earth Analysis, First Edition.
Robert Blair and Joe M. Regenstein.

Genetic modification (GM) can be defined as the manipulation of an organism's genes by introducing, eliminating or rearranging specific genes using the methods of modern molecular biology, particularly those techniques referred to as recombinant deoxyribonucleic acid (rDNA) techniques. This method uses laboratory techniques to introduce specific changes into the genetic code located within the cells so that the succeeding generations possess desired features.

In some cases the inserted genes are derived from another species, in others the genetic change is made by using genes found within the same species or a closely related species. For example, it became possible for the gene responsible for drought tolerance to be identified in one plant, isolated and removed, and inserted into a different plant. The genetically-modified plant then gains drought tolerance as well, and this attribute is passed down to succeeding generations. It also became possible for genes from non-plant organisms to be used with plants. The best known example of this is the use of B.t. genes in maize and other crops. B.t. (*Bacillus thuringiensis*), is a naturally occurring bacterium that produces proteins that are lethal to insect larvae. B.t. protein genes can be transferred into maize, enabling the crop to develop a resistance against insects such as the European corn borer and thus reducing or avoiding the need for spraying with insecticide.

In general, GM has been used to introduce specific traits into plants, such as resistance to insect attack, resistance to virus diseases and tolerance to herbicides.

According to James (2014) the global biotech crop hectarage has increased from 1.7 million hectares in 1996 to over 175 million hectares in 2013, the main crops being maize, soyabeans, rapeseed (canola) and cottonseed. During this 18-year period, more than a 100-fold increase of commercial biotech crop hectarage has been reported. The United States continues to lead global biotech crop plantings at 70.1 million hectares or 40% of total global hectares, but according to the report, more than 90% of farmers (more than 16.5 million) planting biotech crops are small and resource-poor. Of the countries planting biotech crops in 2013, 8 were industrial countries and 19 were developing countries. For the second year in a row, developing countries planted more hectares of biotech crops than industrialised countries. By 2015, it is predicted that more than 120 GM crops (including potatoes and rice) will be cultivated worldwide, with the number of countries growing GM crops and the area planted doubling between 2006 and 2015 (James, 2010).

Although GM has been shown to have important applications with food crops (and animals), the technique is still controversial and continues to raise concerns in several quarters. The main concern is whether genetic modification using rDNA techniques results in harmful attributes in the altered organism, such as allergenicity. A second main concern is whether the nutritional value of the food product is the same.

Food derived from mutation breeding is widely used and accepted since induced mutagenesis is considered a conventional breeding technique. Mutations, defined as any change in the base sequence of DNA, can either occur spontaneously or be induced, and both methods have produced new crop varieties. Crop plants account for 75% of released mutagenic species. In the USA many varieties have been developed using induced mutagenesis, such as lettuce, beans, grapefruit, rice, oats and wheat. Organic farming systems, at least in the USA, permit food from mutated varieties to be sold as organic.

GM technology differs from mutagenesis in that it involves insertion of an alien gene or genes, whereas mutagenesis results in a realignment of the genes contained within the genome of an organism. Mutagenesis also has a longer history of use, although it does involve the use of mutagenic agents such as ionising irradiation. The food regulations in some countries require that only new food products derived using GM techniques, and not mutagenesis, are subject to scientific assessment before being approved for food or feed use. Consequently our review does not include the quality and safety of food products developed by mutagenesis.

Plants and animals may be reproduced by cloning. This technique is not considered to be genetic modification since it does not involve any change in the genetic makeup of the organism. As a result, the evidence relating to the quality and safety of food items produced by cloning is not included in our review.

This book is the first comprehensive text on how GM production methods influence the quality of foods and feeds, based on an unbiased assessment of the scientific findings. Assessments of the religious, ethical and environmental concerns can be found in other publications.

References

James, C. (2010). Global Status of Commercialized Biotech/GM Crops: 2010 ISAAA Brief No. 42-2010. ISAAA, Ithaca, NY.

James, C. (2014). Global Status of Commercialized Biotech/GM Crops: 2013 ISAAA Brief No. 46-2013. ISAAA, Ithaca, NY. http://www.isaaa.org/resources/publications/briefs/46/executivesummary/ (accessed 2 July 2014).

Ronald, P. (2011). Plant genetics, sustainable agriculture and global food security. *Genetics* 188: 11–20.

Royal Society (2009). *Reaping the Benefits: Science and the Sustainable Intensification of Global Agriculture*. The Royal Society, London, UK.

CHAPTER 2

International regulations

Developed countries have government agencies to ensure that the food we buy is safe. All foods, often a specific legal category that excludes other materials such as food additives, are subject to the same regulations, regardless of source and country of origin of the food. The regulations in most such countries include provisions relating to GM foods and also for feed for food-producing animals.

An important feature of the regulatory process relating to the approval of GM foods such as cereal grains is that the approved product is patented and licensed to the companies that developed them. They can then be produced either by the company owning the patent or under license by any other company approved by the patent holder.

The approval process for the assessment of GM products for use in food and feed in most (if not all) jurisdictions involves the submission of extensive supporting documentation by the applicant company to the relevant regulatory agency. Approval follows the assessment and acceptance of the submitted documentation. In many cases, the regulatory agencies will ask for additional information if the material submitted is incomplete. Some commentators dislike this process in that the supporting documentation is not made public. However, it is similar to the approval process for other patented products such as pharmaceuticals. Since the supporting documentation has to be comprehensive, it often requires disclosure of proprietary information that the government agency is required to protect. Thus, companies are confident that information provided to the government will not find its way into the hands of competitors.

Leaders in developing the appropriate regulations for GM foods and feeds are those countries in North America, the European Community (EC) and Australasia. It is clear that not all decisions on food and related issues are reached on the basis of the scientific evidence alone. These other influences can arise

Genetic Modification and Food Quality: A Down to Earth Analysis, First Edition.
Robert Blair and Joe M. Regenstein.

both in the legislative process that defines the framework for how a topic might be handled, and during the regulatory process when the broad scope of the legislation is turned into regulatory language that will be enforced by the appropriate executive branch department or ministry. And sometimes they will even arise thereafter in how the department or ministry actually enforces its own regulations.

Decisions by regulatory authorities and their advisory committees can be swayed by interpretation of safety thresholds, trade benefits and political considerations, which can be influenced by lobbying – the advocacy by different groups for positions that favour their particular interests. Organisations, including religious, charitable, sociable or consumer group, are likely to be involved in lobbying, either formally or informally.

Attempts have been made to develop harmonised international regulations with relevance to GM foods, the goal being to allow goods to move freely between all countries operating with the same rules.

For instance, the Cartagena Protocol on Biosafety to the Convention on Biological Diversity is an international agreement involving 168 parties, including 166 United Nations member states, Niue, and the European Union, which aims to ensure the safe handling, transport and use of living modified organisms (LMO) resulting from modern biotechnology that may have adverse effects on biological diversity, also taking into account any risks to human health. It was adopted in 2000 and entered into force in 2003. Lebanon became the 165th signatory in 2013. Australia, the USA and Canada have yet to become signatories. The Protocol appears to be directed more towards ecological and environmental issues related to genetically modified organisms (GMOs) rather than the quality and safety of GM foods. Additionally, it appears that countries with advanced legislation on food quality and safety are not willing to delegate responsibility for this issue to an international body.

In the following pages, the authors will review some of the key legislation in various countries that specifically relates to GM foods.

North America

USA

Legislation applying to most food products, that is, everything other than meat, poultry and liquid eggs, is governed by the Federal Food, Drug, and Cosmetic Act (FFDCA), and the regulations issued under its authority by the US Food and Drug Administration (FDA). The FFDCA defines food as 'articles used for food or drink for man or other animals'. The FDA Center for Food Safety and Applied Nutrition (CFSAN) is responsible for the regulation of human food products. The FDA Center for Veterinary Medicine (CVM) is responsible for the regulation of animal food (feed) products.

The FDA conducts a rigorous science-based assessment of the safety and quality of GM foods and feeds prior to their entry into the marketplace, unless the food(s) in question have GRAS (Generally Recognised As Safe) status. In addition food manufacturers must ensure the safety and quality of products that contain ingredients derived using GM technology.

The federal government in 1986 developed a Coordinated Framework for the Regulation of Biotechnology to provide for the regulatory oversight of organisms derived through genetic modification. The three principal agencies involved are the US Department of Agriculture (USDA) Animal and Plant Health Inspection Service (APHIS), the Environmental Protection Agency (EPA) and the FDA. The approach taken by the Coordinated Framework is based on the judgement of the National Academy of Sciences, a non-governmental agency set up to provide advice to the government, that the potential risks associated with these organisms fall into the same general categories as those risks created by using traditionally bred organisms. Products are regulated according to their intended use, with some products being regulated by more than one agency depending on use. All the government regulatory agencies involved have a responsibility to ensure that the implementation of regulatory decisions, including approval of field tests of any GM crops and eventual deregulation of approved biotech crops, does not adversely impact human health or the environment. In 1992, the FDA provided a general outline for the safety assessment of GM technology derived food products based on a risk analysis related to the characteristics of the products.

The FDA carries out its responsibility for the regulation of animal feed in cooperation with state and local partners. For instance, FDA cooperates with the Association of American Feed Control Officials (AAFCO) and the states to allow for the implementation of uniform policies in all US jurisdictions for regulating the use of animal feed products. Although FDA has the responsibility for regulating the use of animal feed products, the ultimate responsibility for the production of safe and effective animal feed products lies with the manufacturers and distributors of the products.

The Federal, Food, Drug and Cosmetic Act (FFDCA) requires that any substance that is added to or is expected to become a component of animal feed, either directly or indirectly, must be used in accordance with a food additive regulation unless it is GRAS. Typical feed ingredients – such as forages, grains, and most minerals and vitamins – are GRAS as sources of nutrients.

The FDA regards traditional food and feed products as safe, based on a long history of use. Therefore in assessing the acceptability of a GM food or feed, it uses the concept of substantial equivalence. If it can be demonstrated that the GM product is as safe and nutritious as its conventional equivalent, then the GM product is deemed acceptable. Otherwise the GM product is not approved for food or feed use.

Other agencies outside of the USA also accept the substantial equivalence approach. The Food and Agriculture Organization (FAO) and World Health

Organization (WHO) of the United Nations in 2000 stated that the substantial equivalence approach 'is considered the most appropriate strategy for the safety and nutritional assessment of genetically modified foods'. And while substantial equivalence is not a safety assessment in itself, in that it does not characterise any hazards, 'it is used to structure the safety assessment of a genetically modified food relative to a conventional counterpart'. Similarly, the Organization for Economic Cooperation and Development (OECD) in 2000 concluded: 'Safety assessment based on substantial equivalence is the most practical approach to address the safety of food and food components derived through modern biotechnology.'

Canada

Health Canada is the federal agency responsible for provisions related to public health, food safety and nutrition, under the Food and Drugs Act and the agency's subsequent regulations. This agency conducts pre-market evaluations to assess the safety and nutritional adequacy of foods proposed for sale in Canada as well as health risk assessments related to chemical, physical and microbiological contaminants, toxicants and allergens. GM foods are considered to be a class of 'novel foods', regulated by the same agency. Manufacturers and importers who wish to sell or advertise a GM food in Canada must submit data to Health Canada for a pre-market safety assessment, which involves a rigorous scientific evaluation by Health Canada evaluators with expertise in molecular biology, toxicology, chemistry, immunology, nutritional sciences and microbiology. This safety assessment provides assurance that the food, if acceptable, is safe when prepared or consumed according to its intended use. Once the assessment is complete, a decision is made by this agency on whether or not to approve the product.

The Canadian Food Inspection Agency (CFIA) administers the regulation of novel feeds. Prior to their use in livestock feed in Canada all novel feeds must be authorised based on a safety assessment conducted by the CFIA.

In both the USA and Canada the approval process for GM food products is based on a favourable scientific review of available evidence relating to allergenicity, toxicity and environmental issues, and is similar to the procedure relating to new pharmaceutical drugs. The regulatory authorities in both countries have the power to revoke the approval of a product that subsequent to initial approval is later shown to be deleterious to human or animal health.

Mexico

According to the European website GMO Compass (http://www.gmo-compass.org/eng/news/447.mexico_controlled_cultivation_genetically_modified_maize.html) statutory provisions for the regulation of genetically modified crops were enacted in Mexico in 2009. These allow for the cultivation of genetically modified maize, among others. Farmers or companies wishing to produce GM maize, however, must submit an application for permission. Cultivation of GM maize in areas known as 'diversity centres', in which wild varieties and native breeds of maize grow that exist only in this region of origin, is not permitted. Possible contamination of native strains of maize by cross-breeding with GM cultivars is a main concern of the Mexican regulations.

Europe

The term GMO is defined officially in Europe as 'an organism in which the genetic material has been altered in a way that does not occur naturally through fertilisation and/or natural recombination. GMOs may be plants, animals or micro-organisms, such as bacteria, parasites and fungi'.

GM foods can only be authorised in the European Union (EU) if they have passed a rigorous safety assessment. The procedures for evaluation and authorisation of GM foods and feeds are set out in Regulation (EC) No 1829/2003, which came into force in April 2004 and in Directive 2001/18/EC on the release of GMOs into the environment, which came into force in March 2001.

Procedures for the approval of foods and feeds in the EU differ from those in the USA, in that evaluation and approval are undertaken by two different bodies. The scientific and technical evaluations are undertaken by the European Food Safety Authority (EFSA), which then makes its recommendations to the main governing body for Europe (the European Commission). The Commission then adopts or rejects EFSA's recommendations. EFSA's role with respect to reviewing the safety of GMOs is defined by Regulation (EC) No 1829/2003 and Directive 2001/18/EC.

The regulations relating to GM foods allow individual member states in the EU to adopt additional food legislation and controls beyond those approved by the EC, provided that these do not prevent trade with other member states.

In 2014 a political agreement was reached within the European Council that was aimed at developing legislation designed to allow 'Member States to restrict or prohibit the cultivation, in all or part of their territory, of Genetically Modified Organisms (GMOs) that have been authorised at EU level' (http://ec.europa.eu/food/plant/gmo/legislation/future_rules_en.htm). Currently member states can only restrict or ban the cultivation of GMOs by adopting safeguard clauses where new serious risks to human health, animal health and the environment are identified after the GMO has been authorised.

Russia

According to the website of the agricultural trade organisation AgroCharts, Russia continues the registration of GM crops for imports of food and feed in accordance with existing Russian legislation (http://www.agrochart.com/en/news/news/010814/russia-agricultural-biotechnology-annual-jul-2014-/).

In September 2013, the Russian government initiated legislation for the registration of GM crops for cultivation starting July 2014. Later the government delayed implementation of this legislation to 2017. The Ministry of Agriculture is the main body responsible for seed registration.

Australasia

Food regulations in Australia and New Zealand are administered by a bi-national government agency, Food Standards Australia New Zealand (FSANZ), which is an independent statutory agency established by the Food Standards Australia New Zealand Act (1991) approved by the parliaments of both countries. The agency develops and administers the Australia New Zealand Food Standards Code, which lists requirements for food safety, GM foods, food additives and labelling. Enforcement and interpretation of the Code is the responsibility of state/territory departments and their food agencies within Australia and New Zealand.

All applications for GM food must be assessed on a case-by-case basis by FSANZ. The main objective of the assessment is to identify whether new or altered hazards are present in the food as a result of the genetic modification and, if present, to determine what risk, if any, they may pose to human health.

Currently the only GM food crops produced in Australia are canola and cotton, but a variety of other GM foods can be imported and used as ingredients in packaged foods.

FSANZ allows manufacturers to use a wide range of GM food ingredients imported from overseas. These include specific GM cultivars of soyabeans, maize, rice, potatoes and sugar beet.

GM food products on sale in Australia and New Zealand – either as a whole food or as an ingredient in a processed food – must have their GM status identified if introduced genetic material or protein is present in the final food. However, there are some exceptions, such as foods from animals that have been fed GM feed. The regulations state that highly refined foods such as oils and sugars, that do not normally contain any genetic material or protein, may require labelling if they possess characteristics that are significantly altered from the non-GM counterpart.

China

In 2009 China's National People's Congress (NPC) Standing Committee passed the first comprehensive Food Safety Law (FSL), which created a state-level Food Safety Commission to oversee food-safety monitoring (Petry and Wu, 2009). This Commission is composed of representatives of the Ministry of Health

(MOH), the Ministry of Agriculture (MOA), the General Administration for Quality Supervision, Inspection and Quarantine (AQSIQ), the State (national) Food and Drug Administration (SFDA) and the State Administration for Industry and Commerce (SAIC). Other key provisions of the law deal with supervision, monitoring, enforcement, recall, trace-back, licensing, registration and development of standards and regulations.

The Food and School Hygiene Division of the Bureau of Health and Epidemic Prevention in the Ministry of Public Health (MPH) has overall responsibility for the implementation of national policies on food safety. The Institute of Food Safety Control and Inspection of the MPH serves as a technical and scientific advisory body to the ministry and conducts applied research in the field of food safety. In coordination with the Institute of Food Safety Control and Inspection, the Institute of Nutrition and Food Hygiene which is within the Chinese Academy for Preventive Medicine (CAPM) is also responsible for conducting some of the research work required by the Bureau of Health and Epidemic Prevention.

In addition to the MPH, other ministries with responsibilities for food and feed safety are the ministries of agriculture, commerce, grains, light industry, communications, chemical industry, railways and foreign trade. A national coordination committee for food safety composed of representatives from these ministries has been set up under the chairmanship of the Vice-Minister for Public Health, to promote the exchange of information and facilitate decision-making.

A draft law introduced in 2012 by the Legislative Affairs Office of the State Council (an administrative office within the State Council of the People's Republic of China which assists the Premier in providing legal advice and administrative laws to govern the behaviour of government departments) proposed that no institution or individual would be permitted to apply transgenic technologies to the principal grain cultivars without approval of the relevant departments. The law would require that the research, sale, importation and export of transgenic grain seeds should comply with state regulations.

India

The 1986 Environmental Protection Act (EPA) provides the foundation for India's biotechnology regulatory framework for GM plants, animals and their respective products. According to the current regulations, all biotech food/agricultural products, or products derived from biotech plants and/or other organisms, must receive formal approval from the Genetic Engineering Appraisal Committee (GEAC) prior to commercialisation or importation. In the case of transgenic crops, applicants are also required to seek clearance from the Ministry of Agriculture.

A bill to set up an independent and autonomous national biotech regulatory authority for biosafety clearance of GM products and processes has been proposed by the Indian government but has yet to be implemented.

The Food Safety and Standards Act of 2006 includes specific provisions for regulating GM food products, including processed foods. However, the apex regulatory body under the Act, the Food Safety and Standard Authority of India, is still in the process of formulating specific regulations for overseeing GM food products. Consequently, the GEAC continues to regulate processed food products (containing GM ingredients) under the 1989 Rules.

B.t. cotton is the only GM crop approved for commercial cultivation in India since 2002. The commercial cultivation of B.t. cotton cultivars is approved for seed, fibre and feed production/consumption (http://www.agrochart.com/en/news/news/010814/india-agricultural-biotechnology-annual-jul-2014/). The only GM food product currently authorised for import into India is soyabean oil derived from glyphosate-tolerant soyabeans.

Japan

In 1991 the Ministry of Health and Welfare (now the Ministry of Health, Labor and Welfare (MHLW)) established a Guideline for the Safety Assessment of Foods and Food Additives Produced by Recombinant DNA Techniques (GM foods) based on international guidelines previously prepared by the WHO, the OECD and other organisations. The Food Sanitation Investigation Council of the Ministry assesses GM foods individually based on the Guideline, using substantial equivalence as one of its key provisions.

Under the Food Sanitation Law, the government in 2001 introduced a modified procedure for the safety assessment of GM foods, whereby the MHLW receives applications and its Food Safety Commission (FSCJ) evaluates the product's safety in terms of human health. Environmental assessment of GM crops and the safety assessments of such crops as feedstuffs are carried out by the Ministry of Agriculture, Forestry and Fisheries. Final review and authorisation then come from the MHLW.

Japan is the world's largest per capita importer of foods and feeds that have been produced using GM technology (Sato, 2012). Annually Japan imports about 15 million tonnes of maize and three million tonnes of soyabeans, approximately three quarters of which are produced through biotechnology. Other main imports are processed foods that contain biotech-derived oils, sugars, yeasts, enzymes and other ingredients. By 2012 over 160 products had been approved for food use in Japan. The list included 100 cultivars of 8 crops that were approved for environmental release, which includes commercial cultivation.

Papayas were the first fruit to enter the Japanese market as a GM food product for consumption in the unprocessed form, following a favourable assessment by the Food Safety Commission and clearance by a government panel. The GM papayas were developed in the USA in the 1990s, with resistance to the papaya ringspot virus (PRV).

To date no GM food crops have been grown commercially in Japan (Sato, 2012). A biotech rose introduced in 2009 is the only biotech crop grown commercially in Japan.

Republic of Korea

The regulation of GM foods and food additives in Korea is governed by the Food Sanitation Act, which mandates the safety assessment of food in relation to human health. The current safety assessment system is carried out by the Korean Food and Drug Administration (KFDA), according to the Safety Evaluation Guidelines by Notification 1999-46 regulation. As of December 2008, KFDA had approved 54 types of GM crops (Kim et al., 2010), including soyabeans, corn (maize), canola, cottonseed, potatoes, sugar beet and alfalfa.

Philippines

According to the AgroChart website (http://www.agrochart.com/en/news/news/220914/philippines-agricultural-biotechnology-annual-sep-/), Philippine rules governing biotechnology crop regulations are widely recognised as science-based, thorough and transparent. To ensure human, food, feed and environmental safety, the regulatory regime requires that risk assessments be conducted in accordance with internationally accepted bodies such as the Cartagena Protocol on Biosafety, the Codex Alimentarius Commission, the Organization for Economic Co-operation and Development and the UN Food and Agriculture Organization. GM maize has been on sale in the country since 2003 and comprised 28% of planted maize area in 2013. The Philippine Department of Agriculture's Bureau of Plant Industry recently cited the successful propagation of GM maize over the past decade, with no environmental or health issues reported, as the reason the country is now self-sufficient in the grain. Six transformation events (TEs) in 10 GM maize varieties are approved for commercial production. While B.t. maize is the only GM crop approved for commercial production in the Philippines, Golden Rice and B.t. eggplant have completed most testing requirements.

Africa

In 2010, the Common Market for Eastern and Southern Africa (COMESA) produced a draft policy on GM technology, for discussion by the governments of all 19 member nations. Under the proposed policy, new GM crops would be scientifically assessed by COMESA. If the GM crop was deemed safe for human health then environment clearance would be granted for the crop to be grown in all 19 member countries, subject to the approval of each individual country. The African Biosafety Network of Expertise was established by the African Union in 2010, with funding from the Bill and Melinda Gates Foundation (the founders of Microsoft). Its aim is to train and develop African regulators to assess, approve, monitor and track GM crops.

South Africa (a member of the African Union) is currently the major producer of GM crops in Africa, with smaller amounts grown in Burkina Faso and Egypt. Kenya passed laws in 2011 and Ghana and Nigeria passed laws in 2012 which allowed the production and importation of GM crops. Kenya and Ghana are members of the African Union.

South Africa

As noted above, South Africa is the main African country producing GM crops commercially. The production area of GM crops continued to expand in 2013 to reach 2.9 million hectares, making South Africa the eighth largest producer of biotech crops in the world (James, 2014). The regulations governing GM foods in South Africa are the Genetically Modified Organisms Act of 1999 (under the Ministry of Agriculture and Land Affairs), labelling regulations (under the Ministry of Health) and the Biodiversity Act (under the Ministry of Environmental Affairs and Tourism).

South America

Brazil and Argentina, respectively, are the second and third largest producers of GM foods behind the USA. Honduras, Costa Rica, Colombia, Bolivia, Paraguay, Chile and Uruguay also allow GM crops to be grown. Venezuela and Ecuador have banned the cultivation of GM crops.

Argentina

Argentina was one of the first countries to adopt GM crops and in 2013 had 14% of the global biotech crop area (James, 2014). The Competent Authority for the administration of the regulatory framework of GM crops is the Secretariat of Agriculture, Livestock and Fisheries (SAGyP is the Spanish acronym), within the Ministry of Agriculture, Livestock and Fisheries. Three favourable assessments with respect to the issues below are needed in Argentina for the commercial authorisation of GM crops (Burachik, 2012):

(a) Impact on the agro-ecosystem: evidence that the effects on the agro-ecosystem of the extended cultivation of the GM crop are not significantly different from those of the non-modified counterpart (or the approved available technology).

(b) Food and feed safety: evidence that the food and/or feed derived from the GM crop are as safe as and not less nutritious than the ones derived from their non-modified counterparts.

(c) Impact on trade and production: assurance that the production and commercialisation of the GM crop and of the food or feed derived from it will not encounter any domestic or international commercial restrictions.

A biosafety act is currently under review.

Brazil

In Brazil the National Biosafety Technical Commission (São Paulo) is responsible for making environmental and food safety assessments and prepares guidelines for transport, importation and field experiments involving GM products. The Council of Ministers then evaluates the commercial and economic issues.

Chile

Regulatory actions for GM crops in Chile began shortly after the first GM food product was commercialised internationally. Chile's first GM regulation was a resolution enacted in 1993 by the Ministry of Agriculture's Animal and Livestock Service SAG (Servicio Agrícola y Ganadero), the counterpart of the PPQ (Plant Protection and Quarantine) branch of the USDA. Cultivation of GM seed for export is currently allowed under the strict control of the Agricultural and Livestock Service (Resolution 1523, 2001), which regulates this process. Importation of GM foods for human consumption is permitted, with the approval of the Ministry of Health (Decree 115) and labelling if they are substantially different from the conventional product. Chile does not have a Biosafety Committee at this time.

Food labelling

The labelling of foods is a complex issue that regulatory authorities in each country need to deal with. There are many requests for information about the food supply that consumers would like to have on the food product label. However, the space available for information on the label, the cost of supplying that information (including hidden costs such as how one harvests, stores and transports materials that then may need to be segregated for labelling purposes), the benefit to the consumer of knowing (a very political decision) and the issue of misleading information and information overload all need to be considered.

Many countries try to restrict, from a regulatory point of view, the label requirements to those issues where there is an actual safety or health concern. Consumers then have the necessary information before making a purchasing decision. Thus in practice the issue of labelling for GM crops should be predicated on real safety/health issues that consumers need to be made aware of.

Currently there is a great deal of emotional discussion about GM crops used in the food supply, and it is hoped that this book will provide relevant information to assist this debate.

Clearly a claim that a product is 'GM free' needs to be substantiated and provides an opportunity for marketing such an attribute, possibly at a premium price. Having to label a product as 'containing GM materials' is seen in the marketplace as conveying a negative message. Thus countries that require such labelling often are making a political statement, suggesting to their consumers that these products should not be purchased. Sales of such products in these countries generally reflect this reality even if some crops are legally permitted for sale or planting. Companies providing GM-free products then gain a significant marketing advantage and provide a set of interests that will strongly support such legislation regardless of the costs or the impact on the overall food supply of their country.

Such labels are clearly meant to appeal to consumers on an emotional level. So it is unlikely that consumers can be relied on to deal rationally with label information about GM foods, especially when they have been subjected to so much adverse publicity and misinformation about GM foods. (See Chapter 11 for more comment about this.)

In the EU all food (including processed food) or feed which contains more than 0.9% of approved GMOs must be labelled as such. Australia and New Zealand and two countries in Asia, Saudi Arabia and Kazakhstan, have also established this high level of purity for non-GMO products. This requires the separation and tracking of GMO products, thus leading to their absence in the marketplace even though they are legally permitted in many of the countries. Others among the 64 countries that claim to have GMO labelling have some limits on the amount of GM crops, permitting up to 5% of the final product or up to 5% of certain ingredients to be GM crops before the labelling requirements come into play. Finally, some countries have a requirement for GMO labelling without any real regulations to substantiate the requirement. So the situation remains fluid.

One interesting alternative approach has been that of Bhutan, in Southeast Asia, that has mandated that all food be organic within a specified timeframe. However, this announcement of a future goal does not seem to address the issue of whether organic animals must be fed only organic diets or how imports will be handled. The focus seems to be on decreasing the use of fertiliser and pesticides on farms within the country.

Below are examples of the procedures in a few countries with different policies on the issues involved.

Australia/New Zealand

GM foods, ingredients, additives or processing aids that contain novel DNA or protein must be labelled with the words 'genetically modified'. Novel DNA or protein is defined in the Food Standards Code as DNA or a protein which, as a result of the use of gene technology, is different in chemical sequence or structure from DNA or protein present in counterpart food, which has not been produced using gene technology. Labelling is also required when genetic modification results in an altered characteristic in a food, for example soyabeans with changed nutritional characteristics such as an increase in their oleic acid content.

The Australian/New Zealand regulations state that GM labelling is not about safety. It is about helping consumers make an informed choice about the food they buy.

Canada

Health Canada and the Canadian Food Inspection Agency (CFIA) share the responsibility for food labelling policies under the Food and Drugs Act. Health Canada's responsibilities for food labelling fall within the Department's mandate to safeguard health and safety, while CFIA leads the federal programme to develop general food labelling policies and regulations. In particular, the CFIA is responsible for protecting consumers from misrepresentation and fraud with respect to food labelling, packaging and advertising, and for prescribing basic food labelling and advertising requirements.

Currently, in Canada, labelling is mandatory if there is a health or safety issue with a food, which might be mitigated through labelling. For example, if the nutritional value or composition of the food has been changed, or if there is an allergen present in the food, the food must be labelled as such. In this situation, special labelling is required to alert consumers or susceptible groups in the population. This applies to all foods, including GM foods. Otherwise Canada does not require that GM food products be specially labelled.

The Government of Canada recognises that for many Canadians, labelling of foods derived from biotechnology is an important issue of consumer preference or choice. Under a standards committee established by the Canadian General Standards Board, a Canadian standard for voluntary labelling of GE foods entitled Voluntary Labelling and Advertising of Foods that Are and Are Not Products of Genetic Engineering was developed to address non-health and safety labelling (rather, labelling for method of production, for example, whether a food has or has not been produced through genetic engineering). This committee included a broad range of stakeholders, including consumer groups, food companies, producers, general interest groups, universities and government. In April 2004, the national standard was adopted by the Standards Council of Canada. The objectives of the national standard are to provide meaningful criteria for labelling, understandable messages for consumers, and a consistent policy to verify the truthfulness of labels.

The Canadian government supported the development of a national standard for the voluntary labelling of foods derived from biotechnology. This process was sponsored by the Canadian Council of Grocery Distributors, under the guidance of the Canadian General Standards Board. If marking as non-GMO has economic value, that is, consumers pay a higher price, then it is important that the government ensure that products are properly marked. The practice of defining the standard for a voluntary claim has often been used with other similar types of marketing claims by governments in countries where mandatory labelling is used for health and safety requirements.

Europe

The EU recognises consumers' right to information and labelling as a tool for making an informed choice. Since 1997 Community legislation has made labelling of GM food mandatory for: products that consist of GMOs or contain GMOs; also, products derived from GMOs but no longer containing any GMO if there is still DNA or protein resulting from the genetic modification present. The latest regulation on GM labelling is Regulation (EC) 1830/2003 of the European Parliament and of the Council of 22 September 2003, concerning the traceability and labelling of GMOs and the traceability of food and feed products produced from GMOs and amending Directive 2001/18/EC. Previously, the labelling of GM foods was based on the provisions of Article 8 of Regulation (EC) 258/97 on novel foods and novel foods ingredients. The labelling of GM maize varieties and GM soya varieties that did not fall under Regulation 258/97 are covered by Regulation (EC) 1139/98 concerning the compulsory indication of the labelling of certain foodstuffs produced from genetically modified organisms as amended by Regulation (EC) 49/2000. In addition, all GM additives and GM flavourings have to be labelled according to Regulation (EC) 50/2000 on the labelling of foodstuffs and food ingredients containing additives and flavourings. In accordance with the general labelling rules of Directive 90/220/EEC, the labelling of four out of the eight authorised GMOs for use in feed is mandatory.

Japan

GM foods whose safety has been authorised by the Ministry of Health, Labor and Welfare are required to be labelled such as 'genetically modified' when they are marketed.

The Quality Labelling Standard for Genetically Modified Foods specifies those food items subject to labelling and the method of labelling.

1 Food items for mandatory labelling. Food items subject to mandatory GM labelling are designated as those in which genetically modified DNA or proteins derived from the DNA can be detected even after processing, or those whose compositions or nutritional values differ in comparison with their conventional counterparts.

   As of March 2006, six agricultural products and processed foods made from them have been designated as mandatory labelling items. Taking into account the commercialisation of new GM foods and newly developed detection methods, the items are reviewed every year.

2 Labelling scheme for GM foods. 'In the case that products are ensured to consist of or to contain GM products through identity preserved handling (IP handling), those shall be labelled such as "Soy beans (genetically modified)", while in the case that IP handling has not been undertaken for the product, it shall be labelled for example as "Soy beans (no segregation practice with GM products)".'

'As a voluntary labelling scheme, products that are ensured not to contain GM products though IP handling, can be labelled as "Soy beans (non-GM)," or "Not containing GM products".'

GM papayas are allowed to be offered for sale in Japan. The fruit must be labelled accordingly. The Japanese government has allowed the import of GM papayas from the USA, where a virus-resistant variety has been cultivated in Hawaii since 1999.

Russia

Labelling and information for consumers on the presence of GM ingredients in food products is regulated by the technical regulations of the Customs Union (CU) on safety and labelling of food products. The CU technical regulation for feed has not yet been adopted. However, feed sold in Russia is subject to Russian regulations that do not require labelling of GM feed, but does require registration of GM lines for use in feed, and requires registration of feed as GM feed if the presence of registered lines is over 0.9% and the presence of non-registered lines is over 0.5%.

USA

Currently the USA does not have any national requirements for GMO labelling. A few states have passed such legislation, mainly by referendum, that is, by popular vote rather than through the state legislature. This process of referendum is permitted in some states and tends to be more political than scientific! Other states have rejected the labelling through a referendum. To date, most regulations passed by a legislature have a requirement that a few other states pass similar legislation before the labelling requirement comes into effect. Vermont is currently the only state that has required GMO labelling within a specified period, regardless of legislation introduced by other states. This law is now being challenged in the Federal court system, therefore it remains to be determined whether this law will eventually come into effect. The USA has an amendment to its Constitution (its fundamental governmental document) prohibiting rules that interfere with interstate commerce. This would be one of the legal bases for challenging the law. Rather than have 50 different laws for each of the US states, there has also been a move by the food industry to encourage the development and passage of national legislation. The thinking of the food industry is that one uniform law, even if somewhat onerous, is more satisfactory than 50 different laws that would truly interfere with interstate commerce.

References

Burachik, M. M. (2012). Regulation of GM crops in Argentina. *GM Crops Food* 3: 48–51.

James, C. (2014). Global Status of Commercialized Biotech/GM Crops: 2013 ISAAA Brief No. 46-2013. ISAAA, Ithaca, NY. http://www.isaaa.org/resources/publications/briefs/46/executivesummary/ (accessed 2 July 2014).

Kim, H.-Y., Kim, J.-H. and Oh, M.-H. (2010). Regulation and detection methods for genetically modified foods in Korea. *Pure Applied Chemistry* 82: 129–137.

Petry, M. and Wu, B. (2009). GAIN Report Number: CH9019, US Foreign Agricultural Service, Washington, DC (accessed 4 April 2013).

Sato, S. (2012) GAIN Report Number JA2013. Global Agriculture Information Network (GAIN), United States Foreign Agricultural Service, Washington, DC (accessed 4 April 2014).

CHAPTER 3

Microorganisms

Several genetically modified microorganisms have been approved for use in food and beverage production, although not all are in commercial use at the present time.

Among the products are baked goods such as bread; beer and wines; yoghurt and cheeses; and food processing additives such as amino acids, vitamins, flavours and enzymes. These additives include vitamin B2 (riboflavin, colouring agent), vitamin C (ascorbic acid, preservative), xanthan gum (thickener), citric acid (acidity regulator), natamycin (preservative), nisin (preservative), methionine (amino acid), monosodium glutamate (flavour enhancer) and aspartame (sweetener). Several enzymes derived from microorganisms are used in the production of cheeses and juices, as well as in the production of glucose syrup (corn syrup), glucose and other starch products. The current list of enzymes produced internationally using GM microorganisms includes: acetolactate decarboxylase, aminopeptidase, amylase, asparaginase, catalase, cellulase, cgtase, chymosin, galactosidase, glucanase, glucose isomerase, glucose oxidase, hemicellulase, hexose oxidase, laccase, lactase, lipase, lipoxygenase, mannanase, pectinase, pectinesterase, phytase, protease, pullulanase, transglutaminase and xylanase.

Processes using genetically modified organisms have in many instances replaced more complex chemical production processes.

In general, the genetic manipulation of microorganisms for use in the food industry has involved enhancing the production of a substance produced naturally by the microorganism. A common approach is to alter the regulatory mechanism controlling the gene for production of the substance. The substance can then be produced in much greater quantity. A related approach is to regulate the point at which the microorganism halts production of the substance. Microorganisms usually stop producing a substance when the concentration in

Genetic Modification and Food Quality: A Down to Earth Analysis, First Edition.
Robert Blair and Joe M. Regenstein.

the medium reaches a certain point. Genetic manipulation can be used to regulate the termination point so that the microorganism tolerates a higher concentration of the substance and the substance continues to be produced to a higher concentration.

Some important substances are only produced naturally by microorganisms that are difficult to culture. In this case, the gene involved may be transferred to another microorganism that is easier to grow and is permitted for use in producing food ingredients.

When the growth and production phase ends, the substance of interest is isolated and purified. No traces of the microorganisms are present in the final product, and no microbial DNA is detectable in the substance produced.

In other applications the microorganism may be consumed by humans as a component of the food produced, for example yoghurt and cheese.

As outlined by Schuller and Casal (2005), the approach taken in modifying microorganisms for food production is to avoid genetic sequences from species that are closely related taxonomically to pathogenic species. In addition, the preferred approach is to derive the sequences from within the host species and to utilise organisms with GRAS (Generally Regarded as Safe) status. Genetic modification of microorganisms differs, therefore, from the modification of plants.

In 1958, the United States Food and Drug Administration (FDA) established the Food Additives Amendment, with a list of 700 food substances that were exempt from a new requirement that manufacturers test food additives before placing them on the market. These foods were designated GRAS. The GRAS designation is defined under Sections 201(s) and 409 of the United States Federal Food, Drug, and Cosmetic Act as 'any substance that is intentionally added to food is a food additive, that is subject to premarket review and approval by FDA, unless the substance is generally recognized, among qualified experts, as having been adequately shown to be safe under the conditions of its intended use, or unless the use of the substance is otherwise excluded from the definition of a food additive.'

Under Sections 201(s) and 409 of the above Act, and FDA's implementing regulations in Title 21 of the Code of Federal Regulations Section (21CFR 170.3), the use of a food substance may be GRAS either through scientific procedures or, for a substance used in food before 1958, through experience based on common use in food. General recognition of safety through scientific procedures requires the same quantity and quality of scientific evidence as is required to obtain approval of the substance as a food additive and ordinarily is based upon published studies, which may be corroborated by unpublished studies and other data and information. General recognition of safety through experience based on common use in foods requires a substantial history of consumption for food use by a significant number of consumers.

Manufacturers can petition for GRAS status for new additives if the substances meet the criteria cited above. In addition the regulations allow them to make a

GRAS determination through the scientific evaluation process. According to the FDA, this 'route requires the same quantity and quality of scientific evidence as required to obtain approval of the substance as a food additive and ordinarily is based upon published studies, which may be corroborated by unpublished studies and other data and information'. After a company makes a GRAS determination it may voluntarily submit it for review. The FDA has the authority to review these determinations and advise the company on whether the GRAS determination is appropriate. The FDA also has the authority to de-GRAS a substance, in which case the company must designate the substance as a food additive and obtain pre-market approval from the FDA following the standard procedure.

A complete listing of GRAS substances known to the FDA in the US food regulations is available at http://www.accessdata.fda.gov/scripts/fdcc/?set=GRASNotices. A supplemental list of flavour-type compounds is maintained by the Flavor and Extract Manufacturers Association and updates are routinely published in the US Institute of Food Technologist's monthly journal *Food Technology*. A listing of GRAS food substances that are derived from microorganisms is published in the US Code of Federal Regulations 21 CFR 184 (http://www.fda.gov/Food/IngredientsPackagingLabeling/GRAS/Microorganisms MicrobialDerivedIngredients/default.htm). Conditions for their use are prescribed in the regulations and are predicated on the use of non-pathogenic and non-toxicogenic strains of the respective organisms and on the use of current good manufacturing practice (21 CFR 184.1(b)).

The Canadian Foods Regulations accepts the US GRAS status, provided the food in question also meets the requirements of the Canadian Foods Regulations. Canadian food producers (and those in other countries) seeking to export products to the United States may petition the US FDA for the granting of GRAS status, based on a comprehensive dossier of safety and toxicological data and/or a substantial history of consumption for use of the food in question by a significant number of consumers.

The European Union has adopted a food designation somewhat similar to GRAS. The concept of 'Qualified Presumption of Safety' (QPS) is used as the basis for the safety of the parental organism (EFSA, 2007). The list of QPS organisms is updated annually (EFSA, 2012) and currently includes bacteria, filamentous fungi, viruses and yeasts. If the parental microorganism has QPS status, the risk assessment is focused on the changes introduced (intended and unintended) during the development of a GM microorganism. Where no comparator can be identified for the microorganism and/or its product, a comprehensive safety assessment is carried out. For the toxicological, allergenicity and nutritional assessment, the same basic approach is applied as with GM plants. Although the EU has this legislation in place, currently no GM microorganisms have been approved for use in food. Bacterial biomass from L-lysine production and yeast biomass from processes using GM microorganisms are, however, allowed into the food chain in the form of animal feed.

The Australian/New Zealand approach to the safety assessment of foods derived from GM microorganisms is similar to that of the EU and is based on the Codex Guideline for the Conduct of Food Safety Assessment of Foods Produced Using Recombinant-DNA Micro-organisms (FAO/WHO, 2009). The regulations outline the need to establish whether the GM microorganism itself becomes a component of the food, as in the case of starter cultures and probiotics, or whether the GM microorganism is subsequently removed from the food. The presence or absence of viable organisms in the food determines the range of issues to be considered in the safety assessment. Where viable organisms remain in the final food (e.g. yoghurt) the additional issues to be considered are: (i) possible antibiotic resistance and gene transfer; (ii) possible immunological effects; (iii) potential pathogenicity; and (iv) viability and residence in the digestive tract. As in Europe, these regulations can be regarded as precautionary since currently there is a prohibition in Australia and New Zealand on foods derived from GM microorganisms.

According to Devos et al. (2014) approximately 19 GM microorganism applications have been assessed since 2002, or are under assessment, by the European Food Safety Authority. Two of these applications relate to feed materials derived from biomass of GM bacteria (*Escherichia coli*) and are being considered under the GM Food and Feed Regulation. The others are for products such as enzymes, vitamins or amino acids, obtained by fermentation of GM microorganisms in which no material derived from the production strain is present at detectable levels, or for biodegradable polymers intended to be used in food contact packaging material that are produced by fermentation of GM microorganisms producing the monomers. According to EFSA these products are considered under EU legislation as having been made *with* GM microorganisms in contrast to products made *from* GM microorganisms, in which the GM microorganism or residues of its cells are still present in the product.

This distinction has important consumer implications since products made *from* (but not *with*) GM microorganisms have to be labelled 'either as "genetically modified ...", or as "produced from genetically modified ..."'. 'This information may be indicated, in the same font size, in a footnote to the list of ingredients.' Neither rennet nor chymosin (which are made *with* GM microorganisms are regarded as food ingredients, consequently they do not have to be declared in the list of ingredients.

The vast majority of products made *with* GM microorganisms are feed enzymes, vitamins or amino acids. These products are currently under investigation within the framework of the European Regulations (EC) No 1332/2008 (on food enzymes), 1333/2008 (on food additives), 1334/2008 (on food flavourings) and 1831/2003 (on feed additives). Production organisms encompass GM bacteria (*E. coli, Corynebacterium glutamicum, Bacillus subtilis* and *B. licheniformis*), and GM fungi, including yeast (*Trichoderma reesei, Aspergillus niger, A. oryzae, Penicillium funiculosum, Saccharomyces cerevisiae, Pichia pastoris*). Although these products are not considered GM foods and feeds and are therefore beyond the scope of the

GM Food and Feed Regulation, they still have to undergo a risk assessment prior to authorisation in compliance with existing EU legislation specific for each type of product. In the course of the assessment of GM microorganism applications, EFSA takes into account the characteristics of the production of the GM microorganism and the potential effects that its genetic modification can exert on the safety of the final product (e.g. production of toxic compounds, presence of antibiotic resistance genes). Key aspects considered include the characterisation of the parental microorganism and the inserted sequence(s), the stability of the GM microorganism, and the possible presence of GM microorganism cells or recombinant DNA in the final product. According to EFSA the number of applications for food and feed products made with GM microorganisms is expected to increase significantly from 2013 onwards, following introduction of the EU legislation on food enzymes, which requires that all products already on the market undergo a safety assessment by EFSA.

Bacteria

As indicated above, in the US regulations the conditions for the use of microorganisms in food production are prescribed in the regulations and are predicated on the use of non-pathogenic and non-toxicogenic strains of the respective organisms and on the use of current good manufacturing practice (184.1(b)). Prior sanctions were granted for the use of harmless lactic acid producing bacteria, such as *Lactobacillus acidophilus*, as optional ingredients in specified standardised foods. These bacteria are permitted for use in cultured milk (which includes buttermilk) (21 CFR 131.112), sour cream (131.160), cottage cheese (133.128) and yoghurt (131.200), provided that the mandatory cultures of *Lactobacillus bulgaricus* and *Streptococcus thermophillus* are also used in the yoghurt.

GM bacteria that have been approved (or have approval pending) for food use in the USA include the following.

1 GRN (FDA file number) # 22: α-amylase derived from *Bacillus licheniformis* carrying a gene encoding a modified α-amylase derived from *Bacillus licheniformis* and *Bacillus amyloliquefaciens* (1999).
2 GRN # 79: α-amylase derived from *Bacillus licheniformis* carrying a gene encoding a modified α-amylase from *Bacillus licheniformis* (2001).
3 GRN # 265: Glycerophospholipid cholesterol acyltransferase (GCAT) enzyme preparation from *Bacillus licheniformis* expressing a modified GCAT gene from *Aeromonas salmonicida subsp. salmonicida* (2009).
4 GRN # 277: Maltotetraohydrolase enzyme preparation from *Bacillus licheniformis* expressing a modified maltotetraohydrolase gene from *Pseudomonas stutzeri* (2009).
5 GRN # 476: Asparaginase enzyme preparation produced by genetically modified *Bacillus subtilis* (pending).

6 CFR # 173.150: Milk-clotting enzymes, microbial for use in the production of cheese (milk-clotting enzymes are derived from *Endothia parasitica, Bacillus cereus, Mucor pusillus Lindt* and *Mucor miehei* and *Aspergillus oryzae* modified to contain the gene for aspartic proteinase from *Rhizomucor miehei* var *Cooney et Emerson* (2011)).

Recent publications exemplify the wide range of food and feed products using GM bacterial processes.

Biochemicals, additives

Patent # EP 0036258 A3 was granted in 1982 to researchers Bahl, Rose and White for the production of the artificial sweetener aspartame, using a GM strain of *E. coli*. According to the patent statement this compound, a dipeptide with the formula Asp-Phe-me, is produced and harvested. The free carboxyl group of the large peptide is benzylated and then hydrolysed to benzyl Asp-Phe dipeptide. This dipeptide is methylated and then debenzylated to form aspartame.

Gioia et al. (2011) described a process using genetic engineering of *Pseudomonas fluorescens* for the production of vanillin from ferulic acid. Vanillin is one of the most important flavours in the food industry and there is great interest in its production through biotechnological processes starting from natural substrates such as ferulic acid. Among bacteria, recombinant *Escherichia coli* strains are the most efficient vanillin producers, whereas *Pseudomonas* spp. strains, although possessing a broader metabolic versatility, rapidly metabolise various phenolic compounds including vanillin. In order to develop a robust *Pseudomonas* strain that can produce vanillin in high yields and at high productivity, the vanillin dehydrogenase (vdh)-encoding gene of *Pseudomonas fluorescens* BF13 strain was inactivated via targeted mutagenesis. The results demonstrated that engineered derivatives of strain BF13 accumulate vanillin if inactivation of vdh is associated with concurrent expression of structural genes for feruloyl-CoA synthetase (fcs) and hydratase/aldolase (ech) from a low-copy plasmid (a plasmid is defined as a small DNA molecule that is physically separate from, and can replicate independently of, chromosomal DNA within a cell). The conversion of ferulic acid to vanillin was enhanced by optimisation of growth conditions, growth phase and parameters of the bioconversion process. The developed strain produced up to 8.41 mM vanillin, which is the highest final titer of vanillin produced by a *Pseudomonas* strain to date and presents new opportunities for the use of bacterial biocatalysts for biotechnological production of vanillin from agro-industrial wastes that contain ferulic acid.

EFSA (2013) recently assessed the evidence relating to the efficacy and safety of the amino acid L-methionine produced by *E. coli* (KCCM 11252P) and *E. coli* (KCCM 11340P). This product is used as an additive in animal and poultry feed to improve the amino acid profile. Neither the production strains nor their recombinant DNA were detected in the final product. EFSA concluded, therefore, that L-methionine produced by such *E. coli* strains is safe and efficacious for the target animals when used as a feed additive to meet their requirements. Also, there were no safety concerns for the consumer.

Putignani et al. (2013) reviewed the use of GM *E. coli* as a source of flavonoids and terpenoids. During the last two decades, the engineered biosynthesis of several natural products by microbial sources has made enormous progress. It is noteworthy that *E. coli* has been used as a microbial power plant for the biosynthesis of different types of molecules for both biological and clinical applications. Among these products, flavonoids and terpenoids are particularly attractive because of the high variety of their biochemical activities including anti-oxidant, anti-inflammatory, anti-viral, anti-bacterial, anti-obesity and anti-cancer properties.

Gaspar et al. (2013) outlined developments on the production of biochemicals by lactic acid bacteria. These bacteria comprise a functionally related group of low-G+C Gram-positive bacteria (possessing fewer G and C DNA bases in their DNA bases than A and T bases in comparison with other bacteria) known essentially for their roles in bio-processing of foods and animal feeds. Due to extensive industrial use and high economic value, they have been studied intensively and a large body of comprehensive data on their metabolism and genetics has been generated throughout the years. This knowledge has been instrumental in the implementation of successful applications in the food industry, such as the selection of robust starter cultures with desired phenotypic traits.

Milk products, probiotics

Renault (2002) described how lactic acid bacteria have a long history of use in fermented food products. Progress in gene technology allows their modification by introducing new genes or by modifying their metabolic functions. These modifications may lead to improvements in food technology (bacteria better fitted to technological processes, leading to improved organoleptic properties etc.), or to new applications, including bacteria producing therapeutic molecules that could be delivered by mouth. In the opinion of this researcher, risk assessment and expected benefits will determine the future use of modified bacteria in the field of food technology and health.

An important group of milk-based foods containing live bacteria are yoghurts, which can be classified as probiotics and which have nutritional as well as health benefits (Gill and Guarner, 2004; Namoto, 2005; Steidler and Neirynck, 2005). According to the FAO/WHO Guidelines for the Evaluation of Probiotics in Food (2002), probiotics are defined as 'live microorganisms which when administered in adequate amounts confer a health benefit on the host'. These guidelines specify that prior to being categorised as probiotics, organisms should follow a process of testing, including strain identification by genotype and phenotype, functionalised characterization and safety assessment testing, and double-blind, placebo-controlled human trials to verify their health benefits. Commonly used food-grade probiotics include lactic acid bacteria, such as *Lactobacillus acidophilus*, and *Bifidobacteria*, which have a history of safe use in food products. These bacteria can be added easily to traditionally fermented foods, such as cheeses, yoghurts and other dairy products.

Current research on probiotic bacteria includes the development of GM strains with specific characteristics.

Fungi

Yeasts

Yeasts are unicellular microorganisms classified in the kingdom Fungi, with around 1500 species. Yeasts from the genus *Saccharomyces* display high degrees of fermentation ability and therefore are used in the production of a variety of food products and beverages. Yeasts are used in brewing (beer yeast) and in the production of wine and spirits (wine yeast, distillers' yeast) because of their ability to break down plant starches into simple sugars that are then converted into alcohol (ethanol). Secondary metabolic products help give alcoholic beverages their characteristic aromas and flavours. Yeasts are also used in baking (bakers' yeast) to allow the dough to rise due to the production of carbon dioxide during the fermentation process.

Saccharomyces cerevisiae was the first genetically modified organism (as distinct from a genetically modified product) to be cleared for use in human food production (Hammond, 1995). This yeast has GRAS status, which is an advantage in the production of compounds that are intended for human consumption. To date, more than 700 different strains of *S. cerevisiae* have been identified, allowing food and beverage producers to select appropriate strains for particular food use.

Although *S. cerevisiae* has GRAS status under US regulations, any genetic modification made to the organism requires the approval of the FDA before the new strain is designated as GRAS. The evidence required to be submitted in support of the GRAS designation depends on the nature and extent of the genetic changes made.

GM fungi that have been approved (or have approval pending) for food use in the USA include the following.

1. GRN # 292: Cellulase enzyme preparation derived from a genetically modified strain of *Myceliophthora thermophila* (2009).
2. GRN # 296. Lipase enzyme preparation from a genetically modified strain of *Aspergillus niger* (2009).
3. GRN # 345: Carboxypeptidase enzyme preparation from modified *Aspergillus niger* (2010).
4. GRN # 402: Peroxidase enzyme preparation derived from a genetically modified strain of *Aspergillus niger* (2012).

5 GRN # 422: Modified baker's yeast *Saccharomyces cerevisiae* intended to reduce acrylamide production in a variety of grain-based foods, vegetable-based food (potato), and coffee and coffee substitutes at a level of 0.5 to 5% (2012).
6 GRN # 428: Asparaginase enzyme preparation from genetically modified *Aspergillus niger* (2012).
7 FDA CFR # 173.150: Milk-clotting enzymes, microbial for use in the production of cheese (milk-clotting enzymes are derived from *Endothia parasitica Bacillus cereus, Mucor pusillus* Lindt and *Mucor miehei Cooney et Emerson* and *Aspergillus oryzae* modified to contain the gene for aspartic proteinase from *Rhizomucor miehei* var *Cooney et Emerson* (2011)).

The main areas in which GM fungi are used in food and beverage production – apart from enzyme production – are beer- and wine-making and also to a more limited extent baking.

Dequin (2001) outlined the potential of genetic engineering for improving brewing, wine-making and baking yeasts. The end of the twentieth century was marked by major advances in biotechnology, particularly in areas related to genetics and more recently genomics. Considerable progress was made in the development of genetically improved yeast strains for the wine, brewing and baking industries. Yeast is intensively processed with gene technology. As a rule, genetically modified yeast may be found in all levels of development and experimentation. Goals may include the following:

- Bakers' yeast: reduction of rising times for dough; improvement of dough stability.
- Beer yeast: simplification of the process of fermentation for beers with reduced quantities of alcohol or calories; stability of taste; reduction of lees during filtering.
- Wine yeast: optimisation of the process of fermentation, reduction in the length of fermentation times; stability or improvement of taste; enhancement of fermentation performance, aimed at the complete fermentation of sugar and the elimination of sweetness in wine; reduction of the fermentation of champagne in the bottle; reduction of carcinogenic materials that arise during fermentation.

In the previous decade, recombinant DNA technology widened the possibilities for introducing new properties in beers, wines and baked goods. The most remarkable advances, according to this author, were improved process performance, off-flavour elimination, increased formation of by-products, improved hygienic properties or extension of substrate utilisation. Although the introduction of this technology into traditional industries is currently limited by public perception, the number of potential applications of genetically modified industrial yeast was considered likely to increase in the coming years, as knowledge derived from genomic analyses increases.

Yeast is also available as dry yeast and is a source of yeast extract and nutritional yeast. Yeast dregs, for example as obtained from breweries, are used mainly as animal feed.

Baking

In 1990 the United Kingdom became the first country to permit the use of a live, genetically modified organism in food. It was developed at the National Centre for Biotechnology Education (http://www.ncbe.reading.ac.uk/ncbe/gmfood/yeasts.html). This was a special strain of bakers' yeast engineered to make the bread dough rise faster. Existing genes were placed under the control of stronger, constitutive promoters, which help the yeast to break down the sugar maltose faster than usual. This yeast does not appear, however, to be used commercially by food manufacturers.

Enzymes are often added to baked goods. They can make dough easier to process, make it expand, or provide for an ideal crust.

In 2012 the US FDA granted GRAS status to a company in Vancouver, Canada, for a functionally-enhanced baker's yeast (*S. cerevisiae*) that prevents acrylamide formation in foods and beverages (GRAS Notice No. GRN 000422). This GM strain was approved to reduce acrylamide production in a variety of grain-based foods, vegetable-based food (potato), and coffee and coffee substitutes. Acrylamide is produced during the cooking or processing of certain foods, particularly in foods high in carbohydrates and low in proteins, resulting from the reaction of asparagine with reducing sugars via the Maillard reaction. Acrylamide is categorised as a Group 2A carcinogen by the World Health Organization, and is found in foods such as bread, cookies, crackers, baby food, breakfast cereal, french fries and potato chips. In 2014 Health Canada approved another approach to reducing the level of acrylamide in foods by the addition of asparaginase obtained from *Aspergillus niger* AGN7-41 to the List of Permitted Food Enzymes for addition to bread, flour, whole wheat flour and unstandardised foods. Approval was based on a detailed safety assessment of asparaginase obtained from *Aspergillus niger* AGN7-41 as a food additive. Asparaginase from another strain of *Aspergillus niger* (*A. niger* ASP72), is already permitted for use in Canada as a food enzyme in bread, flour, whole wheat flour and unstandardised foods.

Brewing (beer-making)

In 1994, the Brewing Research Foundation International gained approval (Stewart et al., 2013) for the use of a GM yeast to make low-carbohydrate beer. The yeast *Saccharomyces cerevisiae* var *diastaticus* produces an amylase capable of hydrolysing the starch residues that normally remain in the brew after fermentation. The high-calorie starch residues are thus converted into fermentable sugar. The GM yeast can be used to produce a high alcohol premium product or, alternatively, a greater volume of low calorie 'lite' beer.

Interest has been focused on transferring the gene for glucoamylase from *S. diastaticus* into regular *S. cerevisiae*. Such a yeast would be better able to utilise the carbohydrate present in conventional feedstocks. This would increase the yield of alcohol and enable the production of a full-strength, low-carbohydrate diet beer without the use of extra enzymes after the beer had been brewed (Saerens et al., 2010).

Stewart et al. (2013) outlined developments in brewing and distilling yeast strains over the previous 25 years. The review described differences between ale and lager yeast strains, the uptake and metabolism of wort sugars and amino acids, yeast flocculation, yeast management between fermentations and yeast strain genetic stability.

Steensels et al. (2012) outlined state-of-the-art technologies to select and generate superior yeast strains for the brewing industry. An example of this technology was described by Zhang et al. (2013). Since esters and higher alcohols produced by yeast during the fermentation of wort have the greatest impact on the smell and taste of beer, the objective of their study was to develop GM yeast strains that produce a higher proportion of acetate esters and a lower proportion of higher alcohols in beer produced from wheat. Industrial brewer's yeast (*S. cerevisiae*) strain S5 was used as the parental strain to produce GM strains S5-2 and S5-4. Concentrations of ethyl acetate produced by these GM strains increased to 78.88 and 117.40 mg/l, respectively, representing a 7.7-fold and 11.5-fold increase over that produced by the parental strain. Isoamyl acetate produced by these strains was 5.14 and 9.25 mg/l, respectively; while no isoamyl acetate was detected in the fermentation sample of the parental strain. The researchers also found that the GM strains produced about 65 and 51% of the concentration of higher alcohols, respectively, than were produced by the parental strain.

The successful application of technologies such as these may depend on consumer attitudes towards the use of GM yeasts in beer production. Burton and Pearse (2002) found that Australian respondents were equally averse to first-generation modification of either plants or microorganisms but were willing to pay a premium for a product with positive health benefits.

Wine-making

According to the review of Schuller and Casal (2005), the areas of wine production where GM yeasts have been tested include aroma liberating enzymes, acidity adjustment, glycerol production, volatile phenols, acetate ester formation, hydrogen sulfite reduction, resveratol production, ethyl carbamate elimination, antimicrobial enzymes, stress tolerance, sugar uptake, nitrogen assimilation, agrochemicals resistance, removal of filter clogging polysaccharides and flocculation, generally involving *S. cerevisae*.

GM yeasts have been approved in the USA, Canada, South Africa and Moldavia for use in the production of wine. *S. cerevisae* strain ML01 was developed in Canada to improve the taste and colour stability of wine as well as to avoid the production of undesirable compounds (histamines). Another GM yeast has also been developed in Canada with the aim of reducing the presence of ethylcarbamate that occurs naturally during fermentation and which is suspected of having carcinogenic effects.

The US FDA granted GRAS status to *S. cerevisae* strain ML01 in 2003 (FDA, 2003). Health Canada (2006) published a detailed outline of its assessment in approving the ML01 strain for use in wine-making in Canada. The Food Directorate within Health Canada has a legislated responsibility for pre-market assessment of novel foods and novel food ingredients as detailed in Division 28 of part B of the Food and Drug Regulations ('novel foods'). Food use of ML01 is considered a novel food under the following part of the definition: 'c) a food that is derived from a plant, animal or microorganism that has been genetically modified such that i. the plant, animal or microorganism exhibits characteristics that were not previously observed in that plant, animal or microorganism.' The strain was developed by Dr Hendrik van Vuuren at the University of British Columbia, from *S. cerevisiae* var. *bayanus* S92, by introducing two genes into the yeast genome; mae 1, which allows the transport of malate, malonate and succinate into the yeast cell, and mle A, which produces a malolactic enzyme that converts L-malate to L-lactate. As a result, the alcoholic and malo-lactic fermentations are achieved with one strain of yeast. Also, the wine is produced without the use of starter cultures, thus avoiding the production of allergenic biogenic amines which produce off-flavours and can trigger headaches and migraines in some consumers. In addition, there was evidence that use of the ML01 strain improved the taste and colour stability of the wine. The assessment included the potential for the presence of toxicants, anti-nutrients, allergens, chemical or microbiological contaminants in wine produced using the GM strain. Since wine is not consumed for its nutritional value it was concluded in the assessment that it was not necessary to require compositional data to compare wine produced with the novel yeast ML01 with wine made by traditional methods.

In 2006 a further GM strain (*S. cerevisiae* strain ECMo01) developed by the same researcher for use as a starter culture in alcoholic beverage fermentation was granted GRAS status by the US FDA (GRAS Notice No. GRN 000175). The parent strain (Davis 522) is commonly used in industry to obtain wine active dry yeast and has not been reported to produce allergenic or toxic substances. The aim in developing the strain was to increase the expression of urea amidolyase in *S. cerevisiae*. This enzyme catalyses the hydrolysis of urea which is produced by *S. cerevisiae* during alcoholic fermentation. Urea is a precursor of ethyl carbamate, a suspected carcinogen in humans, which is formed in the wine from the reaction of urea and ethanol. Hydrolysis of urea by urea amidolyase has been shown to significantly reduce the potential for formation and accumulation of ethyl carbamate in the wine.

Later the same year it was announced that the GM strain had met the necessary requirements of Health Canada for use commercially in wine production in Canada.

Evidence submitted to the regulatory authorities in the USA and Canada showed that the clarification and stabilisation processes, which are obligatory in commercial winemaking, remove most of the yeast cells. When the wine is kept on the lees for ageing, the yeast cells undergo autolysis, releasing cellular materials that are degraded through the action of enzymes. Further, white wines can be treated with bentonite to remove all remaining protein fractions. Prior to bottling most wines undergo filtration (e.g. with diatomaceous earth, cellulose filters, or membrane filters) that eliminates any remaining yeast cells. These processes also reduce or eliminate any remaining urea amidolyase in wines at the time the wines are consumed.

In 2011 the FDA granted GRAS status to another yeast developed by a Canadian company, namely *Saccharomyces cerevisiae* strain P1Y0, a variant of *S. cerevisiae* parent strain UCD2034 (FDA, 2011). The strain P1Y0 may be used in the reduction of hydrogen sulfide concentration during the production of red and white wine, champagne, sherry, sake and other rice wines, and beer.

Cheese-making

In the initial stage of hard cheese-making, milk is curdled until it achieves a thick, fluid consistency. This step is achieved using an added enzyme, chymosin (rennin), which breaks down the milk proteins and causes them to coagulate and separate from the watery whey. Rennet is a common source of this enzyme.

Under the US regulations rennet and bovine rennet are commercial extracts containing the active enzyme rennin (CAS Reg. No. 9001-98-3), also known as chymosin (International Union of Biochemistry Enzyme Commission (E.C.) 3.4.23.4). Rennet is the aqueous extract prepared from cleaned, frozen, salted, or dried abomasa of calves, kids or lambs. Bovine rennet is the product from adults of the animals listed above. Both products are called rennet and are available commercially in both liquid and powder forms. Chymosin preparation is a clear solution containing the active enzyme chymosin. It is derived, via fermentation, from a non-pathogenic and non-toxigenic strain of *Aspergillus niger* van Tieghem var *awamori* (Nakazawa) Al-Musallam (synonym *A. awamori* Nakazawa) containing the prochymosin gene. Chymosin is recovered from the fermentation broth, and after extraction and purification yields a product that is 80–90% pure. Natural rennin contains only 4–8% active enzyme. Chymosin is now produced in this way in several countries, including some European countries. As outlined above, no GM microorganisms have been approved for use in food in Europe or Australia/New Zealand. Chymosin and other enzymes are considered under EU and Australian/New Zealand legislation as having been made *with* GM microorganisms in contrast to products made *from* GM microorganisms. Since after purification no traces of GM material remain in the product, chymosin and other enzymes produced in a similar way are permitted as processing agents in food production and products such as cheese are not required to be labelled as either as 'genetically modified ...', or as 'produced from genetically modified ...'. New EU regulations came into effect in 2013. Since chymosin remains in the cheese it has to be declared on the list of ingredients. The use of genetically modified organisms in the production process is not required to be explicitly labelled on the product.

According to the US standards, all materials used in the processing and formulating of chymosin preparation must be either generally recognised as safe (GRAS) or be food additives that have been approved by the Food and Drug Administration for this use. Rennet and chymosin preparation must also meet the general and additional requirements for enzyme preparations of the Food Chemicals Codex (USP, 2014).

Fermentation-Produced Chymosin (FPC) as described above was the first artificially produced enzyme to be registered and approved by the US Food and Drug Administration. It has been on the market since 1990. Currently it may be produced commercially using GM *Aspergillus niger* (as indicated above) or another GM fungus *Kluyveromyces marxianus* var. *lactis* (which also has GRAS status). As indicated above, under FDA 21 CFR 173.150 the milk-clotting enzymes of microbial origin for use in the production of cheese may also be derived from *Endothia parasitica, Bacillus cereus, Mucor pusillus Lindt* and *Mucor miehei* and *Aspergillus oryzae* modified to contain the gene for aspartic proteinase from *Rhizomucor miehei* var *Cooney et Emerson (2011).*

Summary

GM bacteria have been approved for food use in several countries, mainly for the production of enzymes such as milk-clotting enzymes and food additives such as aspartame. An important group of milk-based foods containing live bacteria are yoghurts, which are classified as probiotics. Yeasts from the genus *Saccharomyces*, including GM strains, display high degrees of fermentation ability and therefore are used in the production of a variety of food products and beverages. Their approval is based on their GRAS status by the regulatory authorities. *Saccharomyces cerevisiae* was the first GM organism to be approved for use in food production. GM yeasts are an important source of enzymes used in food production (e.g. cheese-making) and in beer and wine production.

References

Burton, M. and Pearse, D. (2002). Consumer attitudes towards genetic modification, functional foods, and microorganisms: a choice modeling experiment for beer. *AgBioForum* 5: 51–58.

Dequin S. (2001). The potential of genetic engineering for improving brewing, wine-making and baking yeasts. *Applied Microbiology and Biotechnology* 56: 577–588.

Devos, Y., Aguilera, J., Diveki, Z., et al. (2014). EFSA's scientific activities and achievements on the risk assessment of genetically modified organisms (GMOs) during its first decade of existence: looking back and ahead. *GM Research* 23:1–25.

EFSA (2007) Opinion of the Scientific Committee on a request from EFSA on the introduction of a Qualified Presumption of Safety (QPS) approach for assessment of selected microorganisms referred to EFSA. *EFSA Journal* 587: 1–16, http://www.efsa.europa.eu/en/efsajournal/doc/587.pdf (accessed 14 March 2014).

EFSA (2012). Scientific Opinion on the maintenance of the list of QPS biological agents intentionally added to food and feed (2012 update). *EFSA Journal* 10: 3020–3104, http://www.efsa.europa.eu/en/efsajournal/doc/3020.pdf (accessed 14 March 2014).

EFSA (2013). Scientific Opinion on the safety and efficacy of L-methionine produced by *Escherichia coli* (KCCM 11252P) and *Escherichia coli* (KCCM 11340P) for all animal species. *EFSA Journal* 11: 3428–3429, http://www.efsa.europa.eu/en/efsajournal/pub/3428.pdf, (accessed 14 March 2014).

FAO/WHO Guidelines for the Evaluation of Probiotics in Food (2002). Joint FAO/WHO Working Group Report on Drafting Guidelines for the Evaluation of Probiotics in Food. London, Ontario, Canada, April 30 and May 1, 2002.

FAO/WHO (2009). Joint FAO/WHO food standards programme, Codex Alimentarius Commission. *Foods Derived from Modern Biotechnology*, Second Edition, World Health Organization (WHO) and Food and Agriculture Organization of the United Nations (FAO), Rome, Italy.

FDA (2003) Agency Response Letter GRAS Notice No. GRN 000120. CFSAN/Office of Food Additive Safety. http://www.fda.gov/Food/IngredientsPackagingLabeling/GRAS/NoticeInventory/ucm153936.htm (accessed 2 March 2014).

FDA (2006) Agency Response Letter GRAS Notice No. GRN 000175. CFSAN/Office of Food Additive Safety. http://www.fda.gov/Food/IngredientsPackagingLabeling/GRAS/NoticeInventory/ucm154604.htm (accessed 2 March 2014).

FDA (2011). Agency Response Letter GRAS Notice No. GRN 000350. CFSAN/Office of Food Additive Safety. http://www.fda.gov/Food/IngredientsPackagingLabeling/GRAS/NoticeInventory/ucm244717.htm (accessed 2 March 2014).

Gaspar, P., Carvalho, A. L., Vinga, S., et al. (2013). From physiology to systems metabolic engineering for the production of biochemicals by lactic acid bacteria. *Biotechnology Advances* 31: 764–788.

Gill, H. S. and Guarner, F. (2014). Probiotics and human health: a clinical perspective. *Postgraduate Medical Journal* 80: 516–526.

Gioia, D., Luziatelli, F., Negroni, A., et al. (2011). Metabolic engineering of Pseudomonas fluorescens for the production of vanillin from ferulic acid. *Journal of Biotechnology* 156: 309–316.

Hammond, J. R. M. (1995) Genetically-modified brewing yeasts for the 21st century. Progress to date. Yeast 11: 1613–1627.

Health Canada (2006). Wines derived from the genetically modified wine yeast Saccharomyces cerevisiae ML01. http://www.hc-sc.gc.ca/fn-an/gmf-agm/appro/nf-an107decdoc-eng.php (accessed 2 March 2014).

Health Canada (2014). Notice of Modification to the List of Permitted Food Enzymes to Enable the Use of Asparaginase Obtained from *Aspergillus niger* AGN7-41 as a Food Enzyme in Bread, Flour, Whole Wheat Flour, and Unstandardized Foods. http://www.hc-sc.gc.ca/fn-an/consult/nom-adm-0023/index-eng.php (accessed 2 March 2014).

Namoto, K. (2005). Review: prevention of infections by probiotics. *Journal of Bioscience and Engineering* 100: 583–592.

Putignani, L., Massa, O. and Alisi, A. (2013). Engineered Escherichia coli as new source of flavonoids and terpenoids. http://www.sciencedirect.com/science/journal/09639 (accessed 21 February 2014).

Renault, P. (2002). Genetically modified lactic acid bacteria: applications to food or health and risk assessment. *Biochimie* 84: 1073–1087.

Saerens, S. M., Duong, C. T. and Nevoigt, E. (2010). Genetic improvement of brewer's yeast: current state, perspectives and limits. *Applied Microbiology and Biotechnology* 86: 1195–1212.

Schuller, D. and Casal, M. (2005). The use of genetically modified *Saccharomyces cerevisiae* strains in the wine industry. *Applied Microbiology and Biotechnology* 68: 292–304.

Steensels, J., Snoek, T., Meersman, E., et al. (2012). Selecting and generating superior yeasts for the brewing industry. *Cerevisia* 37: 63–67.

Steidler, L. and Neirynck, S. (2005). *Genetically Modified Probiotics. Probiotics and Probiotics: Scientific Aspects.* Ed. Tannock, G. Caister Academic Press, Poole.

Stewart, G. G., Hill, A. E. and Russell, I. (2013). 125th Anniversary review: developments in brewing and distilling yeast strains. *Journal of the Institute of Brewing* 119: 202–220.

USP (2014). *Food Chemicals Codex*, 9th Edition. United States Pharmacopeial Convention (USP), Rockville, Maryland 20852-1790.

Zhang, C., Liu, Y., Qi, Y., et al. (2013). Increased esters and decreased higher alcohols production by engineered brewer's yeast strains. *European Food Research and Technology* 236: 1009–1014.

CHAPTER 4
Cereals

The use of biotechnology to improve genetic makeup has become widely adopted in breeding new and improved cereal crops. However, to date most of the attention has been concentrated on maize. According to statistics published by the Global GMO Compass (2014) the proportion of genetically modified (GM) maize rose globally to 32% in 2013. The proportion was 80–86% in Argentina, Brazil and South Africa and was highest at 90–98% in the USA and Canada.

Brookes (2008) reviewed the impact of using GM insect-resistant corn (maize) in Europe since 1998. In 2006 there were plantings of this crop in seven EU member states. An economic analysis showed an improvement in profitability of between 12 and 21%. Insecticidal spraying had also been reduced and the grain quality was improved as a result of lower mycotoxin contamination.

Some consumers are concerned that the cereal grains they buy may have come from GM crops, preferring instead food from these crops to be unaltered by biotechnology. Organic producers have a declared objection to the technology and consumers who have a similar objection often opt for organic grains. Objection to GM grains is especially strong in Europe where the majority of consumers appear to prefer traditional types of cereal grains.

Not all societies that are considered very traditional reject crops derived by biotechnology. For example, the Muslim Council of Indonesia (Indonesian Ulemas Council) approved GM foods in 2003. Also, GM maize is being grown by Amish farmers in the USA, a group that follows a very traditional lifestyle.

Consumer attitudes to grain crop improvement by GM technology may be dependent, therefore, on the type and accuracy of the information available to them in addition to their general view on the sanctity of certain foods.

Genetic Modification and Food Quality: A Down to Earth Analysis, First Edition.
Robert Blair and Joe M. Regenstein.

Table 4.1 Global area of biotech crops in 2013 (ISAAA, 2014a). Reproduced with permission of ISAAA.

Rank	Country	Area, million hectares	Crops
1	USA	70.1	Maize, soyabeans, cotton, canola, sugar beet, alfalfa, papaya, squash
2	Brazil	40.3	Maize, soyabeans, cotton
3	Argentina	24.4	Maize, soyabeans, cotton
4	India	11.0	Cotton
5	Canada	10.8	Maize, soyabeans, canola, sugar beet
6	China	4.2	Cotton, papaya, poplar, tomato, sweet pepper
7	Paraguay	3.6	Maize, soyabeans, cotton
8	South Africa	2.9	Maize, soyabeans, cotton
9	Pakistan	2.8	Cotton
10	Uruguay	1.5	Maize, soyabeans
11	Bolivia	1.0	Soyabeans
12	Philippines	0.8	Maize
13	Australia	0.6	Cotton, canola
14	Burkina Faso	0.5	Cotton
15	Myanmar	0.3	Cotton
16	Mexico	0.1	Cotton, soyabeans
17	Spain	0.1	Maize
18	Colombia	0.1	Maize, cotton
19	Sudan	0.1	Cotton
20	Chile	<0.1	Maize, soyabeans, canola
21	Honduras	<0.1	Maize
22	Portugal	<0.1	Maize
23	Czech Republic	<0.1	Maize
24	Cuba	<0.1	Maize
25	Costa Rica	<0.1	Cotton, soyabeans
26	Romania	<0.1	Maize
27	Slovakia	<0.1	Maize

As shown in Table 4.1 (International Service for the Acquisition of Agri-biotech Applications (ISAAA), 2014a) the only GM cereal crop grown internationally in 2013 in commercially significant quantities was maize.

Maize (*Zea mays* L.)

Several transgenic maize (corn) cultivars have been developed by different companies and approved in several countries.

The ISAAA database contains 136 approved maize cultivars that have been approved in various countries (Appendix Table 4.1A). Not all of the countries listed have approved these cultivars for domestic cultivation. For instance, only one GM maize cultivar (MON810) is approved for cultivation in the EU, with another maize cultivar (T25) in process of being re-approved for cultivation. The other GM maize cultivars that have been approved for use in the EU have to be imported.

An example of the genetic modification introduced into maize is MON810 (YieldGard: MaisGard in Europe) which was developed by Monsanto USA, approved in the United States in 1996, and cultivated in 1997 on approximately 1 million hectares. YieldGard maize is resistant to lepidopteran insects, particularly European corn borers, due to an insertion into the genotype of the maize that allows it to express a gene coding for the production of a synthetic cryIA(b) δ-endotoxin derived from *Bacillus thuringiensis* ssp. *Kurstaki* which is toxic to these types of insects. The insects are then repulsed from eating the modified maize, avoiding the need for spraying with a chemical insecticide.

Maize lines tolerant to imidazoline herbicides have been developed by Pioneer Hi-Bred using tissue culture selection and the chemical mutagen ethyl methanesulfonate, using a process known as mutagenesis, which is distinct from genetic engineering. Mutations, defined as any change in the base sequence of DNA, can either occur spontaneously or be induced, and both methods have produced new crop varieties. Induced mutagenesis is considered a traditional breeding technique. In this process cells are cultured in the laboratory and subjected to the influence of a strong mutagen. The Joint FAO/IAEA Mutant Variety Database (2014) lists more than 3218 cultivars of various plant species worldwide that have been developed using induced mutagenesis agents, including ionising irradiation and ethyl methane sulfonate. China and Europe account for most of the registered cultivars. Crop plants account for 75% of released mutagenic species.

There is some concern within the scientific community that the likelihood of unintended changes with mutagenesis is possibly higher than with modern, highly targeted gene manipulations. In addition to the change being sought, it is possible that other genes might be changed in an unforeseen – possibly harmful – way, yet the comprehensive approval process required of a GM maize cultivar does not always apply to cultivars produced by mutagenesis. Canada is one country that subjects new cultivars to the standard approval process whether developed by genetic modification or mutagenesis.

According to the Biotech Annual Update compiled by ISAAA (ISAAA, 2014b), GM maize represented 32% of the total area of maize grown globally in 2013. GM maize was planted in 57.4 million hectares, up from 55.1 million hectares in 2012. Seventeen countries grew GM maize in 2013, those with the highest hectarages being: USA (35.6 million), Brazil (12.9 million), Argentina (3.2 million), South Africa (2.4 million) and Canada (1.7 million hectares).

Five EU countries (Spain, Portugal, Czechia, Romania and Slovakia) planted a total of 148,000 hectares of GM maize in 2013, up from 129,000 hectares in five countries in 2012.

Globally, most of the GM maize was of 'stacked trait' type, that is cultivars containing more than one gene manipulation in the genotype so that multiple beneficial traits can be found in a single cultivar. The stacked maize category includes three combinations of traits: a double stack with insect resistance (B.t.) and herbicide tolerance (HT) (i.e. B.t./HT), a double stack with two traits for insect resistance (i.e. B.t./B.t.) and a triple stack with two types of insect resistance plus herbicide tolerance (i.e. B.t./B.t./HT).

Stacked traits continue to be an important and growing feature of biotech crops. Thirteen countries planted GM crops with two or more traits in 2013, of which 10 were developing countries. About 47 million hectares (equivalent to 27% of the 175 million hectares) were of stacked type in 2013, up from 43.7 million hectares in 2012. This trend of additional stacked traits is expected to continue.

Safety aspects

All cereal grains, whether grown conventionally, organically or using biotechnology, must meet the mandated standards of the countries in which they are consumed. It is important, therefore, to review the current evidence related to the relative safety and nutritional content of grains grown using biotechnological methods.

Currently 136 cultivars of GM maize are approved for use in the USA and 71 cultivars in the EU, following reviews and assessments of research findings related to their use in human and/or livestock diets. As noted in Chapter 2, the regulations require that any cultivar which is found to be hazardous to human health subsequent to approval and use in the marketplace has to be withdrawn from the market.

An example of a maize line that was withdrawn is StarLink, a line of B.t. (*Bacillus thuringiensis* insect resistant maize discussed above) which was developed by Aventis Crop Sciences for use in animal feed. During the mid 1990s the company submitted data to the US Environmental Agency (EPA) on the safety of StarLink and applied for approvals under the Federal Insecticide, Fungicide, and Rodenticide Act (FIFRA) and the Federal Food, Drug, and Cosmetic Act (FFDCA). After reviewing the available data, the EPA was unable to determine whether the Cry9C protein that this cultivar created to provide insect resistance was a potential human allergen. All of the other information submitted indicated that Cry9C would not pose any other types of risks to human health or the environment. Accordingly the EPA in 1998 approved StarLink for commercial use, provided that all grain derived from StarLink corn was directed to domestic animal feed or to industrial uses (e.g. biofuels) so that none would be directly consumed by humans. This was based on the possibility that some consumers might develop an allergic reaction to the B.t. protein contained in StarLink, since the protein is less rapidly digested than the protein contained in conventional maize lines. In September 2000, StarLink maize was found in food that had been prepared and distributed for human consumption, with a total of 28 consumers reporting apparent allergic reactions. The US Centers for Disease Control and Prevention (CDC) investigated the situation and concluded there was no evidence that the symptoms reported were associated with hypersensitivity to the Starlink B.t. protein (CDC, 2001). Nevertheless, the registration for the Starlink varieties of crops was voluntarily withdrawn by Aventis in 2000 and the FDA recommended that dry grain mills processing yellow maize should test for the presence of Cry9C. No verified positive test for the protein in maize has been reported since 2003.

Cancer induction

The most serious health concern related to GM-maize is the claim it causes cancer in mammalian species (Séralini et al., 2012a). The report in the journal *Food and Chemical Toxicology* resulted in a considerable amount of controversy in the scientific community and in the media, which publicised the findings worldwide. Specifically, the paper reported that male and female laboratory rats fed diets containing Roundup Ready (GM) maize and the herbicide Roundup for a prolonged period (two years) developed cancer and died at higher rates, particularly females, than rats fed a control diet. The paper suggested a link between exposure to GM maize and Roundup and an increased incidence of tumours in rats. The aim of the study was to test the hypothesis that prolonged feeding of the GM maize resulted in cancer. The data that had been used in the studies submitted for official approval of the product were from previous studies that were of a shorter duration (~90 days).

The GM maize used in the study was reported to be MON-00603-6 (a glyphosate-tolerant 'Round-up' maize named NK603) that had been authorised for food and feed use in the EU, but not for cultivation.

The GM maize and an isogenic (i.e. the same cultivar without the extra gene), non-transgenic control maize were grown under similar normal conditions in the same location, spaced at a sufficient distance to avoid cross-contamination. One field of NK603 was treated with Roundup (R) at 3l/ha (Weather-MAX, 540 g/l of glyphosate, EPA Reg. 524-537 to provide 1720 g glyphosate/hectare: a single treatment of 1800 g/hectare is used commonly on Round-up Ready maize in North America) and another field of NK603 was not treated with Roundup. Laboratory rat chow was made to contain 11, 22 or 33% of GM maize, cultivated either with or without Roundup, or 33% of the non-transgenic control line.

The Roundup added to the drinking water was the commercial formulation (GT Plus, 450 g/l of glyphosate), to provide the concentrations of Roundup shown in Table 4.2 (50 ng/l, stated to be the contaminating level of some regular tap waters; 400 mg/kg, stated to be the FDA Maximum Residue Limit of glyphosate in some GM feed) and 2.25 g/l, described as half of the minimal agricultural working concentration.

Virgin albino Sprague–Dawley rats at 5 weeks of age were used, with free access to food and water. After 20 days of acclimatisation, 100 male and 100 female animals were randomly assigned on a weight basis to provide 10 groups of equivalent starting weight. The experiment was then conducted according to guidelines approved by the Organisation for Economic Co-operation and Development, for a period of two years. The treatment layout was as shown in Table 4.2.

For each sex, one control group had access to plain water and a standard diet based on the control maize; six groups were fed the diets with 11, 22 and 33% of GM NK603 maize either treated or not with Roundup. The final three groups were fed on the control diet and had access to water supplemented with Roundup at the concentrations described above and in Table 4.2.

The animals were examined clinically twice weekly, when feed and water consumption and individual body weights were recorded. Any animals removed for ethical reasons and animals remaining at the end of the study were euthanised and their organs collected for histopathological examination.

The published data on mortality were not set out in tabular form by Séralini et al. (2012a), making interpretation difficult. Therefore we constructed Table 4.3

Table 4.2 Treatments used by Séralini et al. (2012a).

	Control	Dietary treatment, content of maize %						Water treatment, content of R			Total, *n*
		Untreated maize, g/kg			Maize treated with R, g/kg						
		110	220	330	110	220	330	50 ng/l	400 mg/kg	2.25 g/l	
Males, *n*	10	10	10	10	10	10	10	10	10	10	100
Females, *n*	10	10	10	10	10	10	10	10	10	10	100
Total	20	20	20	20	20	20	20	20	20	20	200

Table 4.3 Total mortality reported in the study conducted by Séralini et al. (2012a).

	Control	Dietary treatment, content of maize %						Water treatment, content of R				Total, *n*
		Untreated maize, g/kg			Maize treated with R, g/kg							
		110	220	330	110	220	330	0	50 ng/l	400 mg/kg	2.25 g/l	
Males, *n*	3	5	1	1	4	5	3	3	3	4	1	30
Females, *n*	2	3	7	4	4	7	4	2	5	5	4	45
Total, *n*	5	8	8	5	8	12	7	5	8	9	5	75

from the data presented, showing total mortality over the entire experimental period. Dr Séralini was contacted by us and asked to verify the table but he declined, stating that the mortality data in Table 4.3 do not show when in the two-year period each rat died. We were unable to do this based on the data presented in the Séralini et al. (2012a) report. Our constructed data, therefore, show only one aspect of mortality, that is the total number in each group that died or had to be euthanised.

Table 4.3 shows that male and female rats in all treatment groups had higher mortality than in the control group, but not in a dose-related pattern. For example, males fed the diet with the lowest content of NK603 grain which had not been treated with Roundup had higher overall mortality (5/10), while males fed the diets with the higher contents had a lower level of mortality than the controls. In males fed the diets containing NK603 sprayed with Roundup, those fed the diet containing the highest level showed the lowest mortality, similar to that of the controls. In males given Roundup in drinking water, the group given the highest level showed the lowest mortality, similar to that of the controls.

A higher level of mortality was recorded in females than in males. Mortality of females was higher in all treatment groups than in the control group, but not following a dose-related pattern. There was no evidence of increasing mortality as the concentration of Roundup was raised in the drinking water, even to a concentration that was described as half of the minimal agricultural working concentration.

Publication of the report resulted in a large number of letters to the editor of the journal in question. A summary of the points made by the authors, some of whom wrote on behalf of official societies, is shown in Table 4.4. All comments were critical of the paper, with one exception.

A common criticism of the paper was that the study had a poor design that did not permit a proper statistical analysis; an inadequate total number of animals; an inadequate number of control animals; a poor choice of animal; and an inadequate description of the dietary regimen. The reported effects of the GM maize were not dose-related, an unusual finding if the GM maize was indeed toxic. The ethical supervision of the animals was criticised in that the prolonged duration of the trial caused unnecessary suffering for publicity purposes, and was contrary to accepted guidelines in that a different set of protocols should have been used to investigate possible long-term effects. Another claim made was that the researchers displayed a bias against GM foods in the design and conduct of the study and that funding support had been received from an anti-GM agency (CRIIGEN: Committee for Research and Independent Information on Genetic Engineering), of which the lead researcher is the president, which many saw as a potential conflict of interest. Several letters requested a full disclosure of the data. Some called for the paper to be withdrawn, on the grounds that it did not meet acceptable standards for publication in a scientific journal. The journal subsequently agreed.

Table 4.4 Summary of points made in letters to the Elsevier journal *Food and Chemical Toxicology* following the publication of the paper by Séralini et al. (2012a).

Author	Critical comments	Supportive comments	Journal reference volume:page(s)
Barale-Thomas, E. (on behalf of French Society of Toxicologic Pathology, SFPT)	Conflict of interest statement incorrect: not disclosed that Séralini is President of the Scientific Board of CRIIGEN, a 'major support' of the study. Paper written in poor English. A low number of animals per group, published guidelines not followed. Critical failure in ethical supervision. Readers lack a clear understanding of the procedure followed for the pathologic examination of tissues. Data cannot be analysed adequately. Study has serious deficiencies in the protocol, the procedures and the interpretation of the results: SFPT cannot support any of the scientific claims made or any relevance to human risk assessment.		53: 473–474
Berry, C.	Design flaw: not a formal carcinogenicity study. A tumour-prone strain of rat was used, with an inadequate number of control sets. Groups too small to detect meaningful differences or to draw conclusions about tumour incidence. No comparisons of relative frequencies and total tumour burden, no adjustment for survival times, no analysis of cumulative tumour risks relative to survival duration, no analysis of time to onset of tumours and no discussion of test facility's historical control tumour incidence data. No images of tumours in control animals are shown and the illustrations in test animals appear typical for aged SD rats (though not normally allowed to live that long, on ethical grounds). The tumour histopathological data are poorly and incompletely reported, preventing serious analysis.		53: 445–446.

Dung and Ham.	The strain of rat used is known to be prone to cancer (incidence 70–95.8% during lifetime of 89–105 weeks). Thus a comparison of tumour incidence is difficult and unclear. Inappropriate experimental design: rat numbers too small for a reliable comparison; essential to have a control group for each experimental treatment. No details on composition of the different diets used. No details on levels of Roundup (R) residue in the GMO-containing diets and whether the maize used was free from mycotoxins. Data contradictory and in relation to previous findings: toxicity of a poison is concentration-dependent; however, the data indicated that lower levels of GM corn (with or without R) in the diets were more toxic than higher levels. Rats given the highest concentration of R had longer lifespan than controls and those given lower concentrations of R. Conclusions not supported by the data. Complete release/publication of data requested, for a rigorous re-evaluation. Without supporting evidence, an editorial retraction must be made.	53: 443–444
Grunewald, W. and Bury, J.	Too few negative controls, the first fundamental flaw of the study. The second and third fundamental flaws in the experimental design are: far too few animals per treatment, no correct internal standard to calibrate the number of spontaneous tumors. From 90 days onwards Sprague–Dawley rats (those used) have a high propensity to develop tumours spontaneously, also incidence of tumours increases with the age of the rats. The design needs to permit a distinction between spontaneous and treatment-related tumour incidence. Mortality was not dose-related. This paper should not have been accepted for publication in a peer-reviewed scientific journal. Public debate about the health and safety of genetically modified crops is very sensitive: the public has the right to be provided with correct scientific information.	53: 447–448.

Continued

Table 4.4 *Continued*

Author	Critical comments	Supportive comments	Journal reference volume:page(s)
Hammond et al.	Authors did not follow OECD GLP guidelines or OECD testing guideline (TG) 453 (2009). Inadequate animal numbers. Authors failed to acknowledge that mammary and pituitary tumors observed in the study are very common in untreated female SD rats fed ad libitum for 2 years. Numerous problems in the statistical analysis of this data. The authors have a history of inappropriate application of statistical methods to analyse toxicology data. Mortality was not dose related. Conclusions not supported by the data presented.		53: 459–464
Heinemann, J.A.		The study is a valuable contribution to the scientific literature, debate and process of evaluating technologies. 'Most criticisms were of a general nature and without substance worthy of entering the scientific debate. Some criticisms were specific, referring to the type of rat used, the kind of statistical analysis, and the interpretation of the response to increasing concentrations of the agrichemicals, Roundup, or genetically modified plant ingredient.' This author reported finding 7 other papers published between 2004 and 2013 in *Food and Chemical Toxicology* in which Sprague–Dawley rats were fed diets supplemented with material from GM plants. All of these papers were authored by scientists associated with the companies that developed the GM plant used in the study. None of the studies extended beyond (about) 90 days. These studies used approximately the same number of rats as in the study by Séralini et al. All of them used the same kind of rat as Séralini et al. The statistics used in these other studies passed anonymous peer-review, and this author found the statistical analysis used by Séralini et al. equally valid.	53:442

		'Where the Séralini et al. study has no peer in this group of papers is in its duration. No number of 90 day feeding studies can refute the findings of a long term study when the effects are largely those that appear after 90 days.'	
Langridge, P.	Major problems with this manuscript: absence of any statistical analysis of the data, the peculiar treatment of controls, the biased interpretation of the results (for example, the males feed high doses of GM maize appeared to show reduced tumour rates but this is ignored), failure to adequately reference other studies that have reached different conclusions. The comment 'All data cannot be shown in one report, and the most relevant are described here' is clearly unacceptable. 'I find it hard to believe that your journal is unable to support supplementary data files, particularly since these are unlikely to be substantial relative to other studies. I have joined colleagues in signing a petition to urge Dr Séralini to make all of the data publically available. It is unfortunate that the scientific community should need to initiate such a petition. Full release of data should have been a requirement of your journal.'		53: 441
Ollivier, L.	A classical test for comparing mortality rates among treatments is the Chi-square test. Use of this test on the mortality data reported shows no differences statistically significant at the 5% level. Lack of any statistical test on the tumour data.		53: 458
Panchin, A.Y.	Using a one-tailed Fishers exact test there is no statistically significant increase in mortality rates or the number of tumors in rats fed GM maize. None of the reported differences between the number of rats with specific organ pathologies in control and GM-fed groups are statistically significant. Reported tumours may occur in rats not fed GM maize: no statistical evidence that the incidence of tumours in general or any specific type of tumour is increased with GM maize.		53: 475

Continued

Table 4.4 Continued

Author	Critical comments	Supportive comments	Journal reference volume:page(s)
Pilu, R.	This author questioned whether the mycotoxin content of the diets had been monitored. Contamination of grains with mycotoxins such as aflatoxins and fumonisins is associated with serious animal and human diseases, in particular cancer.		53: 454
Sanders et al.	'We are concerned that the normally robust scientific process has been used to publicise an agenda rather than evidence – as part a campaign against GM technology.' Inadequate justification for the study in question. Length of study was inappropriate for the strain of rats used. Selected images and data only presented. Flaws found in the statistical analysis and study design. Anomalous findings. Lack of rigour in the peer-review process.		53: 450–453
Schorsch, F. (on behalf of the European Society of Toxicologic Pathology)	Methodology did not concur with the established protocols for interpreting rodent carcinogenicity studies and their relevance for human risk assessment. Mammary and pituitary tumors are common background lesions in the Sprague–Dawley rat strain after 2 years. The pathology description and conclusions of this study are unprofessional. Misinterpretations of tumours and related biological processes, misuse of diagnostic terminology; pictures are not informative and presented changes do not correspond to the narrative. The scientific rationale for acceptance of the paper for publication is questionable.		53: 465–466
de Sousa, L.	Flaws in the experimental design, statistical analysis and presentation of the data. Some crucial information for the proper understanding and full assessment of the work is lacking. The paper 'presents poor quality science and dubious ethics'.		53: 440

Tester, M	Results are refuted by numerous studies in the peer-reviewed scientific literature. Similar feeding studies by Séralini have been found to be deficient by EFSA and by the Public Research and Regulation Initiative (PRRI). Poor work. Paper should be withdrawn.	53: 457
Trewavas, A.	The company supplying the rats reported that the incidence of spontaneous tumours in the strain of rats used and receiving no special diet or treatment are identical with those reported by Séralini et al. Mortality differences are small and appear related to natural variation rather than induced response. Results reported depend totally on an inadequate number of control rodents and reflect random error. Unethical treatment of animals and use of visual presentations. Paper should be withdrawn by the authors, with an apology for misleading the public and the scientific community.	53: 449
Tribe, D	Use of Chi square analysis shows that random variation is sufficient to account for the mortality reported in this trial. Tumour numbers can similarly be assessed. Call made for a full conventional statistical assessment of the paper, and made available in the journal. Urge also that a full description of the experiments be provided. In support, this author presented an analysis of the comments of several relevant agencies and scientific societies.	53: 467–472.
Wager, R.	Design flawed: too few animals per treatment, strain of rats and feeding protocol inappropriate. Insufficient control animals. Lack of a random, unrelated diet. Inappropriate and non-standard statistical tests used. Critical details on control diet and on individual food consumption lacking. Ethical guidelines not followed. Incomplete data presented.	53:455–456

A logical interpretation of the Séralini et al. (2012a) findings is that much more detailed research is required before the claim that GM maize causes cancer can be substantiated. While the study conducted by these researchers reported cancer induction in rats fed GM maize, the incidence may have been spontaneous and unrelated to treatment. The study was too small to differentiate these separate effects, a point conceded by Séralini et al. (2012b) in a rebuttal to the criticisms. ('The variability in rates of mortality can indeed, if looked at in isolation, arise in principle by chance.'). Additionally, the toxic effects claimed were not dose-related, unusual if the GM maize (and Roundup) were indeed toxic. Another conclusion is that Séralini et al. (2012a) may have a point in suggesting that the testing period for GM foods should be longer than 90 days, but that any study designed to investigate long-term effects would have to follow an approved protocol and involve a suitable species of test animal.

EFSA (2012) issued an official response to the paper by Séralini et al. (2012a) stating that it did not meet acceptable scientific standards and there was no need to re-examine previous safety evaluations of genetically modified maize NK603. This followed separate and independent assessments carried out by EFSA and by six EU member states (Belgium, Denmark, France, Germany, Italy and the Netherlands). Member states also identified many of the same weaknesses in the methodology and design of the paper as raised by EFSA: unclear study objectives, the low number of rats used in each treatment group, a lack of detail on the feed and treatment formulation, key information missing on the statistical methods employed and incomplete endpoint reporting.

In the course of the review process, EFSA had requested that Séralini et al. (2012a) provide further information on their study documentation. No such material had reached the Authority before publication of the EFSA statement. However, on 9 November 2012, Séralini et al. (2012b) published a general reply to the reactions from across the globe to their paper. After carefully examining this reply, EFSA concluded it provided only a limited amount of relevant information that failed to address the majority of the outstanding questions raised in the Authority's first statement. In their 'Answer to critics' document', Séralini et al. (2012b) conceded the sample size of their treatment groups was too small to allow them to draw conclusions with regard to long-term carcinogenicity and mortality. EFSA noted this acknowledgement from the authors but pointed out that it was inconsistent with the overall conclusions regarding the tumours and mortality made in the original paper.

Similar reviews and rejection of the claims made by Séralini et al. (2012a) and a decision to make no change to the existing authorisation for the use of Roundup Ready Maize NK603 and the herbicide glyphosate in food production were made by Health Canada and the Canadian Food Inspection Agency (CFIA); the German Federal Institute for Risk Assessment (BfR); Food Standards Australia New Zealand (FSANZ) and the French Agence nationale de sécurité sanitaire de l'alimentation, de l'environnement et du travail (Anses). These agencies reviewed information submitted previously to the regulatory agencies as well as published scientific literature on Roundup Ready Maize NK603 and Roundup in addition to the Séralini et al. (2012a) data.

It should be noted that Séralini and co-workers claimed in previous publications (Séralini et al., 2007; Spiroux de Vendomois et al., 2009) that GM crops caused liver and kidney pathologies. The studies involved three cultivars of GM maize (NK 603, MON 810 and MON 863). However the claims have not been supported by regulatory agencies and other scientists (EFSA, 2007, 2010; FSANZ, 2009a,b; HCB, 2009; Doull et al., 2007) MacKenzie et al. (2007). He et al. (2008) and Appenzeller et al. (2009a) found no evidence of toxicity when GM maize 1507 or a related hybrid (1507 × 59122) were fed to rats.

In its response to the Séralini et al. (2012a) claims, EFSA indicated that one of the datasets used in the initial approval process for Roundup Ready Maize NK603 was described by Hammond et al. (2004). This study involved a 13-week feeding study in which Sprague–Dawley rats were fed diets containing Roundup Ready or a commercial line of isogenic (not GM) maize at a concentration of 11 or 33% (w/w). There were 400 rats in the study, divided into 10 groups of 20 rats/sex/group. Overall health, body weight, food consumption, clinical pathology parameters (hematology, blood chemistry, urinalysis), organ weights and histopathology of tissues were monitored. No significant differences in the study parameters were recorded. No health problems were recorded and all 400 animals appeared normal throughout the study. One male fed the diet containing 33% GM maize died unexpectedly on day 82, from an undiscovered cause: the animal had previously appeared healthy. No apparent cause of death was determined at necropsy, following microscopic examination of tissues. One male from a reference control group was euthanised on day 86 following an accident. Neither of these deaths was considered to be related to treatment.

More recent findings do not confirm the claim made by Séralini et al. (2012a). Hardisty et al. (2013) conducted a 13-week study in which young adult male and female Sprague–Dawley rats were fed diets containing grain from GM maize (DP-004114-3) that was either untreated (4114) or treated in the field with glufosinate ammonium (4114GLU). Control rats were fed diets containing the same concentration of near isogenic, non-GM maize grain (091) or one of three types of commercially available non-GM maize. At the end of the test period renal tubule tumours were reported in two male rats consuming diets containing the 4114 maize. Histopathological examination of the kidney tumours found that these were characteristic of amphophilic-vacuolar (AV) tumours. AV atypical tubular hyperplasia is reported to occur sporadically in young Sprague–Dawley rats. No degenerative/regenerative or cytotoxic changes consistent with nephrotoxicity leading to tumour induction were observed in the rats, supporting the conclusion that the tumours were unrelated to the test maize.

Delaney et al. (2013) conducted a similar study with GM maize 4114 (a stacked-trait cultivar with tolerance to glufosinate herbicide and resistance to insects) that was included at a level of 320 g/kg in diets fed to Sprague–Dawley rats. A control diet contained an isogenic line (091 maize). The GM maize was from crops either untreated or treated with glufosinate ammonium. This study was conducted in accordance with OECD (1998) guidelines, for a total of 91 days. No significant ($p < 0.05$) treatment differences in body weight, food consumption or clinical pathology parameters (hematology, blood chemistry, urinalysis, or organ weight) were observed.

Zhu et al. (2013) fed diets containing 12.5, 25 and 50% GM (glyphosate-tolerant) or a non-GM isogenic control in a 90-day feeding study with Sprague–Dawley rats. An additional group of rats was fed a commercial diet as a control. Body weights, food consumption, serum biochemistry, hematology and absolute and relative organ weights were compared, and gross and microscopic examination of tissues was conducted. No adverse effects related to GM maize were detected.

Thus, other than the Séralini work, all similar studies reported in peer-reviewed publications show no safety issues. No other investigators appear to have replicated the findings reported by Séralini et al. (2012a). Nevertheless, the issue of conducting a proper long-term study remains.

In an interesting development Séralini et al. (2014) republished the retracted paper in an open-access environmental journal.

Allergenicity

Several proteins in maize have been characterised as being allergens (Fonseca et al., 2012): a lipid transport protein, a thioredoxin, two glutelins and one trypsin inhibitor. Fonseca et al. (2012) used a quantitative real time RT-PCR (reverse transcription polymerase chain reaction) methodology to evaluate the expression of these five allergens during seed development in MON 810 and a non-GM counterpart. No statistically significant differences were found in the expression of the five allergens between the two maize types.

Batista et al. (2005) tested the potential allergenicity of protein extracts prepared from GM maize (MON810, B.t.11, T25 and B.t.176) and soya (Roundup Ready) samples and from non-GM control samples in two sensitive groups: children with food and inhalant allergy and individuals with asthma-rhinitis. The test used was a skin prick test. In addition the IgE immunoblot reactivity of sera from patients with a food allergy to soyabeans (Roundup Ready) and maize (MON810, B.t.11, B.t.176) samples, as well as to the pure transgenic proteins (CryIA[b] and CP4 5-enolpyruvylshikimate-3-phosphate synthase), was tested. No adverse reactions were observed and none of the individuals tested showed any detectable IgE antibodies against the pure transgenic proteins. It was concluded that the GM products tested appeared safe in terms of allergenic potential.

Residues of herbicides and insecticides

A particular concern is that the harvested grain from GM maize cultivars that are resistant to herbicides such as Roundup may contain excessive amounts of residual glyphosate. On the other hand, low (or zero) levels of insecticidal residues should be found in grain from GM crops with resistance to insect attack.

Although there is a lack of evidence to support the claim that Roundup Ready maize is more contaminated with glyphosate residues than conventional crops, the issue of possible glyphosate contamination is important to consumers.

Ragab et al. (1981) studied the levels of glyphosate and its metabolites in crops following their pre-emergence treatment with Roundup. Glyphosate at 4 kg/ha was applied on a sandy loam soil the day after sowing sweet corn, oats, beans (?*Phaseolus* sp.), peas, red beets and carrots. Analysis of the harvested crop by gas chromatography showed no residues of glyphosate within the limit of detection of 0.05 mg/kg.

Ruhland et al. (2004) studied the distribution of glufosinate and its metabolites in transgenic, glufosinate-tolerant crops of maize (*Zea mays* L. ssp. *mays*) and oilseed rape (*Brassica napus* L. var *napus*) and found no indication of an accumulation of total residues or of residue levels above the official tolerances for glufosinate.

A related concern with all cereal grains is that growing cereals might take up chemicals in the environment and deposit them in the harvested grain. These chemicals include some metals and some persistent pesticides remaining in the environment as a result of past use. It is not known whether GM maize is more or less likely to take these up than conventional maize since no comparative data appear to be available in the published literature.

The available data from agencies such as the USDA Pesticide Data Program (USDA, 2007, 2010, 2013) indicate that grains may contain pesticide residues, but that the residues are too small to be of concern in relation to human health. Many of the residues found are at the limit of detection. More studies need to be carried out to provide comparable data on residue levels in GM and conventionally grown grains.

The fact that no regulatory agency worldwide that routinely monitors pesticide levels in cereal grains (or other foods and feeds) reports their findings by genetic type, that is GM or conventional, can be interpreted as meaning that a possibly higher content of pesticide residues in GM products is not considered a concern.

Mycotoxin contamination

Mycotoxins are toxins produced by moulds. The mould infection may occur in the field or during storage of the crop and can result in the development of mycotoxins in the grain. Some mycotoxins are poisonous to humans and farm stock and have caused serious problems worldwide.

Flachowsky et al. (2007) reported a lower content of the mycotoxins zearalenone and deoxynivalenol in B.t. maize than in conventional maize, an important finding in relation to food and feed safety. Neither mycotoxin could be detected in the GM maize whereas zearalenone was found at a concentration of 3.0 μg/kg and deoxynivalenol at a concentration of 343 μg/kg in the conventional maize. These are well below the upper limit (tolerance level) set by the United States EPA. Munkvold et al. (1999) reported a lower concentration of fumonisin mycotoxin in transgenic B.t. maize than in conventional maize. These results confirm findings made by other researchers (Bakan et al., 2002; Cahagnier and Melcion, 2000; Pietri and Piva, 2000). The explanation for the lower mycotoxin contamination in B.t. maize is a greater resistance to insect attack and the subsequent field infections by *Fusarium* spp. that exploit sites created by the insect invasion (Dowd, 2000).

In a 2003 review, Munkvold concluded that the presence of transgenic insect protection in maize is more effective in reducing ear rot infection and fumonisin contamination than pesticide control. As a result B.t. maize hybrids consistently showed a much lower rate of fumonisin contamination than conventional hybrids. Some reports indicated concentrations of fumonisins as much as ten times higher in conventional than in B.t. hybrids. Other reports showed that fumonisins, Fusarium infections and ergosterol content (a measure of overall fungal invasion) in kernels were consistently and significantly lower for B.t. hybrids than conventional hybrids. Some experiments reported significantly lower concentrations of deoxynivalenol and nivalenol (another trichothecene mycotoxin) in B.t. hybrids. The efficacy of using B.t. maize hybrids to indirectly reduce aflatoxin accumulation has been investigated, but consistent reductions have not been reported.

Bowers et al. (2014) conducted field trials from 2007 to 2010 to compare grain fumonisin levels among non-B.t. maize hybrids and B.t. hybrids with transgenic protection against manual infestations of European corn borer and Western bean cutworm. They reported that B.t. hybrids experienced less insect injury, Fusarium ear rot and fumonisin contamination compared to non-B.t. hybrids and concluded that transgenically expressed B.t. proteins active against multiple lepidopteran pests can provide broad, consistent reductions in the risk of fumonisin contamination.

The available findings support the conclusion of Monkvold (2003) that the greatest potential for managing the mycotoxin problem in maize, either directly or indirectly, is by using genetic resistance. Managing the mycotoxin problem in this way should also reduce (or eliminate) the need for fungicidal treatment of the growing and stored grain.

Nutritional and organoleptic qualities

Most of the available data on cereal grains and grain products has been published from animal feeding studies since detailed information on nutrient composition is required before feeds can be formulated correctly to provide optimum productivity in the animals being fed.

There is a preponderance of evidence from these studies indicating that the approved transgenic and related non-transgenic cereal crops and their products (e.g. grain and silage) are of equivalent nutritive value (Table 4.5, 4.6).

Table 4.5 Examples of reports on the nutrient composition of GM and non-GM cultivars of maize grain.

Nutrient component g/kg unless stated	Maize type Reuter et al. (2001b) Dry matter basis		Maize type Ridley et al. (2002) Dry matter basis		Maize type Jacobs et al. (2008) As received basis			Maize type Rehout et al. (2009) Dry matter basis	
	Control Isogenic	B.t. maize	Conventional	NK603	Control isogenic	59122	Commercial	Control	MON 810
Dry matter	(870)	(872.4)	(882)	(889)	885	883.3	887.1	(909.8)	(911.4)
Crude protein	115.9	110.6	126.0	122.0	77.1	84.9	71.8	84.1	84
Crude fat	23.3	23.7	36.7	36.1	31.8	30.7	30.7	42.5	43.9
Gross energy (kcal/kg)	–	–	–	–	3923	3906	3912	–	–
Crude fibre	22.7	20.1			19.0	18.0	18.0	28.7	31.6
Ash	13.6	14.3	14.9	14.5	13.3	12.5	12.6	13.4	18.9
Organic matter	986.4	985.7	–	–	–	–	–	–	–
Nitrogen-free extractives	824.6	831.4	–	–	–	–	–	831.3	821.6
Acid detergent fibre	–	–	36.0	37.2	–	–	–	33.4	35.6
Neutral detergent fibre	–	–	100	100.6	–	–	–	125.3	133.8
Starch	720.1	733.5	–	–	–	–	–	700.3	690.5
Sugars	18.6	19.2	822.9	827.6	–	–	–	39.1	39
Non-starch polysaccharides	65.1	63.2	–	–	–	–	–	–	–
ME pig MJ/kg	15.9	15.9	–	–	–	–	–	–	–
AME poultry MJ/kg	–	–	–	–	–	–	–	14.96	14.84
Minerals									
Calcium	0.09	0.08	0.046	0.047	0.1	0	0	0.28	0.4
Phosphorus	3.11	3.13	3.6	3.6	2.4	2.7	2.3	2.4	3.38
Magnesium	1.09	1.06	1.2	1.2	–	–	–	–	–
Manganese mg/kg	–	–	65.6	64.7	–	–	–	–	–
Potassium	3.4.0	3.39	3.6	3.6	–	–	–	–	–
Copper mg/kg	–	–	19.0	17.9	–	–	–	–	–
Iron mg/kg	–	–	229.5	227.1	–	–	–	–	–
Zinc mg/kg	–	–	287.2	283.6	–	–	–	–	–

Continued

Table 4.5 Continued

Nutrient component g/kg unless stated	Maize type Reuter et al. (2001b) Dry matter basis		Maize type Ridley et al. (2002) Dry matter basis		Maize type Jacobs et al. (2008) As received basis			Maize type Rehout et al. (2009) Dry matter basis	
	Control Isogenic	B.t. maize	Conventional	NK603	Control isogenic	59122	Commercial	Control	MON 810
Amino acids (AA)									
Arginine	4.0	4.0	4.24% total AA	4.16% total AA	3.1	3.7	3.1	3.54	3.52
Cystine	2.3	2.2	2.0% total AA	2.0% total AA	1.6	1.7	1.5	1.4	1.53
Histidine	3.2	3	2.74% total AA	2.72% total AA	2.1	2.4	2.0	–	–
Isoleucine	3.9	3.8	3.8% total AA	3.87% total AA	2.6	3.0	2.3	–	–
Leucine	15.8	14.9	14.07% total AA	14.2% total AA	9.3	10.6	8.1	–	–
Lysine	2.6	2.8	2.67% total AA	2.69% total AA	2.3	2.5	2.4	2.3	2.3
Methionine	1.9	1.8	2.03% total AA	1.94% total AA	1.6	1.6	1.6	1.8	1.7
Phenyalanine	5.8	5.5	5.24% total AA	5.32% total AA	3.5	4.2	3.3	–	–
Threonine	4.1	3.9	3.33% total AA	3.37% total AA	2.6	2.8	2.5	2.75	2.73
Tryptophan	–	–	0.54% total AA	0.53% total AA	0.7	0.7	0.7	–	–
Tyrosine	4.4	4.4	3.25% total AA	3.02% total AA	1.4	2.2	1.8	–	–
Valine	5.6	5.2	4.71% total AA	4.74% total AA	3.7	4.1	3.4	–	–
Alanine	8.9	8.5	7.89% total AA	7.93% total AA	5.8	6.5	5.2	–	–
Aspartic acid	7.5	7.2	6.4% total AA	6.45% total AA	5.4	6.0	5.2	–	–
Glutamic acid	24.7	22.7	19.81% total AA	19.84% total AA	14.5	16.5	13.2	–	–
Glycine	3.7	3.5	3.51% total AA	3.49% total AA	2.8	3.1	2.7	–	–
Proline	9.8	9.4	8.96% total AA	8.88% total AA	6.8	8.1	6.7	–	–
Serine	5.6	5.2	4.86% total AA	4.87% total AA	3.4	3.9	3.2	–	–
Fatty acids (FA)									
Arachidic	–	–	0.37% total FA	0.36% total FA	–	–	–	–	–
Behenic	–	–	0.16% total FA	0.16% total FA	–	–	–	–	–
Eicosenoic	–	–	0.3% total FA	0.29% total FA	–	–	–	–	–
Linoleic	50.2% total FA	50.9% total FA	64.26% total FA	64.62% total FA	–	–	–	–	–
Linolenic	1.0% total FA	1.0% total FA	1.11% total FA	1.11% total FA	–	–	–	–	–
Oleic	30.4% total FA	29.6% total FA	23.06% total FA	22.4% total FA	–	–	–	–	–
Palmitic	13.2% total FA	12.7% total FA	8.89% total FA	9.13% total FA	–	–	–	–	–
Stearic	2.2% total FA	2.0% total FA	1.83% total FA	1.92% total FA	–	–	–	–	–

Others									
Arabinose	16.0	14.4	–	–	–	–	–	–	–
Glucose	20.6	21.7	–	–	–	–	–	–	–
Xylose	20.2	19.1	–	–	–	–	–	–	–
Phytic acid	–	–	10	9.7	–	–	–	–	–
Trypsin inhibitor TIU/mg	–	–	2.67	3.16	–	–	–	–	–
Vitamin E mg/g	–	–	0.009	0.0088	–	–	–	–	–
Ferulic acid	–	–	2.0	2.0	–	–	–	–	–
p-coumaric acid	–	–	0.15	0.16	–	–	–	–	–
Raffinose	–	–	1.3	1.3	–	–	–	–	–

Table 4.6 Summary of reported findings on the nutritional composition of GM and non-GM cultivars of cereal grains.

Food/Feedstuff	Reference	Reported findings
		Maize
B.t. maize	Aeschbacher et al., 2002	Nutrient composition of the two cultivars similar.
B.t. maize	Aeschbacher et al., 2005	Nutrient composition of the two cultivars similar.
B.t. maize	Aulrich et al., 2001	Nutrient composition of two grain cultivars and silage samples similar in all investigated components including proximate nutrients, amino acids, fatty acids, minerals and non-starch polysaccharides.
B.t. maize, maize silage	Barrière et al., 2001	No significant difference in the analytical composition of the two cultivars of maize.
Glyphosate-tolerant maize	Berger et al., 2002	Nutrient composition of the two cultivars similar.
Glufosinate-tolerant maize	Böhme et al., 2001	Nutrient composition of the two cultivars similar.
B.t. maize	Brake and Vlachos, 1998	No significant difference in the analytical composition of the two cultivars of maize.
B.t. maize	Brake et al., 2003	Nutrient composition of the two cultivars similar.
VIP3 Maize	Brake et al., 2005	Higher protein content in the control cultivar.
B.t. maize	Bressner et al., 2003	No significant difference in the analytical composition of the two cultivars of maize.
B.t. maize	Chrastinová et al., 2002	No significant difference in the nutritional composition or utilization of the two cultivars of maize.
Glyphosate resistant maize	Chrenková et al., 2002	Nutritional composition of the two cultivars similar.
B.t. maize, maize silage	Daenicke et al., 1999	No significant differences in nutrient composition between silage sources or between grain sources.
B.t. maize silage	Donkin et al., 2000	No significant differences in nutrient composition between silage sources.
B.t. maize silage	Donkin et al., 2003	No significant differences in nutrient composition between silage sources.
B.t. maize grain, glyphosate-tolerant maize	Donkin et al., 2003	No significant differences in nutrient composition between silage sources or between grain sources within an experiment.
Glyphosate-tolerant maize	Erickson et al., 3003	No significant differences in nutrient composition.
Glyphosate-tolerant maize	Fischer et al., 2002	No significant differences in nutrient composition.
B.t. maize	Fischer et al., 2003	No significant differences in nutrient composition.
B.t. maize, glyphosate-tolerant maize	Gaines et al., 2001a	No significant differences in nutrient composition between GM and conventional maize types.

Table 4.6 Continued

Food/Feedstuff	Reference	Reported findings
	Maize	
B.t. maize, glyphosate-tolerant maize	Gaines et al. 2001b	No significant differences in digestible energy between GM and conventional maize types.
B.t. maize, glyphosate-tolerant maize	Gaines et al. 2001c	No significant differences in nutritive value between GM and conventional maize types.
Glyphosate-tolerant maize-silage and grain	Grant et al., 2002	No significant differences in nutritive value between GM and conventional maize types.
B.t. + glyphosate-tolerant maize grain, silage	Grant et al., 2003	No significant differences in nutritive value between GM and conventional maize types.
Glyphosate-tolerant maize silage and grain	Grant et al., 2003	No significant differences in nutritive value between GM and conventional maize types.
B.t. maize	Halle et al., 1998	No significant differences in nutritive value between GM and conventional maize types.
B.t., glufosinate-resistant maize.	He et al., 2008	No significant differences in nutritive value between GM and conventional maize types.
B.t.-resistant, glufosinate-resistant maize.	Herman et al., 2007	No significant differences in nutritive value between GM and conventional maize forage and grain in terms of proximate constituents, fibre and minerals, amino acids, fatty acids, vitamins, secondary metabolites and anti-nutrients.
Glyphosate-tolerant maize	Hyun et al., 2004	No significant differences in nutritive value.
Glyphosate-tolerant maize, maize silage	Ipharraguerre et al., 2002	No significant differences in nutritive value between GM and conventional maize types.
B.t. maize	Masoero et al., 1999	Reduced fibre content in GM maize, levels of other nutrients not significantly different between maize types.
B.t. maize	Mireles et al., 2000	No significant differences in nutritive value between maize types.
B.t. maize	Petty et al., 2001a	No significant differences in nutritive value between maize types.
B.t. maize	Piva et al., 2001	No significant differences in nutritive value between maize types.

Continued

Table 4.6 Continued

Food/Feedstuff	Reference	Reported findings
	Maize	
B.t. maize	Řehout et al., 2009	No significant differences in nutritional composition between maize types.
B.t. maize	Reuter et al., 2001a	Levels of starch, sugar, non-starch polysaccharides (NSP), amino acids, fatty acids, and selected minerals not significantly different.
B.t. maize	Rossi et al., 2003	No significant differences in nutritive value between maize types, although level of phenylalanine higher in GM maize.
B.t. maize	Russell et al., 2000	No significant differences in nutritive value between maize types.
Glyphosate-tolerant maize	Sidhu et al., 2000	No significant differences in nutritive value between maize types.
Glyphosate-tolerant maize	Simon et al., 2002.	No significant differences in nutritional composition between maize types.
Glyphosate-tolerant maize	Stanisiewski et al., 2001	No significant differences in nutritive value.
B.t., glufosinate-resistant maize.	Stein et al., 2009	Nutritional value of the GM and non-GM near-isoline control maize similar.
B.t.-maize	Steinke et al. 2010	No significant differences in the chemical composition and estimated net energy content of the two cultivars.
Glyphosate-tolerant maize	Taylor et al., 2001a	No significant differences in nutritional composition between maize types.
Glyphosate-tolerant maize	Taylor et al., 2001b	No significant differences in nutritional composition between maize types.
B.t. maize	Taylor et al., 2002a	No significant differences in nutritional composition between maize types.
Glyphosate-tolerant maize	Taylor et al., 2002b	No significant differences in nutritional composition between maize types.
B.t. maize, glyphosate-tolerant maize	Taylor et al., 2003a	No significant differences between GM and conventional cultivars.
B.t. maize, glyphosate-tolerant maize	Taylor et al., 2003b	No significant differences between GM and conventional cultivars.
B.t. maize, glyphosate-tolerant maize	Taylor et al., 2003c	No significant differences between GM and conventional cultivars.
B.t.-maize	Tony et al., 2003	No significant differences in nutritive value between GM and conventional cultivars.
B.t.-maize	Venneria et al., 2008	No significant differences in fatty acid content, unsaponifiable fraction of antioxidants, total phenols, polyphenols, carotenoids, vitamin C, total antioxidant activity, and mineral composition.

Table 4.6 Continued

Food/Feedstuff	Reference	Reported findings
		Maize
Rice		
Glufosinate-tolerant rice	Choi et al. 2012	Two GM cultivars compared with conventional rice. No significant differences in levels of proximates, fatty acids, amino acids, minerals, vitamins, and antinutrients. Some small differences in a few fatty acids, minerals, and trypsin inhibitor content.
Glufosinate-tolerant rice	Cromwell et al., 2005.	GM and non-GM cultivars had similar nutritional composition.
Glufosinate-tolerant rice	Oberdoerfer et al., 2005	Two cultivars had similar nutritional composition. Most values for nutritional components fell within the range of values reported for commercial rice.
Wheat		
Wheat (various)	Baker et al., 2006	The metabolome was found to be affected by environmental conditions. Differences between the control and GM lines were generally within the same range as the differences observed between the control lines grown at different sites and in different years.
Glyphosate-tolerant wheat	Kan and Hartnell, 2004	Two cultivars had similar nutritional composition.
Glyphosate-tolerant wheat	Obert et al., 2004	Levels of proximates, amino acids, fatty acids, minerals, vitamins, secondary metabolites, and antinutrients in two types of wheat forage similar, grain also equivalent to commercial wheat cultivars.
Glyphosate-tolerant wheat	Peterson et al., 2003	Two cultivars had similar nutritional composition.

These studies have included maize, rice and wheat. The data presented in several of the reports are quite detailed.

The nutritional data are not exactly similar for GM and non-GM crops since environmental and soil conditions have an important influence on composition (Baker et al., 2006). However the data on comparability are within an acceptable biological range.

Confirmatory data on nutritional equivalence are shown in Table 4.7, which summarises the results of feeding diets containing transgenic and related non-transgenic food and feed products on the growth and productivity of farm stock and laboratory animals. In general, these results indicate similar growth and productivity when fed transgenic or non-transgenic food and feed sources.

Table 4.7 Reported effects of dietary inclusion of conventional or GM cereal feedstuffs on the productivity of animals, poultry and fish.

Animal species	Reference	GM feedstuff	Reported effects
		Meat/lab animals	
Beef cattle	Aulrich et al., 2001	B.t. maize	No significant difference in nutrient digestibility.
Beef cattle	Berger et al., 2002	Herbicide-tolerant maize	No significant differences in growth performance.
Beef cattle	Berger et al., 2003	B.t. maize	No significant differences in growth performance or body score.
Beef cattle	Daenicke et al., 1999	B.t. maize silage	No significant differences in digestibility, weight gain or dry matter intake.
Beef cattle	Erickson et al., 2003	Glyphosate-tolerant maize	No significant differences in growth performance.
Beef cattle	Folmer et al., 2002a	B.t. maize, maize residues, maize silage	No significant differences in growth performance.
Beef cattle	Hendrix et al., 2000	B.t. maize silage, maize residues	No significant differences in growth performance of cows, improved growth performance of steer calves.
Beef cattle	Kerley et al., 2001	B.t. maize	No significant differences in growth performance.
Beef cattle	Petty et al., 2001b	Glyphosate-tolerant maize, silage	No significant differences in growth performance.
Beef cattle	Russell et al., 2000	B.t. maize, crop residues	No significant differences in growth performance.
Beef cattle	Russell et al., 2002	B.t. maize, crop residues	No significant differences in growth performance.
Beef cattle	Simon et al., 2002.	Herbicide-tolerant maize	No significant differences in growth performance.
Beef cattle	Van der Pol et al., 2002	Corn rootworm protected maize	No significant differences in growth performance.
Sheep	Aulrich et al., 2001	B.t. maize	No significant difference in nutrient digestibility.
Sheep	Barriere et al., 2001	B.t. maize, maize silage	No significant differences in growth performance.
Sheep	Daenicke et al., 1999	B.t. maize silage	No significant differences in digestibility, weight gain or dry matter intake.
Pigs	Böhme and Alrich, 1999	Herbicide-tolerant maize	No significant effects on nutrient digestibility.
Pigs	Bressner et al., 2003	B.t. maize	No significant difference in growth performance.
Pigs	Cromwell et al., 2005	Glufosinate-tolerant rice	No significant differences in feed intake or growth performance.

Table 4.7 Continued

Animal species	Reference	GM feedstuff	Reported effects
		Meat/lab animals	
Pigs	Custodio et al., 2006	B.t. maize	No significant difference in growth rate but feed conversion efficiency better with control maize. P digestibility higher with control maize.
Pigs	Fischer et al., 2002	Herbicide-tolerant maize	No significant difference in growth performance.
Pigs	Fischer et al., 2003	B.t. maize	No significant differences in growth performance.
Pigs	Gaines et al., 2001b	Herbicide-tolerant maize	No significant difference in energy digestibility.
Pigs	Hyun et al., 2004	Herbicide-tolerant maize	No significant difference in growth performance.
Pigs	Peterson et al., 2008	Glyphosate-tolerant wheat	No significant effect of GM or near isogenic cultivar or 4 commercial types on growth performance.
Pigs	Piva et al., 2001	B.t. maize	Significantly higher weight gain with B.t. maize (lower mycotoxin contamination), no significant effect on feed:gain ratio.
Pigs	Reuter et al., 2001b	B.t. maize	No significant difference in nutrient digestibility, dry matter intake or weight gain.
Pigs	Reuter et al., 2002a	B.t. maize	Digestibility of all measured parameters similar. Metabolizable energy contents similar.
Pigs	Stanisiewski et al., 2001	B.t. maize	No significant effect on growth performance.
Pigs	Stein et al., 2004	B.t. maize	No significant difference in growth performance.
Pigs	Stein et al., 2009	Double-stacked maize	No significant difference in growth performance.
Pigs	Świątkiewicz et al., 2013	Glyphosate-tolerant soyabean meal and B.t. maize	No significant effects on growth or reproductive performance.
Pigs	Weber and Reichert, 2001	B.t. maize	No significant differences in dry matter intake or weight gain. No significant difference in growth performance.
Broilers	Aeschbacher et al., 2001	B.t. maize	No significant difference in growth performance or ME value.

Continued

Table 4.7 Continued

Animal species	Reference	GM feedstuff	Reported effects
		Meat/lab animals	
Broilers, laying hens	Aeschbacher et al., 2005	B.t. maize	No significant effects of diet on metabolic data or growth or laying performance.
Broilers	Aulrich et al., 2001	B.t. maize	No significant difference in digestibility.
Broilers	Brake and Vlachos, 1998	B.t. maize	No significant differences in growth performance or livability. Feed:gain ratio significantly lower with B.t. maize.
Broilers	Brake et al., 2003	B.t. maize	No significant difference in growth performance.
Broilers	Gaines et al., 2001a	B.t. maize	No significant difference in growth performance.
Broilers	Gaines et al., 2001b	Glyphosate-tolerant maize	No significant difference in growth performance.
Broilers	Halle et al., 1998	B.t. maize	No significant difference in growth performance
Broilers	Kan and Hartnell, 2004	Glyphosate-tolerant wheat	No significant difference in growth performance, but lower carcass yield with control wheat.
Broilers	Mireles et al., 2000	B.t. maize	No significant difference in growth performance or ME value.
Broilers	Piva et al., 2001	B.t. maize	Significantly higher weight gain with B.t. maize (lower mycotoxin contamination), no significant effect on feed:gain ratio.
Broilers	Reuter et al., 2001b	B.t. maize	Improved growth rate with B.t. maize.
Broilers	Řehout et al., 2009	B.t. maize	No significant effect on growth performance.
Broilers	Sidhu et al., 2000	Glyphosate-tolerant maize	No significant effect on growth performance.
Broilers	Taylor et al., 2001a	B.t. maize, glyphosate-tolerant maize	No significant difference in growth performance.
Broilers	Taylor et al., 2001a	Glyphosate-tolerant maize	No significant difference in growth performance.
Broilers	Taylor et al., 2002a	B.t. maize	No significant difference in growth performance.
Broilers	Taylor et al., 2002b	Glyphosate-tolerant maize	No significant difference in growth performance.
Broilers	Taylor et al., 2003a	B.t. maize, glyphosate-tolerant maize	No significant effect on growth performance.

Table 4.7 Continued

Animal species	Reference	GM feedstuff	Reported effects
		Meat/lab animals	
Broilers	Taylor et al., 2003b	B.t. maize, glyphosate-tolerant maize	No significant effect on growth performance.
Broilers	Taylor et al., 2003c	B.t. maize, glyphosate-tolerant maize	No significant effect on growth performance.
Broilers	Taylor et al., 2004	B.t. maize	No significant differences in growth performance.
Broilers	Taylor et al., 2005	B.t. maize + Gt maize	No significant differences in growth performance.
Broilers	Tony et al., 2003	B.t. maize	No significant differences in growth performance.
Broilers	Yonemochi et al., 2002	B.t. maize	Growth performance in starter phase better with GM cultivar.
Laying hens	Daenicke et al., 1999	B.t. maize	No significant difference in laying performance.
Laying hens	Aeschbacher et al., 2001	B.t. maize	No significant difference in laying performance or ME value.
Laying hens	Aulrich et al., 2001	B.t. maize	No significant difference in nutrient digestibility.
Quail	Halle et al., 2004	B.t. maize	No significant difference in growth or layer performance.
Rabbits	Chrastinová et al., 2002	B.t. maize	No significant difference in growth performance.
Rats	Appenzeller et al., 2009	1507x59122 maize, insect-resistant and tolerant to glufosinate.	No toxicologically significant differences observed in nutritional performance variables.
Rats	Chrenková et al., 2002	Glyphosate-tolerant maize	Similar values for protein efficiency ratio, nitrogen digestibility and net protein utilisation.
Rats	Delaney et al. 2013	4114 maize insect-resistant and tolerant to glufosinate.	No biologically significant, treatment related differences in body weight, feed consumption, clinical pathology parameters (hematology, blood chemistry, urinalysis, or organ weight) were observed.
Rats	He et al., 2008	Maize tolerant to glufosinate and insects.	No significant differences in body weight and feed utilisation were observed between rats consuming diets formulated with 59122 and 091 Control corn. Statistical differences were observed in certain hematology and serum chemistry response variables between rats consuming diets formulated with 59122 or 091.

Continued

Table 4.7 Continued

Animal species	Reference	GM feedstuff	Reported effects
		Meat/lab animals	
Rats	MacKenzie et al., 2007	B.t. maize grain with pat gene	No significant differences observed in nutritional performance variables.
Milk			
Dairy cows	Calsalmiglia et al., 2007	B.t., herb tolerant maize silage	No significant difference in milk production.
Dairy cows	Donkin et al., 2000	Herbicide-tolerant maize, maize silage	No significant differences in dry matter intake or milk production.
Dairy cows	Donkin et al., 2003	B.t. maize, glyphosate tolerant maize silage	No significant differences in dry matter intake or milk production.
Dairy cows	Faust and Miller, 1997	B.t. maize chopped	No significant differences in dry matter intake or milk production.
Dairy cows	Faust and Spangler, 2000	B.t. maize silage, maize silage	No significant difference in milk production.
Dairy cows	Folmer et al., 2000	B.t. maize silage, maize silage	No significant differences in dry matter intake or milk production.
Dairy cows	Grant et al., 2002	Herbicide-tolerant maize, maize silage	Intake of maize silage and milk production reduced significantly (ascribed to silage quality).
Dairy cows	Grant et al., 2002	Glyphosate-tolerant maize -silage and grain	Lower dry matter intake of cows fed GM diet, resulting in a lower milk production.
Dairy cows	Grant et al., 2003	Corn rootworm protected corn -silage and grain	No significant difference in milk production.
Dairy cows	Ipharraguerre et al., 2002	Herbicide-tolerant maize, maize silage	No significant difference in milk production.
Dairy cows	Rutzmoser et al., 1999	B.t. maize silage	No significant differences in dry matter intake or milk production.
Dairy cows	Steinke et al. 2010	B.t.-maize, maize silage	Similar dry matter intake and milk yield.

Several studies indicate better growth and productivity with diets containing insect-tolerant GM grain than non-GM grain and this has been attributed to a lower content of mycotoxins in the insect-protected grain. Mycotoxin contamination of the feed is known to have a deleterious effect on animal productivity (Flachowsky et al., 2005).

Cultivars of maize with improved nutritional characteristics have been developed. For instance, Tang et al. (2013) conducted an assessment of transgenic lysine-rich maize (Y642) compared to conventional maize (Nongda 108). The Y642 line contained the sb401 gene, encoding a lysine-rich protein designed to improve the nutritional value of the maize. Its expression resulted in increased levels of lysine and total protein in the maize kernels. The contents of total protein, lysine, some other amino acids, several minerals and riboflavin in Y642 were significantly higher than those in conventional maize (Table 4.8). The content of water-soluble protein was increased by 60% in the GM maize and the salt-soluble protein fraction was about 20% of that in the conventional maize. The content of G2-glutelin protein was also significantly higher in the GM maize. In other respects the grains were similar in nutritive value.

The overall conclusion that can be drawn from the available data is that GM and conventional maize lines are of similar nutritive value, except in those cases where the GM line has an enhanced nutritional feature such as increased lysine content.

No peer-reviewed studies appear to have been conducted on the organoleptic qualities of GM maize relative to conventionally grown maize.

Other cereal crops

Rice (*Oryza sativa*)

Rice plants can be modified genetically to be herbicide resistant, resist pests, improve grain size, and to have to have an enhanced content of nutrients (Sharma et al., 2009). A GM rice cultivar (LL62, herbicide-resistant) has been approved in the USA, but is not yet in commercial production. An approval application for the food and feed use of LL62 rice has been submitted to the EU.

Currently, seven lines of GM rice have been approved internationally; however, no large-scale commercial production of GM rice is taking place in any country. Reports on the comparative nutritional composition of GM and conventional cultivars are summarised in Tables 4.6 and 4.7.

Other GM lines are being developed, with the aim of enhancing the nutritional value of rice. Rice is known to contain insufficient amounts of iron and vitamin A. To the extent that populations that are heavily reliant on rice as a staple suffer from deficiencies of these nutrients, improvements in the content of iron and vitamin A would be expected to be beneficial and rice might be the easiest carrier for these components in these countries. A GM rice with an enhanced content of beta-carotene, a precursor of vitamin A in the diet, has been developed. It also has an enhanced content of iron, of potential importance since the World Health Organization has stated that iron deficiency affects 30% of the world's population. Due to its yellow colour, it has been called Golden

Table 4.8 Mean composition of GM (high-lysine) and conventional maize, dry-matter basis (from Tang et al., 2013).

	Maize cultivar	
Component, g/kg Unless stated	**GM (Y642)**	**Conventional (Nongda 108)**
Dry matter	896.0	898.0
Crude protein	107.8.*	90.6*
Crude fibre	6.7	5.6
Ether extract	53.3	49.3
Ash	13.6	13.8
Carbohydrate**	714.6	737.1
Minerals		
Calcium	0.131*	0.253*
Copper	0.0131	0.0140
Iron	0.165	0.125
Magnesium	9.40	8.90
Manganese	0.0345*	0.0376*
Phosphorus (total)	21.2	20.5
Potassium	26.3	25.6
Sodium	0.265	0.200
Zinc	0.154*	0.125*
Amino acids g/kg		
Histidine	2.98	2.66
Isoleucine	4.34	3.46
Leucine	8.83	7.25
Lysine	3.40*	2.50*
Methionine	1.65	1.47
Cystine	3.87	3.54
Phenylalanine	9.67	8.05
Tyrosine	4.74	4.34
Threonine	8.53	7.07
Valine	5.22	4.42
Vitamins mg/kg		
Carotene (total)	2.1	3.1
Thiamin	0.739	0.792
Riboflavin	1.41*	1.23*
Vitamin E	23.2	24.1

*Significantly different at $p < 0.05$.

**Carbohydrate calculated as 100 – (water + ash + fat + crude protein).

Rice. The plan is to make it available free of charge to small-scale farmers in developing countries. Current research is focused on obtaining levels of beta-carotene and iron to be sufficiently high to have a real impact on the diet of people consuming the GM rice instead of traditional white rice.

Wheat (*Triticum aestivum*)

The first herbicide tolerant wheat produced through genetic engineering was developed by Monsanto, MON 71800 known commercially as Roundup Ready wheat. A gene from a common soil bacterium *Agrobacterium tumefaciens* strain CP4 was introduced into the wheat to produce a glyphosate-tolerant wheat line. This line has been approved for production in the USA but is currently not being cultivated.

BASF released the first herbicide-tolerant wheat in 2007 in Canada (Clearfield wheat), the product of non-GM mutation breeding developed to survive the presence of the imidazolinone herbicides that block the activity of acetohydroxyacid synthase (AHAS). AHAS is the first enzyme in the biosynthetic pathway of the branched amino acids essential for plant growth. Reports on the comparative nutritional composition of GM and conventional cultivars of wheat are summarised in Tables 4.6 and 4.7.

Several companies have plans to develop wheat lines with features such as resistance to fungal diseases; increased lysine content; lower incidence of mycotoxins; resistance to viruses; and salt-tolerance; drought tolerance; and reduction in gluten content.

Summary

Several GM cultivars of cereal crops have been developed, but maize is the only GM cereal crop being grown internationally in commercially significant quantities. The main producing countries are the USA, Brazil, Argentina, Canada and India. Some countries have approved GM cultivars of maize for importation only and not for cultivation. All of the approved cultivars have received clearance by regulatory authorities, based on safety tests and on substantial equivalence with conventional lines of maize. Globally, about 47 million hectares (equivalent to 27% of the 175 million hectares) were of stacked type in 2013.

A large body of data published in peer-reviewed scientific journals shows that no safety concerns are evident when GM lines of maize are included in the diet of test animals. Reports from one laboratory of the possibility of cancer induction as a result of consuming GM maize have been discredited.

The available data indicate that cereal grains contain chemical and pesticide residues, but that the amounts are too small to be of concern in relation to human health. Many of the residues found are at the limit of detection. No separate datasets on residue levels in GM grains are currently publically available: more studies need to be carried out to provide comparable data on GM and conventionally grown grains. The published data indicate a lower level of mycotoxins in GM maize possessing any of the traits for insect resistance.

The overall conclusion that can be drawn from the available data is that GM and conventional cereal crops are of similar nutritive value, except in those cases where the GM cultivar has an enhanced nutritional feature such as increased lysine content. No studies appear to have been conducted on the organoleptic qualities of GM maize relative to conventionally grown maize or other cereal crops.

References

Aeschbacher, K., Meile, L., Messikommer, R. and Wenk, C. (2001). Genetically modified maize in diets for chickens and laying hens: influence on performance and product quality. *Proceedings: International Symposium on Genetically Modified Crops and Co-products as Feeds for Livestock*, Nitra, Slovak Republic: pp. 41–42.

Aeschbacher, K., Meile, L., Messikommer, R. and Wenk, C. (2002). Influence of genetically modified maize on performance and product quality of chickens. *Proceedings Society of Nutritional Physiology* 11: 196.

Aeschbacher, K., Messikommer, R., Meile, L. and Wenk, C. (2005). B.t.176 corn in poultry nutrition: physiological characteristics and fate of recombinant plant DNA in chickens. *Poultry Science* 84: 385–394

Appenzeller, L. M., Malley, L., Mackenzie, S. A., et al. (2009). Subchronic feeding study with genetically modified stacked trait lepidopteran and coleopteran resistant (DAS-Ø15Ø7-1xDAS-59122-7) maize grain in Sprague-Dawley rats. *Food and Chemical Toxicology* 47: 1512–1520.

Aulrich, K., Böhme, H., Daenicke, R., Halle, I. and Flachowsky, G. (2001). Genetically modified feeds in animal nutrition. 1st communication: Bacillus thuringiensis (B.t.) corn in poultry, pig and ruminant nutrition. *Archives of Animal Nutrition* 54: 183–195.

Bakan, B., Melcion, D., Richard-Molard, C. and Cahagnier, B. (2002). Fungal growth and fusarium mycotoxin content in isogenic traditional maize and genetically modified maize grown in France and Spain. *Journal of Agricultural and Food Chemistry* 50: 728–731.

Baker, J. M., Hawkins, N. D., Ward, J. L., et al. (2006). A metabolomic study of substantial equivalence of field-grown genetically modified wheat. *Plant Biotechnology Journal* 4: 381–392.

Barrière, Y., Verite, R., Brunschwig, P., et al. (2001) Feeding value of corn silage estimated with sheep and dairy cows is not altered by genetic incorporation of B.t.176 resistance to Ostrinia nubilalis. *Journal of Dairy Science* 84: 1863–1871.

Batista, R., Nunes, B., Carmo, M., et al. (2005). Lack of detectable allergenicity of transgenic maize and soya samples. *Journal of Allergy and Clinical Immunology* 116: 403–410.

Berger, L. L., Robbins, N. D., Sewell, J. R. and Stanisiewski, E. P. (2002). Effect of feeding diets containing corn grain with Roundup (event GA21 or NK603), control, or conventional varieties on steer feedlot performance and carcass characteristics. *Journal of Animal Science* 80 (Suppl. 1): 270.

Berger, L.L., Robbins, N.D., Sewell, J.R., et al. (2003). Effect of feeding diets containing corn grain with corn rootworm protection (event MON 863), control, or conventional varieties on steer feedlot performance and carcass characteristics. *Journal of Animal Science* 81 (Suppl.1): 214–215 Abstr.M150.

Böhme, H. and Aulrich, K. (1999). Ingredients and digestibility of transgenic sugar beets and corn in comparision to the isogenic varieties in the case of pigs. *VDLUFA Conference Proceedings 1999, 111th VDLUFA Conference*, 13–17 September 1999, Halle/Saale: pp. 289–292.

Böhme, H., Aulrich, K., Daenicke, R. and G. Flachowsky (2001). Genetically modified feeds in animal nutrition. Second communication: Glufosinate tolerant sugar beets (roots and silage) and maize grains for ruminants and pigs. *Archives of Animal Nutrition* 54: 197–207.

Bowers, E., Hellmich, R. and Munkvold, G. (2014). Comparison of fumonisin contamination using HPLC and ELISA methods in B.t. and near-isogenic maize hybrids infested with European corn borer or western bean cutworm. *Journal of Agricultural and Food Chemistry* 62: 6463–6472.

Brake, J. and Vlachos, D. (1998). Evaluation of transgenic event 176 "B.t." corn in broiler chickens. *Poultry Science* 77: 648–653.

Brake, J., Faust, M. and Stein, J. (2003). Evaluation of transgenic event B.t.11 hybrid corn in broiler chickens. *Poultry Science* 82: 551–559.

Brake, J., Faust, M. and Stein, J. (2005). Evaluation of transgenic hybrid corn (VIP3a) in broiler chickens. *Poultry Science* 84: 503–512.

Bressner, G. E., Hyun, Y., Stanisiewski, E. P., et al. (2003). Performance comparison of growing-finishing pigs fed diets containing Corn Root Worm Protected corn (Event MON 863) or conventional corn hybrids. *Journal of Animal Science* 81 (Suppl. 1): 207 Abstract M119.

Brookes, G. (2008). The impact of using genetically modified insect resistant maize in Europe since 1998. *International Journal of Biotechnology* 10: 148–166.

Cahagnier, B. and Melcion, D. (2000): Mycotoxins de Fusarium dans les mais – grains a la recolte: relation entre la presence d'insects (pyrale, sesamie) et la teneur en mycotoxins. *Proceedings 6th International Feed Production Conference*, 27–28 November, Piacenca, Italy: pp. 237–324.

Calsamiglia, S., Hernandez, B., Hartnell, G. F. and Phipps, R. (2007). Effects of corn silage derived from a genetically modified variety containing two transgenes on feed intake, milk production, and composition, and the absence of detectable transgenic deoxyribonucleic acid in milk in Holstein dairy cows. *Journal of Dairy Science* 90: 4718–4723.

CDC, National Center for Environmental Health (2001). Investigation of Human Health Effects Associated with Potential Exposure to Genetically Modified Corn: A Report to the U.S. Food and Drug Administration from the Centers for Disease Control and Prevention. Atlanta, GA: Centers for Disease Control and Prevention.

Choi, H., Moon, J.-K., Park, B.-S., et al. (2012). Comparative nutritional analysis for genetically modified rice, Iksan483 and Milyang204, and nontransgenic counterparts. *Journal of the Korean Society for Applied Biological Chemistry* 55: 19–26.

Chrastinová, L., Sommer, A., Rafay, J., et al. (2002). Genetically modified maize in diets for rabbits – influence on performance and product quality. *Proceedings of Society of Nutrition Physiology* 11: 195.

Chrenková, M., Sommer, A., Čerešňáková, Z., et al. (2002). Nutritional evaluation of genetically modified maize corn performed on rats. *Archives of Animal Nutrition (Archiv fur Tiernahrung)* 56: 229–235.

Cromwell, G., Henry, B., Scott, A., et al. (2005). Glufosinate herbicide-tolerant -liberty link- rice vs. conventional rice in diets for growing-finishing swine. *Journal of Animal Science* 83: 1068–1074.

Custodio, M., Powers, W., Huff-Lonergan, E., et al. (2006). Growth, pork quality, and excretion characteristics of pigs fed B.t. corn or non-transgenic corn. *Canadian Journal of Animal Science* 86 (4): 461–469.

Daenicke, R., Aulrich, K. and Flachowsky, G. (1999). GMO in animal feedstuffs: nutritional properties of B.t.-maize unaffected. *Mais* 27 (September): 135–137.

Delaney, B., Karaman, S., Roper, J., et al. (2013). Thirteen week rodent feeding study with grain from molecular stacked trait lepidopteran and coleopteran protected (DP-ØØ4114-3) maize. *Food and Chemical Toxicology* 53: 417–427.

Donkin, S., Velez, J., Stanisiewski, E. and Hartnell, G. (2000). Effect of feeding Roundup Ready® corn silage and grain on feed intake, milk production and milk composition in lactating dairy cattle. *Journal of Dairy Science* 83 (Supplement 1): 273.

Donkin, S., Velez, J., Stanisiewski, E. and Hartnell, G. (2003). Effects of feeding silage and grain from glyphosate-tolerant or insect-protected corn hybrids on feed intake, ruminal digestion, and milk production in dairy cattle. *Journal of Dairy Science* 86: 1780–1788.

Doull, J., Gaylor, D., Greim, H. A., et al. (2007). Report of an Expert Panel on the reanalysis by Séralini et al. of a 90-day study conducted by Monsanto in support of the safety of a genetically modified corn variety (MON 863). *Food and Chemical Toxicology* 45: 2073–2085.

Dowd, P. F. (2000). Indirect reduction of ear moulds and associated mycotoxins in Bacillus thuringiensis corn under controlled and open field conditions: utility and limitations. *Journal of Economic Entomology* 93: 1669–1679.

EFSA (2007). EFSA reaffirms its risk assessment of genetically modified maize MON863. Available from: http://www.efsa.europa.eu/en/press/news/gmo070628.htm (accessed February 2015).

EFSA (2010). EFSA response to de Vendomois et al. Available from: http://www.efsa.europa.eu/en/events/event/gmo100127-m.pdf (accessed February 2015).

EFSA (2012). Review of the Séralini et al. (2012) publication on a 2-year rodent feeding study with glyphosate formulations and GM maize NK603 as published online on 19 September 2012 in Food and Chemical Toxicology. *EFSA Journal* 10: 2910.

Erickson, G. E., Robbins, N. D., Simon, J. J., et al. (2003). Effect of feeding glyphosate-tolerant (Roundup-Ready events GA21 or nk603) corn compared with reference hybrids on feedlot steer performance and carcass characteristics. *Journal of Animal Science* 81: 2600–2608.

Faust, M. A. and Spangler, S. M. (2000). Nutritive value of silages from MON810 B.t. and non-B.t. near-isogenic corn hybrids. *Journal of Dairy Science* 83: 1184 Abstract.

Fischer, R. L., Lewis, A. J., Miller, P. S., et al. (2002). Comparison of swine performance when fed diets containing Roundup Ready corn (event NK603), control or conventional corn grown during 2000 in Nebraska. *Journal of Animal Science* 80 (Suppl.1): 224.

Fischer, R. L., Miller, P. S., Hyun, Y., et al. (2003). Comparison of swine performance when fed diets containing corn root worm protected corn, parental line corn, or conventional corn grown during 2000 in Nebraska. *Journal of Animal Science* 81 (Suppl. 1): 207 Abstract M118.

Flachowsky, G., Chesson A., and Aulrich, K. (2005) Animal nutrition with feeds from genetically modified plants. *Archives of Animal Nutrition* 59: 1–40.

Flachowsky G., Aulrich, K, Bohme, H. and Halle, I. (2007). Studies on feeds from genetically modified plants (GMP) – Contributions to nutritional and safety assessment. *Animal Feed Science and Technology* 133: 2–30.

Folmer, J. D., Grant, R. J., Milton, C. T. Beck, J. F. (2000). Effect of B.t. corn silage on short term lactational performance and ruminal fermentation in dairy cows. *Journal of Dairy Science* 83: 1182 Abstract 272.

Fonseca, C., Planchon, S., Renaut, J., et al. (2012). Characterization of maize allergens—MON810 vs. its non-transgenic counterpart. *Journal of Proteomics* 75: 2027–2037.

Faust, M. and Miller, L. (1997). Study finds no B.t. in milk. Fall Special Livestock Edition, Iowa State University Extension, Ames, Iowa: 6–7.

FSANZ (2009a). Feeding studies and GM corn MON863. Available from: http://www.foodstandards.gov.au/scienceandeducation/factsheets/factsheets/feedingstudiesandgmc5604.cfm (accessed February 2015).

FSANZ (2009b). A comparison of the effects of three GM corn varieties on mammalian health. *International Journal of Biological Science* 5: 706–726.

Gaines, A. M., Allee, G. L. and Ratliff, B. W. (2001a). Nutritional evaluation of B.t. (MON810) and Roundup Ready® corn compared with commercial hybrids in broilers. *Poultry Science* 80 Suppl. 1: 51 Abstract 214.

Gaines, A. M., Allee, G. L. and Ratliff, B. W. (2001b). Swine digestible energy evaluations of B.t. (MON810) and Roundup Ready corn compared to commercial varieties. *Journal of Animal Science* 79 (Suppl.1)/*Journal of Dairy Science* 84 (Suppl. 10): 109.

Gaines, A. M., Allee, G. L. and Ratliff, B. W. (2001c). Nutritional evaluation of B.t. (MON810) and Roundup Ready corn compared to commercial hybrids in broilers. *Journal of Animal Science* 79 (Suppl.1)/*Journal of Dairy Science* 84 (Suppl. 10): 51.

GMO Compass (2014). Global cultivation area: Maize. http://www.gmo-compass.org/eng/agri_biotechnology/gmo_planting/341.genetically_modified_maize_global_area_under_cultivation.html (accessed 12 June 2014).

Grant, R. J., Kleinschmit, D., Sparks, A. L., et al. (2002). Influence of glyphosate-tolerant (trait NK603) corn silage and grain on feed consumption and milk production in Holstein dairy cattle. *Journal of Dairy Science* 85(Suppl. 1): 384 Abstract 1540.

Grant, R., Fanning, K., Kleinschmit, D., et al. (2003). Influence of glyphosate-tolerant -event NK603- and corn rootworm protected -event MON863- corn silage and grain on feed consumption and milk production in holstein cattle. *Journal of Dairy Science* 86: 1707–1715.

Halle, I., Aulrich, K. and Flachowsky, G. (1998). Einsatz von Maiskörnen der Sorte Cesar und des gentechnisch veränderten B.t.-Hybriden in der Broiler mast. *Proceedings 5. Tagung, Schweine- und Geflhgelernährung*, Wittenberg, Deutschland: pp. 265–267.

Hammond, B., Dudek, R., Lemen, J. and Nemeth, M. (2004). Results of a 13 week safety assurance study with rats fed grain from glyphosate tolerant corn. *Food and Chemical Toxicology* 42: 1003–1014.

Hardisty, J. F., Banas, D. A., Gopinath, C., et al. (2013). Spontaneous renal tumors in two rats from a thirteen week rodent feeding study with grain from molecular stacked trait lepidopteran and coleopteran resistant (DP-ØØ4114-3) maize. *Food and Chemical Toxicology* 53: 428–431.

HCB (2009). French High Council of Biotechnologies response to de Vendomois as translated by UK ACNFP (Advisory Committee on Novel Foods and Processes). Available from: http://www.food.gov.uk/multimedia/pdfs/acnfp9612a2.

He, X.Y., Huang, K.L., Li, X., et al. (2008). Comparison of grain from corn rootworm resistant transgenic DAS-59122-7 maize with non-transgenic maize grain in a 90-day feeding study in Sprague-Dawley rats. *Food and Chemical Toxicology* 46: 1994–2002.

Hendrix, K. S., Petty, A. T. and Lotgreen, D. L. (2000). Feeding value of whole plant silage and crop residues from B.t. or normal corns. *Journal of Animal Science* 78 (Suppl.1): 273.

Herman, R.A., Scherer, P.N., Phillips, A.M., et al. (2010). Safe composition levels of transgenic crops assessed via a clinical medicine model. *Biotechnology Journal* 5: 172–182.

Herman, R. A., Storer, N. P., Phillips, A. M., et al. (2007). Compositional assessment of event DAS-59122-7 maize using substantial equivalence. *Regulatory Toxicology and Pharmacology* 47: 37–47.

Hyun, Y., Bressner, G. E., Ellis, M., et al. (2004). Performance of growing-finishing pigs fed diets containing Roundup Ready corn (event NK603), a nontransgenic genetically similar corn, or conventional corn lines. *Journal of Animal Science* 82: 571–580.

Ipharraguerre, I., Younker, R., Clark, J., et al. (2002). Performance of lactating dairy cows fed glyphosate-tolerant corn -event NK603. *Journal of Dairy Science* 85 (Supplement 1): 358.

ISAAA (2014a). Global Status of Commercialized Biotech/GM Crops: 2013. ISAAA Brief No. 46-2013. ISAAA: Ithaca, New York. http://www.isaaa.org/resources/publications/briefs/46/executivesummary/ (accessed 2 May 2014).

ISAAA (2014b). Biotech Crop Annual Updates: Maize. ISAAA: Ithaca, New York. www.isaaa.org/resources/publications/biotech_crop_annual_update/download/biotech-crop-annual-update-corn.pdf (accessed February 2015).

ISAAA (2014c). GM approval database. ISAAA: Ithaca, New York. http://www.isaaa.org/gmapprovaldatabase/default.asp (accessed 27 October 2014).

Jacobs, C. M., Utterback, P. L., Parsons, C. M., et al. (2008). Performance of Laying hens fed diets containing DAS-59122-7 maize grain compared with diets containing nontransgenic maize grain. *Poultry Science* 87: 475–479.

Kan, C. and Hartnell, G. (2004). Evaluation of broiler performance when fed roundup-ready wheat (event mon71800), control, and commercial wheat varieties. *Poultry Science* 83: 1325–1334.

Kerley, M. S., Felton, E. E. D., Lehmkuhler, J. W. and Shillito, R. (2001). B.t. corn that is genetically modified to prevent insect damage is equal to conventional corn in feeding value for beef cattle. *Journal of Animal Science* 7 (Supplement 2): 98 Abstract 301.

MacKenzie, S. A., Lamb, I., Schmidt, J., et al. (2007). Thirteen week feeding study with transgenic maize grain containing event DAS-Ø15Ø7-1 in Sprague-Dawley rats. *Food and Chemical Toxicology* 45: 551–562.

Masoero, F., Moschini, M., Rossi, F., et al. (1999). Nutritive value, mycotoxin contamination and in vitro rumen fermentation of normal and genetically modified corn (Cry1A(b)) grown in Northern Italy. *Maydica* 44: 205-209.

Mireles, A., Kim, S., Thompson, R. and Amudsen, B. (2000). GMO (B.t.) corn is similar in composition and nutrient availability to broilers as non-GMO corn. *Poultry Science* 79 (Suppl.1): 65.

Munkvold, G. P. (2003). Cultural and genetic approaches to managing mycotoxins in maize. *Annual Review of Phytopathology* 41: 99–116.

Munkvold, G. P., Hellmich, R. L., Rice, L. G. (1999). Comparison of fumonisin concentrations in kernels of transgenic B.t.-maize hybrids and nontransgenic hybrids. *Plant Diseases* 83: 130–138.

Oberdoerfer, R. B., Shillito, R. D., de Beuckeleer, M. and Mitten, D. H. (2005). Rice (Oryza sativa L.) containing the bar gene is compositionally equivalent to the nontransgenic counterpart. *Journal of Agricultural and Food Chemistry* 53: 1457–1465.

Obert, J. C., Ridley, W. P., Schneider, R. W., et al. (2004). The composition of grain and forage from glyphosate tolerant wheat MON 71800 is equivalent to that of conventional wheat (*Triticum aestivum* L.). *Journal of Agricultural and Food Chemistry* 52: 1375–1384.

Peterson, B. A., Hyun, Y., Stanisiewski, E. P., et al. (2003). A comparison of swine performance when fed diets containing Roundup Ready® wheat (event MON 71800) and conventional wheat varieties. *Journal of Animal Science* 81 (Suppl. 1): 207–208 Abstract M120.

Peterson, B. A., Hyun, Y., Stanisiewski, E. P., et al. (2008). Performance of growing-finishing pigs fed diets containing Roundup Ready wheat (MON 71800), a non-transgenic genetically similar wheat, or conventional wheat varieties. *Animal* 2: 1602–1609.

Petty, A. T., Hendrix, K. S., Stanisiewski, E. P. and Hartnell, G. F. (2001a). Feeding value of B.t. corn grain compared with its parental hybrid when fed in beef cattle finishing diets. *Journal of Animal Science* 79 (Suppl. 2):102 Abstract 320.

Petty, A.T., Hendrix, K. S., Stanisiewski, E. P. and Hartnell, G. F. (2001b). Performance of beef cattle fed Roundup Ready corn harvested as whole plant silage or grain. *Journal of Animal Science* 79 (Suppl. 2):102 Abstract 321.

Pietri, A. and Piva, G. (2000). Occurrence and control of mycotoxins in maize grown in Italy. In: *Proceedings of the Sixth International Feed Production Conference*, November 27–28, Piacenza, Italy: pp. 226–236.

Piva, G., Morlacchini, M., Pietri, A., et al. (2001). Performance of broilers and piglets Fed B.t. corn. *Proceedings International Symposium on Genetically Modified Crops and Co-products as Feeds for Livestock*, September 2001, Nitra, Slovak Republic: pp. 27–30.

Ragab, M. T. H., Abdelkader, M. H. T. and Stiles, D. A. (1981). Glyphosate and its metabolite residues in crops following their pre-emergence treatment with Roundup. Annual report 1980, Agriculture Canada Research Station, Kentville, Nova Scotia: pp. 233–235.

Řehout, V., Kadlec, J., Čítek, J., et al. (2009). The influence of genetically modified B.t. maize MON 810 in feed mixtures on slaughter, haematological and biochemical indices of broiler chickens. *Journal of Animal and Feed Sciences* 18: 490–498.

Reuter, T., Aulrich, K., Berk, A. and Flachowsky, G. (2001a). Nutritional Evaluation of B.t.-Corn in Pigs. *Journal of Animal Science* 79: 1073.

Reuter, T., Aulrich, K., Berk. A. and Flachowsky, G. (2001b). Influence of B.t.-corn on growing and slaughtering results of pigs. *Proceedings of the Society of Nutrition Physiology* 10: 150.

Reuter, T., Aulrich, K., Berk, A. and Flachowsky, G. (2002a). Investigations on Genetically Modified Maize (B.t.-Maize) in Pig Nutrition: Chemical Composition and Nutritional Evaluation. *Archives of Animal Nutrition* (Archiv Fur Tierernahrung) 56: 23–31.

Reuter, T., and Aulrich, K. (2003). Investigations on genetically modified maize (B.t.-maize) in pig nutrition: fate of feed ingested foreign dna in pig bodies. *European Food Research and Technology* 216: 185–192.

Ridley, W. P., Sidhu, R. S., Pyla, P. D., et al. (2002). Comparison of the nutritional profile of glyphosate-tolerant corn event nk603 with that of conventional corn (*Zea mays L.*). *Journal of Agricultural and Food Chemistry* 50: 7235–7243.

Rossi, F., Moschini, M., Fiorentini, L., et al. (2003). Analytical composition and rumen degradability of isogenic and transgenic corn varieties. *Journal of the Science of Food and Agriculture* 83: 1337–1341.

Ruhland, M., Engelhardt, G. and Pawlizki, K. (2004). Distribution and metabolism of D/L-, L- and D-glufosinate in transgenic, glufosinate-tolerant crops of maize (*Zea mays* L. ssp. *mays*) and oilseed rape (*Brassica napus* L. var *napus*). *Pest Management Science* 60: 691–696.

Russell, J. R., Farnham, D., Berryman, R. K., et al. (2000). Nutritive value of the crop residues from bt-corn hybrids and their effects on performance of grazing beef cows. 2000 Beef Research Report, Iowa State University, Ames: pp. 56–61.

Russell, J. R., Hersom, M. J., Haan, M. M., et al. (2002). Effects of Grazing Crop Residues of bt-Corn Hybrids on Performance of Pregnant Beef Cows (2002). Beef Research Report, 2001 Paper 11. http://lib.dr.iastate.edu/beefreports_2001/11.

Rutzmoser, K., Meyer, J. and Obermaier, A. (1999). Verfütterung von Silomais der Sorten "Pactol" und "Pactol CB" (gentechnisch veränderte B.t.-Hybride) an Milchkühe. Bodenkultur und Pflanzenbau, Schriftenreihe der Bayer. Landesanstalt für Bodenkultur und Pflanzenbau 3: pp. 25–34.

Séralini, G-E., Cellier, D. and de Vendômois, J.S. (2007). New analysis of a rat feeding study with a genetically modified maize reveals signs of hepatorenal toxicity. *Archives of Environmental Contamination and Toxicology* 52: 596–602.

Séralini, G-E., Clair, E., Mesnage, R., et al. (2012a) Long term toxicity of a Roundup herbicide and a Roundup-tolerant genetically modified maize. *Food and Chemical Toxicology* 50: 4221–4231. (paper retracted by the journal).

Séralini, G-E., Mesnage, R., Defarge, N., et al. (2012b). Answers to critics: Why there is a long term toxicity due to NK603 Roundup-tolerant genetically modified maize and to a Roundup herbicide. *Food and Chemical Toxicology* 53: 476–483.

Séralini, G. E., Clair, E., Mesnage, R., et al. (2014). Republished study: long-term toxicity of a Roundup herbicide and a Roundup-tolerant genetically modified maize. *Environmental Sciences Europe* 26: 14.

Sharma, A. K. and Sharma, M. K. (2009). Plants as bioreactors: recent developments and emerging opportunities. *Biotechnology Advances* 27: 811–832.

Sidhu, R. S., Hammond, B. G., Fuchs, R. L., et al. (2000). Glyphosate-tolerant corn: the composition and feeding value of grain from glyphosate-tolerant corn is equivalent to that of conventional corn (*Zea mays* L.). *Journal of Agricultural and Food Chemistry* 48: 2305–2312.

Simon, J. J., Vander Pol, K. J., Erickson, G. E., et al. (2002). Effect of Roundup Ready® corn (event NK603) on performance in beef feedlot diets. *Journal of Animal Science* 80 (Suppl. 1): 46 Abstract 179.

Stanisiewski, E. P., Hartnell, G. F. and Cook, D. R. (2001). Comparison of swine performance when fed diets containing Roundup Ready® corn (GA21), parental line or conventional corn. *Journal of Animal Science* 79 (Suppl. 1): 319-320 Abstract 1322.

Stein, H. H., Rice, D. W., Smith, B. L., et al. (2009). Evaluation of corn grain with the genetically modified input trait DAS-59122-7 fed to growing-finishing pigs. *Journal of Animal Science* 87: 1254–1260.

Stein, H. H., Sauber, T., Rice, D., et al. (2004). Comparison of corn grain from biotech and non-biotech counterparts for grow-finish pig performance. *Journal of Animal Science* 82 (Suppl. 1): 328–329 Abstract W65.

Steinke, K., Guertler, P., Paul, V., et al. (2010). Effects of long-term feeding of genetically modified corn (event MON810) on the performance of lactating dairy cows. *Journal of Animal Physiology and Animal Nutrition* 94: 185–193.

Świątkiewicz, M., Bednarek, D., Markowski, J., et al. (2013). Effect of feeding genetically modified maize and soybean meal to sows on their reproductive traits, haematological indices and offspring performance. *Bulletin Veterinary Institute, Pulawy, Poland* 57: 413–418.

Tang, M., He, X., Luo, Y., et al. (2013). Nutritional assessment of transgenic lysine-rich maize compared with conventional quality protein maize. *Journal of the Science of Food and Agriculture* 93: 1049–1054.

Taylor, M. L., Hartnell, G. F., Nemeth, M. A., et al. (2001a). Comparison of broiler performance when fed diets containing YieldGard® and Roundup Ready® corn, parental lines, or commercial corn. *Poultry Science* 80 (Suppl. 1): 319 Abstract 1321.

Taylor, M. L., Hartnell, G. F., Nemeth, M. A., et al. (2001b). Comparison of broiler performance when fed diets containing Roundup Ready® corn event NK603, parental line, or commercial corn. *Poultry Science* 80 (Suppl. 1): 320 Abstract 1323.

Taylor, M. L., Hartnell, G. F., Astwood, J. D., et al. (2002a). Comparison of broiler performance when diets containing MON863 (Corn Rootworm Protection), non-transgenic control, or commercial corn. *Poultry Science* 81 (Suppl. 1): 95 Abstract 406.

Taylor, M. L., Hartnell, G. F., Astwood, J. D., et al. (2002b). Comparison of broiler performance when fed diets containing YieldGard® (MON810) x Roundup Ready® (NK603), nontransgenic control, or commercial corn. *Poultry Science* 81 (Suppl. 1): 95 Abstract 405.

Taylor, M., Hartnell, G., Riordan, S., et al. (2003a). Comparison of broiler performance when fed diets containing grain from Roundup Ready® - nk603, Yieldgard® x Roundup Ready® - Mon810 x NK603, non-transgenic control, or commercial corn. *Poultry Science* 82: 443–453.

Taylor, M., Hartnell, G., Riordan, S., et al. (2003b). Comparison of broiler performance when fed diets containing grain from Yieldgard® MON810, Yieldgard® X Roundup Ready® - GA21, nontransgenic control, or commercial corn. *Poultry Science* 82: 823–830.

Taylor, M., Hyun, Y., Hartnell, G., et al. (2003c). Comparison of broiler performance when fed diets containing grain from YieldGard® Rootworm - MON863, YieldGard® Plus - MON810 x MON863, nontransgenic control, or commercial reference corn hybrids. *Poultry Science* 82: 1948–1956.

Taylor, M. L., George, B., Hyun, Y., et al. (2004). Comparison of broiler performance when fed diets containing insect-protected (MON 88017 or MON 88017 x MON 810), control, or commercial corn. *Poultry Science* 83 (Suppl. 1): 322 Abstract W39.

Taylor, M., Hartnell, G., Nemeth, M., et al. (2005). Comparison of broiler performance when fed diets containing corn grain with insect-protected (corn rootworm and European corn borer) and herbicide-tolerant (glyphosate) traits, control corn, or commercial reference corn – revisited. *Poultry Science* 84: 1893–1899.

Tony, M., Butschke, A., Broll, H., et al. (2003). Fate of DNA from isogenic and transgenic (B.t. 176) maize after feeding to broilers (Abbau von DNA aus isogenem und transgenem (B.t. 176) Mais in Broilern). *Archives of Animal Nutrition* 57: 235–252.

Vander Pol, K. J., Erickson, G. E., Macken, C. N., et al. (2002). Effect of corn root worm protected corn (event MON863) on performance in beef feedlot diets. *Journal of Animal Science* 80 (Suppl. 1): 46 Abstract 180.

Venneria, E., Fanasca, S., Monastra, G., et al. (2008). Assessment of the nutritional values of genetically modified wheat, corn, and tomato crops. *Journal of Agricultural and Food Chemistry* 56: 9206–9214.

USDA (2007). US Department of Agriculture Pesticide Data Program. Annual Summary, Calendar Year 2006. http://www.ams.usda.gov/pdp (accessed 28 April 2013).

USDA (2010). US Department of Agriculture Pesticide Data Program. Annual Summary, Calendar Year 2008. http://www.ams.usda.gov/pdp (accessed 28 April 2013).

USDA (2013). US Department of Agriculture Pesticide Data Program. Annual Summary, Calendar Year 2011. http://www.ams.usda.gov/pdp (accessed 28 April 2013).

de Vendômois, J. S., Roullier, F., Cellier D. and Séralini, G. E. (2009). A comparison of the effects of three GM corn varieties on mammalian health. *International Journal of Biological Sciences* 5: 706–726.

Weber, T. E. and Richert, B. T. (2001). Grower-finisher growth performance and carcass characteristics including attempts to detect transgenic plant DNA and protein in muscle from pigs fed genetically modified "BT" corn. *Journal of Animal Science* 79 (Suppl. 2): 67 (Abstract).

Yonemochi, C., Fujisaki, H., Harada, C., et al. (2002). Evaluation of transgenic event CBH 351 (Starlink) corn in broiler chicks. *Animal Science Journal* 73: 221–228.

Zhu, Y., He, X., Luo, Y., et al. (2013). A 90-day feeding study of glyphosate-tolerant maize with the G2-aroA gene in Sprague-Dawley rats. *Food and Chemical Toxicology* 51: 280–287.

Appendix Table 4.1A Approved lines of GM maize (total 136), rice (total 7) and wheat (total 1), all countries (ISAAA 2014c). Reproduced with permission of ISAAA.

Maize		
Name	**Code**	**Trade name**
32138	DP-32138-1	32138 SPT maintainer
3272	SYN-E3272-5	Enogen
3272 x B.t.11	SYN-E3272-5 x SYN-BT011-1	not available
3272 x B.t.11 x GA21	SYN-E3272-5 x SYN-BT011-1 x MON-00021-9	not available
3272 x B.t.11 x MIR604	SYN-E3272-5 x SYN-BT011-1 x SYN-IR604-5	not available
3272 x BT11 x MIR604 x GA21	SYN-E3272-5 x SYN-BT011-1 x SYN-IR604-5 x MON-00021-9	not available
3272 x GA21	SYN-E3272-5 x MON-00021-9	not available
3272 x MIR604	SYN-E3272-5 x SYN-IR604-5	not available
3272 x MIR604 x GA21	SYN-E3272-5 x SYN-IR604-5 x MON-00021-9	not available
33121	DP-033121-3	not available
4114	DP-004114-3	not available
5307	SYN-05307-1	Agrisure Duracade
5307 x MIR604 x B.t.11 x TC1507 x GA21	SYN-05307-1 x SYN-IR604-5 x SYN-BT011-1 x DAS-01507-1 x MON-00021-9	Agrisure Duracade 5122
5307 x MIR604 x B.t.11 x TC1507 x GA21 x MIR162	SYN-05307-1 x SYN-IR604-5 x SYN-BT011-1 x DAS-01507-1 x MON-00021-9 x SYN-IR162-4	Agrisure Duracade 5222
59122	DAS-59122-7	Herculex RW
59122 x GA21	DAS-59122-7 x MON-00021-9	not available
59122 x MIR604	DAS-59122-7 x SYN-IR604-5	not available
59122 x MIR604 x GA21	DAS-59122-7 x SYN-IR604-5 x MON-00021-9	not available
59122 x MIR604 x TC1507	DAS-59122-7 x SYN-IR604-5 x DAS-01507-1	not available
59122 x MIR604 x TC1507 x GA21	DAS-59122-7 x SYN-IR604-5 x DAS-01507-1 x MON-00021-9	not available
59122 x MON810	DAS-59122-7 x MON-00810-6	not available
59122 x MON810 x NK603	DAS-59122-7 x MON-00810-6 x MON-00603-6	not available
59122 x MON88017	DAS-59122-7 x MON-88017-3	not available
59122 x NK603	DAS-59122-7 x MON-00603-6	Herculex RW Roundup Ready
59122 x TC1507 x GA21	DAS-59122-7 x DAS-01507-1 x MON-00021-9	not available
676	PH-000676-7	not available
678	PH-000678-9	not available
680	PH-000680-2	not available
98140	DP-098140-6	Optimum GAT

Appendix Table 4.1A Continued

Maize		
Name	**Code**	**Trade name**
98140 x 59122	DP-098140-6 x DAS-59122-7	not available
98140 x TC1507	DP-098140-6 x DAS-01507-1	not available
98140 x TC1507 x 59122	DP-098140-6 x DAS-01507-1 x DAS-59122-7	not available
B.t.10	not available	B.t.10
B.t.11 (X4334CBR, X4734CBR)	SYN-BT011-1	Agrisure CB/LL
B.t.11 x 59122	SYN-BT011-1 x DAS-59122-7	not available
B.t.11 x 59122 x GA21	SYN-BT011-1 x DAS-59122-7 x MON-00021-9	not available
B.t.11 x 59122 x MIR604	SYN-BT011-1 x DAS-59122-7 x SYN-IR604-5	not available
B.t.11 x 59122 x MIR604 x GA21	SYN-BT011-1 x DAS-59122-7 x SYN-IR604-5 x MON-00021-9	not available
B.t.11 x 59122 x MIR604 x TC1507	SYN-BT011-1 x DAS-59122-7 x SYN-IR604-5 x DAS-01507-1	not available
BT11 x 59122 x MIR604 x TC1507 x GA21	SYN-BT011-1 x DAS-59122-7 x SYN-IR604-5 x DAS-01507-1 x MON-00021-9	Agrisure 3122
B.t.11 x 59122 x TC1507	SYN-BT011-1 x DAS-59122-7 x DAS-01507-1	not available
B.t.11 x 59122 x TC1507 x GA21	SYN-BT011-1 x DAS-59122-7 x DAS-01507-1 x MON-00021-9	not available
B.t.11 x GA21	SYN-BT011-1 x MON-00021-9	Agrisure GT/CB/LL
B.t.11 x MIR162	SYN-BT011-1 x SYN-IR162-4	Agrisure Viptera 2100
B.t.11 x MIR162 x GA21	SYN-BT011-1 x SYN-IR162-4 x MON-00021-9	Agrisure Viptera 3110
BT11 x MIR162 x MIR604	SYN-BT011-1 x SYN-IR162-4 x SYN-IR604-5	Agrisure Viptera 3100
B.t.11 x MIR162 x MIR604 x GA21	SYN-BT011-1 x SYN-IR162-4 x SYN-IR604-5 x MON-00021-9	Agrisure Viptera 3111, Agrisure Viptera 4
B.t.11 x MIR162 x TC1507	SYN-BT011-1 x SYN-IR162-4 x DAS-01507-1	not available
B.t.11 x MIR162 x TC1507 x GA21	SYN-BT011-1 x SYN-IR162-4 x DAS-01507-1 x MON-00021-9	Agrisure Viptera 3220
B.t.11 x MIR604	SYN-BT011-1 x SYN-IR604-5	Agrisure CB/LL/RW
BT11 x MIR604 x GA21	SYN-BT011-1 x SYN-IR604-5 x MON-00021-9	Agrisure 3000GT
B.t.11 x MIR604 x TC1507	SYN-BT011-1 x SYN-IR604-5 x DAS-01507-1	not available
B.t.11 x TC1507	SYN-BT011-1 x DAS-01507-1	not available
B.t.11 x TC1507 x GA21	SYN-BT011-1 x DAS-01507-1 x MON-00021-9	not available

Continued

Appendix Table 4.1A Continued

Maize		
Name	**Code**	**Trade name**
B.t.176 (176)	SYN-EV176-9	NaturGard KnockOut, Maximizer
BVLA430101	not available	not available
CBH-351	ACS-ZM004-3	Starlink Maize
DAS40278	DAS-40278-9	Enlist Maize
DAS40278 x NK603	DAS-40278-9 x MON-00603-6	not available
DBT418	DKB-89614-9	B.t. Xtra Maize
DLL25 (B16)	DKB-89790-5	not available
GA21	MON-00021-9	Roundup Ready Maize, Agrisure GT
GA21 x MON810	MON-00021-9 x MON-00810-6	Roundup Ready YieldGard Maize
GA21 x T25	MON-00021-9 x ACS-ZM003-2	not available
HCEM485	HCEM485	not available
LY038	REN-00038-3	Mavera Maize
LY038 x MON810	REN-00038-3 x MON-00810-6	Mavera YieldGard Maize
MIR162	SYN-IR162-4	Agrisure Viptera
MIR162 x GA21	SYN-IR162-4 x MON-00021-9	not available
MIR162 x MIR604	SYN-IR162-4 x SYN-IR604-5	not available
MIR162 x MIR604 x GA21	SYN-IR162-4 x SYN-IR604-5 x MON-00021-9	not available
MIR162 x TC1507	SYN-IR162-4 x DAS-01507-1	not available
MIR162 x TC1507 x GA21	SYN-IR162-4 x DAS-01507-1 x MON-00021-9	not available
MIR604	SYN-IR604-5	Agrisure RW
MIR604 x GA21	SYN-IR604-5 x MON-00021-9	Agrisure GT/RW
MIR604 x NK603	SYN-IR604-5 x MON-00603-6	not available
MIR604 x TC1507	SYN-IR604-5 x DAS-01507-1	not available
MON801 (MON80100)	MON801	not available
MON802	MON-80200-7	not available
MON809	PH-MON-809-2	not available
MON810	MON-00810-6	YieldGard, MaizeGard
MON810 x MON88017	MON-00810-6 x MON-88017-3	YieldGard VT Triple
MON832	not available	Roundup Ready Maize
MON863	MON-00863-5	YieldGard Rootworm RW, MaxGard
MON863 x MON810	MON-00863-5 x MON-00810-6	YieldGard Plus
MON863 x MON810 x NK603	MON-00603-6 x MON-00810-6 x MON-00863-5	YieldGard Plus with RR
MON863 x NK603	MON-00863-5 x MON-00603-6	YieldGard RW + RR

Appendix Table 4.1A Continued

Maize		
Name	**Code**	**Trade name**
MON87411	MON-87411-9	not available
MON87427	MON-87427-7	Roundup Ready Maize
MON87427 x MON89034 x MON88017	MON-87427-7 x MON-89034-3 x MON-88017-3	not available
MON87427 x MON89034 x NK603	MON-87427-7 x MON-89034-3 x MON-00603-6	not available
MON87427 x MON89034 x TC1507 x MON88017 x 59122	MON-87427-7 x MON-89034-3 x DAS-01507-1 x MON-88017-3 x DAS-59122-7	not available
MON87460	MON-87460-4	Genuity DroughtGard
MON87460 x MON89034 x MON88017	MON-87460-4 x MON-89034-3 x MON-88017-3	not available
MON87460 x MON89034 x NK603	MON-87460-4 x MON-89034-3 x MON-00603-6	not available
MON87460 x NK603	MON-87460-4 x MON-00603-6	not available
MON88017	MON-88017-3	YieldGard VT Rootworm RR2
MON89034	MON-89034-3	YieldGard VT Pro
MON89034 x 59122	MON-89034-3 x DAS-59122-7	not available
MON89034 x 59122 x MON88017	MON-89034-3 x DAS-59122-7 x MON-88017-3	not available
MON89034 x MON88017	MON-89034-3 x MON-88017-3	Genuity VT Triple Pro
MON89034 x NK603	MON-89034-3 x MON-00603-6	Genuity VT Double Pro
MON89034 x TC1507	MON-89034-3 x DAS-01507-1	not available
MON89034 x TC1507 x 59122	MON-89034-3 x DAS- 01507-1 x DAS-59122-7	not available
MON89034 x TC1507 x MON88017	MON-89034-3 x DAS-01507-1 x MON-88017-3	not available
MON89034 x TC1507 x MON88017 x 59122	MON-89034-3 x DAS-01507-1 x MON-88017-3 x DAS-59122-7	Genuity SmartStax
MON89034 x TC1507 x MON88017 x 59122 x DAS40278	MON-89Ø34-3 x DAS-Ø15Ø7-1 x MON-88017-3 x DAS-59122-7 x DAS-40278-9	not available
MON89034 x TC1507 x MON88017 x DAS40278	MON-89034-3 x DAS-01507-1 x MON-88017-3 x DAS-59122-7 x DAS-40278-9	not available
MON89034 x TC1507 x NK603	MON-89034-3 x DAS-01507-1 x MON-00603-6	Power Core
MON89034 x TC1507 x NK603 x DAS40278	MON-89034-3 x DAS-01507-1 x MON-00603-6 x DAS-40278-9	not available
MS3	ACS-ZM001-9	InVigor Maize
MS6	ACS-ZM005-4	InVigor Maize
NK603	MON-00603-6	Roundup Ready 2 Maize

Continued

Appendix Table 4.1A Continued

Name	Code	Trade name
Maize		
NK603 x MON810 x 4114 x MIR 604	MON-00603-6 x MON-00810-6 x DP004114-3 x SYN-IR604-4	not available
NK603 x MON810	MON-00603-6 x MON-00810-6	YieldGard CB + RR
NK603 x T25	MON-00603-6 x ACS-ZM003-2	Roundup Ready Liberty Link Maize
T14	ACS-ZM002-1	Liberty Link Maize
T25	ACS-ZM003-2	Liberty Link Maize
T25 x MON810	ACS-ZM003-2 x MON-00810-6	Liberty Link Yieldgard Maize
TC1507	DAS-01507-1	Herculex I, Herculex CB
TC1507 x 59122 x MON810 x MIR604 x NK603	DAS-01507-1 x DAS-59122-7 x MON-00810-6 x SYN-IR604-5 x MON-00603-6	Optimum Intrasect Xtreme
TC1507 × MON810 × MIR604 × NK603	DAS-01507-1 × MON-00810-6 × SYN-IR604-5 × MON-00603-6	not available
TC1507 x 59122	DAS-01507-1 x DAS-59122-7	Herculex XTRA
TC1507 x 59122 x MON810	DAS-01507-1 x DAS-59122-7 x MON-00810-6	not available
TC1507 x 59122 x MON810 x NK603	DAS-01507-1 x DAS-59122-7 x MON-00810-6 x MON-00603-6	Optimum Intrasect XTRA*
TC1507 x 59122 x MON88017	DAS-01507-1 x DAS-59122-7 x MON-88017-3	not available
TC1507 x 59122 x NK603	DAS-01507-1 x DAS-59122-7 x MON-00603-6	Herculex XTRA RR
TC1507 x GA21	DAS-01507-1 x MON-00021-9	not available
TC1507 x MIR604 x NK603	DAS-01507-1 x SYN-IR604-5 x MON-00603-6	Optimum TRIsect
TC1507 x MON810	DAS-01507-1 x MON-00810-6	not available
TC1507 x MON810 x MIR162 x NK603	DAS-01507-1 x MON-00810-6 x SYN-IR162-4 x MON-00603-6	not available
TC1507 x MON810 x NK603	DAS-01507-1 x MON-00810-6 x MON-00603-6	Optimum Intrasect
TC1507 x MON88017	DAS-01507-1 x MON-88017-3	not available
TC1507 x NK603	DAS-01507-1 x MON-00603-6	Herculex I RR
TC6275	DAS-06275-8	not available
VCO-01981-5	VCO-Ø1981-5	not available
Rice		
7Crp#10	not available	not available
GM Shanyou 63	not available	BT Shanyou 63
Huahui-1/TT51-1	not available	Huahui-1
LLRICE06	ACS-OS001-4	Liberty Link™ rice
LLRICE601	BCS-OS003-7	Liberty Link™ rice
LLRICE62	ACS-OS002-5	Liberty Link™ rice
Tarom molaii + cry1Ab	not available	not available
Wheat		
MON71800	MON-71800-3	Roundup Ready wheat

CHAPTER 5
Oilseed crops

Oilseeds are important crops that provide both oil and protein for human and animal feeding. Some crops provide oils for industrial use, for example as a lubricant for machinery.

The US Department of Agriculture publishes data annually on worldwide oilseed production. The most recent data (USDA, 2014) are shown in Table 5.1. As can be seen from these data, the three most important oilseed crops for food/feed use in terms of amount produced are soyabeans (commonly called soybean), rapeseed (canola) and cottonseed. However, the data do not include figures for olive oil, palm oil and palm kernel oil.

Palm oil is extracted from the fleshy portion of the fruit (mesocarp), whereas palm kernel oil is extracted from the kernel or the seed of the fruit. These two oils have different fatty acid compositions. Palm oil contains about 50% saturated fat and about 50% unsaturated fat. The fatty acid composition of palm kernel oil resembles that of coconut oil, with about 82% saturated fat and about 18% unsaturated fat (Chow, 2007).

Based only on the amount of oil produced, the most important crops are then palm oil, soyabean oil and rapeseed (canola) oil in that order (Table 5.2).

The average distribution of fatty acids by category of the important vegetable oils is shown in Table 5.3 (Dubois et al., 2007). The main categorisation of fatty acids is based on whether the fatty acid is saturated (containing only C–C bonds) or unsaturated (containing C=C bonds). Within the unsaturated, there are those with only one unsaturated bond, that is, monounsaturated, or those with more than one, that is, polyunsaturated. The first double bond from the methyl end (CH_3) determines whether it is an omega 3 (n-3) or an omega 6 (n-6) polyunsaturated fatty acid. (There are other fatty acids such as omega 7 and omega 9

Genetic Modification and Food Quality: A Down to Earth Analysis, First Edition.
Robert Blair and Joe M. Regenstein.

Table 5.1 World supply of major oilseeds, 2010/11–2013/14 (USDA, 2014).

	Year			
Crop	**2010/11**	**2011/12**	**2012/13**	**2013/14**
		Oilseed, million tonnes		
Production				
Copra	5.89	5.56	5.80	5.58
Cottonseed	44.26	47.98	46.31	45.10
Palm kernel	12.92	13.77	14.79	15.76
Peanut	39.84	38.33	40.12	39.81
Rapeseed/Canola	60.57	61.49	63.67	71.09
Soyabean	263.89	239.69	267.83	285.01
Sunflower seed	33.36	39.68	36.00	42.87
Total	460.72	446.5	474.52	505.22

Table 5.2 World vegetable oil supply, 2010/11–2013/14 (USDA, 2014).

	Year			
Crop	**2010/11**	**2011/12**	**2012/13**	**2013/14**
		Oil, million tonnes		
Production				
Coconut	3.71	3.41	3.66	3.45
Cottonseed	4.96	5.21	5.21	5.14
Olive	3.25	3.24	2.38	3.03
Palm	48.84	52.11	55.97	59.60
Palm kernel	5.73	6.13	6.52	6.96
Peanut	5.31	5.30	5.49	5.58
Rapeseed/Canola	23.47	24.12	24.98	26.05
Soyabean	41.29	42.60	42.89	44.77
Sunflower	12.43	14.87	13.48	15.59
Total	148.97	156.99	160.56	170.15

found naturally.) These characteristics help determine the relative value of the oils for both human and animal use.

An important point to note is that the oil extracted from any GM oilseeds does not contain any genetically modified material. All of the proteins and DNA remain with the oilseed residue (termed a 'meal') after oil extraction. Annual estimates of the world production of major protein meals are shown in Table 5.4 (USDA, 2014).

Several GM (transgenic) cultivars of oilseed crops have been introduced, primarily designed to provide resistance to insects and/or herbicides. According to James (2013) soyabean was the principal GM oilseed crop in 2012 (amounting to 84.5 million hectares out of a total of 107 million hectares of soyabeans),

Table 5.3 Fatty acid composition of important vegetable oils (Dubois et al., 2007).

Vegetable Oil	Fatty acid (FA) profile, %								
	Canola	Coconut	Cottonseed	Maize	Olive	Palm	Peanut	Soyabean	Sunflower
Saturated	8.0	62.9	27.8	14.8	15.3	50.4	18.3	15.7	12.8
Unsaturated									
MUFA	62.4	34.1	18.2	28.1	73.8	39.4	49.6	24.2	22.4
PUFA	31.5	3.0	53.4	57.1	10.0	10.5	30.8	59.8	66.0
n-6	21.6	2.8	53.2	56.1	9.4	10.2	30.3	52.1	65.6
n-3	9.9	0.2	0.2	1.0	0.6	0.3	0.4	7.8	0.5

Table 5.4 World supply of major protein meals, 2010/11–2013/14 (USDA, 2014).

	Year			
Crop	**2010/11**	**2011/12**	**2012/13**	**2013/14**
	Million tonnes			
Production				
Coconut	1.97	1.82	1.95	1.85
Cottonseed	14.83	15.64	15.64	15.51
Fish	4.89	3.96	4.18	4.16
Palm kernel	6.77	7.22	7.76	8.29
Peanut	6.47	6.46	6.73	6.84
Rapeseed/Canola	34.60	35.65	36.95	38.62
Soyabean	174.63	180.42	180.95	188.15
Sunflower	13.24	15.82	14.29	16.39
Total	257.40	266.99	268.43	279.81

Table 5.5 GM share of crop plantings in 2010 by country, percentage of total plantings. Brookes and Barfoot (2012). Reproduced with permission of PG Economics Ltd, UK.

Country	**Soyabeans % GM**	**Cotton % GM**	**Canola % GM**	**Maize % GM**
USA	93	93	88	86
Canada	70	N/A	93	94
Argentina	99	98	N/A	86
South Africa	85	100	N/A	69
Australia	N/A	99	8	N/A
China	N/A	67	N/A	N/A
Philippines	N/A	N/A	N/A	22
Paraguay	96	N/A	N/A	N/A
Brazil	76	27	N/A	55
Uruguay	94	N/A	N/A	95
India	N/A	85	N/A	N/A
Colombia	N/A	87	N/A	8
Mexico	24	47	N/A	N/A
Bolivia	79	N/A	N/A	N/A

N/A: not applicable.

followed by GM cotton (23.9 million hectares out of a total of 34 million hectares of cotton) and GM rapeseed/canola (8.2 million hectares out of a total of 34 million hectares of rapeseed/canola).

Brookes and Barfoot (2012) published similar data, showing the percentage of total crop plantings represented by GM cultivars of oilseeds in the main producing countries in 2010 (Table 5.5). The percentage of GM cultivars planted is now quite high in several countries, particularly the USA.

The International Service for the Acquisition of Agri-biotech Applications (ISAAA, 2014) lists the following in its current database of approved GM oilseed crops that have been approved for at least one cultivar in at least one country. The ISAAA is a not-for-profit organisation whose main mission is to foster knowledge-based and transparent decision-making on crop biotechnology.

- Argentine Canola (*Brassica napus*)
- cotton (*Gossypium hirsutum* L.)
- Polish canola (*Brassica rapa*)
- soyabeans (*Glycine max* L.).

The issue of the scope of the patentability of GM seed cultivars was clarified by the US Supreme Court in 2013. The case involved an Indiana farmer (Vernon Bowman) who purchased Monsanto Roundup Ready soyabean seeds for his first crop and initially followed the terms of the licensing agreement with Monsanto. That agreement permits a grower to plant the purchased seeds in one season. The farmer can consume the resulting crop or sell it as a commodity, usually to a grain elevator or agricultural processor. But under the agreement, the farmer may not save any of the harvested soyabeans for replanting, nor may the farmer supply them to anyone else for that purpose. In accordance with the agreement, he used all of the purchased seed for planting and sold the entire harvested crop to a grain elevator.

But he then took another approach for his second crop of the season. Claiming that late-season planting was more risky, he decided not to pay the premium price for Monsanto Roundup Ready seed. Instead he purchased 'commodity soybeans' intended for human or animal consumption from a grain elevator and planted them in his fields. Those soyabeans came from local harvested crops. Use of a glyphosate-based herbicide during the growth period confirmed that his crop had the Roundup Ready trait. Mr Bowman then saved the seeds from that crop to use in his late-season planting the next year and continued doing so (sometimes adding more soyabeans bought from a grain elevator). He did this for at least eight years.

After discovering this practice, Monsanto sued the farmer for patent infringement. The farmer claimed 'patent exhaustion' as a defence, arguing that Monsanto could not control his use of the soyabeans because they were the subject of a prior authorised sale (i.e. they were purchased legally from local farmers by the grain elevator operator). The District Court rejected that argument, and awarded damages to Monsanto of $84,456. The Federal Circuit Court of Appeals affirmed the decision. It reasoned that patent exhaustion did not protect Mr Bowman because he had 'created a newly infringing article'. The Supreme Court then held that patent exhaustion does not permit a farmer to reproduce patented seeds through planting and harvesting without the patent holder's permission.

Soyabeans (soybeans) (*Glycine max* L.)

The ISAAA database contains 30 approved soyabean cultivars (Appendix Table 5.1A). None have been approved for cultivation in the EU.

Examples of approved cultivars are MON 40-3-2 (Monsanto), which has been modified to provide tolerance to glyphosate-containing herbicides, and MON 87705 (Monsanto), which has been modified to provide tolerance to glyphosate-containing herbicides and also has an increased oleic acid content. Oleic acid is considered desirable because this MUFA improves the nutritional characteristics of the oil and improves its oxidative stability.

Roundup Ready soyabeans have been genetically modified to tolerate the applications of the Monsanto herbicide Roundup while the crop is growing, by insertion of a single gene derived from a common soil bacterium into the genome of the soyabean. The gene produces an altered version of the plant enzyme that is normally blocked by glyphosate, the active ingredient in Roundup herbicides. The bacterial version of the enzyme is naturally tolerant of glyphosate so that the altered soyabeans are unaffected by spraying with the herbicide, while weeds growing in the same field are killed. The bacterial enzyme is a protein that like most proteins is readily digested by both humans and animals.

Among other government agencies, Health Canada has determined that glyphosate-tolerant soyabeans are as safe and nutritious as other commercially available soyabean varieties, and are suitable for food use. Consequently GM soyabeans have been designated 'substantially equivalent' to other non-GM soyabean cultivars. This decision followed an extensive and thorough assessment of data relating to the GM soyabeans, according to the Canadian Guidelines for Safety Assessment of Novel Foods (September 1994). These guidelines are based on internationally accepted principles for establishing the safety of foods derived from genetically modified organisms, and were developed in consultation with other government agencies, consumers and industry.

The assessment compared glyphosate-tolerant soyabeans with the commercial variety from which they were derived, as well as with other commercial varieties of soyabeans. It determined that, apart from minute levels of the introduced enzyme (protein), there was no difference in composition or nutritional quality between the genetically altered soyabeans and other commercial varieties. This assessment applied only to the food use of these soyabeans. Issues related to the cultivation of these soyabeans in Canada and to their use as animal feeds have been addressed separately by another Canadian government agency, Agriculture & Agri-Food Canada.

In 2011 the European Food Safety Authority (EFSA, 2011a) published its scientific assessment of the proposal by Pioneer to permit the use of GM herbicide-tolerant soyabean 356043 for food and feed uses including import and processing in Europe. A review of the genetic modification procedure used and data on the levels of newly expressed proteins did not raise a safety concern. Comparative analysis of phenotypic and agronomic characteristics indicated that soyabean 356043 is not different from its conventional counterpart. In terms of composition, differences were found between 356043 soyabean and its conventional counterpart as increases in the newly expressed proteins Glycine max-HRA and GAT4601, and the levels of the fatty acids heptadecanoic, heptadecenoic and heptadecadienoic acid and the acetylated amino acids *N*-acetylaspartate (NAA) and *N*-acetylglutamate (NAG). As observed by EFSA, the levels of these acetylated amino acids and odd-chain fatty acids fall outside the natural ranges observed for other commercial non-GM soyabean varieties. The overall level of NAA and NAG (taken together) in soyabean 356043 was found to be less than 0.15% of the total amino acids. The total level of odd-chain fatty acids amounts to less than 1% of total fatty acids. No statistically significant differences in total amino acid contents in seed were observed between the 356043 soybean and its conventional counterpart. Levels of major fatty acids in 356043 soyabean seed were found to be comparable to those observed in the conventional counterpart.

The safety assessment of the newly expressed proteins Glycine max-HRA and GAT4601 identified no concerns regarding any potential toxicity or allergenicity issues with these proteins. Heptadecanoic, heptadecenoic and heptadecadienoic acid are present in the human diet and the intake of small amounts of these fatty acids via food or feed is not expected to produce adverse effects. NAA and NAG are normal components of mammalian metabolism and the estimated increases in their intake are considered low when related to the normal intake of L-aspartic acid and L-glutamic acid.

Further toxicological, allergenicity and nutritional analyses provided no indications of adverse effects. The EFSA GMO Panel considered that the information available for soyabean 356043 addresses the scientific comments raised by the various EU member states. EFSA then stated that soyabean 356043, as described in the application, is as safe as its conventional counterpart with respect to potential effects on human and animal health and on the environment in the context of its intended uses.

EFSA (2012a) published an updated scientific assessment of the available data relating to the cultivation and marketing of MON 40-3-2 (glyphosate-tolerant soyabeans) in Europe. The agency determined that this cultivar is as safe as its conventional counterpart and commercial non-GM soyabean varieties with respect to potential adverse effects on human and animal health. The agency therefore recommended managing the use of glyphosate herbicide on MON 40-3-2 in a manner that resulted in similar or reduced environmental impacts compared with conventional soyabean cultivation. There was no evidence of adverse effects on non-target organisms (including animal pollinators) and there were no indications of the occurrence of adverse effects on non-target predators, herbivores and decomposers. The agency concluded that, if subjected to appropriate management measures, the cultivation of MON 40-3-2 is unlikely to have environmental effects any more adverse than those associated with conventional soyabean cultivation. In spite of these scientific recommendations, the EU Commission approved the cultivar for importation and for food and feed use only, but not for cultivation.

EFSA (2012b) published a scientific assessment of the available data relating to the cultivation and marketing of MON 87705 in Europe. The safety assessment identified no concerns regarding potential toxicity and allergenicity related to this cultivar. The altered fatty acid profile did not raise concerns related to toxicity, and the overall allergenicity of the whole plant was not changed by the genetic modification. A feeding study with broiler chickens confirmed that the defatted meal was as nutritious as meals produced from its conventional counterpart and non-GM reference varieties. The agency therefore concluded that the MON 87705 cultivar is as safe as its conventional counterpart with respect to potential effects on human and animal health, and the environment in the context of its intended use as food and feed, import and processing. In spite of these scientific recommendations, the EU legislators approved the cultivar for importation and for food and feed use only, not for cultivation.

EFSA has published similar scientific assessments of other approved GM oilseeds (and other approved GM crops). In general, though, the EU Commission has been unwilling to permit cultivation of GM crops in Europe.

Safety aspects

As with cereal grains, all oilseed crops whether grown conventionally, organically or using biotechnology must meet the mandated standards of the countries in which they are both grown and/or consumed. It is important therefore to review the current evidence related to the relative safety and nutritional content of oilseeds grown using biotechnological methods.

The main concerns about potentially adverse effects of GM foods on health are allergenicity and toxicity. To some extent the possibility of allergenicity in an altered protein can be predicted using bioinformatic comparisons of the amino acid sequence of the proteins added to the crop with reference to databases of known allergenic sequences (Lack, 2002), but the allergenicity still needs to be tested and proven to be absent both experimentally and with respect to sequence matches with known allergens in the database prior to approval of the product.

Soyabeans are an important food used by humans and animals, and it has had proteins identified that have been characterised as major allergens. Soyabeans are considered to be among the top eight allergens in the USA.

The International Food Biotechnology Committee (Delaney et al., 2008a) developed a scientifically based two-tiered, weight-of-evidence strategy to assess the safety of novel proteins used in the context of agricultural biotechnology. Recommendations draw upon knowledge of the biological and chemical characteristics of proteins and the testing methods used for evaluating the potential intrinsic hazards of chemicals. Studies in Tier I (potential hazard identification) include an assessment of the biological function or mode of action, and the intended application of the protein, its history of safe use, a comparison of the amino acid sequence of the protein to those of other proteins, as well as the biochemical and physico-chemical properties of the proteins. Studies outlined in

Tier II (hazard characterisation) are conducted when the results from Tier I are not sufficient to allow a determination of safety (using the standard of reasonable certainty of no harm) on a case-by-case basis. These Tier II studies may include acute and repeated dose toxicology studies and hypothesis-based testing. Application of these guidelines has included examples of their application to transgenic proteins as agricultural inputs and as food and feed output traits in GM modified crops. The guidelines also provide recommendations for future research considerations related to protein safety assessment.

An important criticism of the current situation regarding the official approval process for GM foods is that the relevant data are not published in peer-reviewed journals, thus they are not available for review by other scientists. For instance, Domingo (2007) claimed in a review of relevant studies 'where is the scientific evidence showing that GM plants/food are toxicologically safe?'

One explanation for this apparent lack of openness is that the policy for GM plants is similar to that which has existed for many years for drug approvals: a positive decision based on submitted data (again not necessarily peer-reviewed) results in the granting of a patent and gives the company the right to market the product in question for its intended use.

Some critics go further and view the apparent lack of openness as being suggestive of collusion between the applicant companies and the governmental agencies to allow onto the market products that have not been tested adequately. Such a claim is somewhat questionable, given that no reputable company or government agency would knowingly allow any suspect product to enter the human or animal food market. The subsequent negative publicity could be devastating.

In spite of the claim about a lack of data on the safety of GM oilseeds, reports providing peer-reviewed data have been published in the scientific literature.

Allergenicity

Batista et al. (2005) tested the potential allergenicity of protein extracts prepared from transgenic maize (MON810, Bt11, T25 and Bt176) and soyabeans (glyphosate-tolerant) and from non-transgenic control samples in two sensitive groups: children with food and inhalant allergies and individuals with asthma-rhinitis. The test used was a skin prick test. In addition the IgE immunoblot reactivity of sera from patients with food allergy to soyabeans (glyphosate-tolerant) and maize (MON810, Bt11, Bt176) samples, as well as to the pure transgenic proteins (CryIA[b] and CP4 5-enolpyruvylshikimate-3-phosphate synthase) was tested. No adverse reactions were observed and no individuals tested showed any detectable IgE antibodies against the pure transgenic proteins. It was concluded that the transgenic products tested were safe in terms of their allergenic potential.

The allergenicity and IgE-reactive components of conventional and GM soyabeans (containing the EPSPS transgene) extracts were compared in allergic adults who had been sensitised to soyabeans (Kim et al., 2006). The study involved 1716 adult allergy patients and 40 healthy, non-atopic controls. Skin prick tests and IgE enzyme linked immunosorbent assays (ELISAs: Engvall and Perlmann, 1971) were done using non-transgenic and GM soyabean extracts, in addition to other common inhaled allergens. The specificities of serum IgE antibodies from allergic patients and the identities of the IgE-reactive components of the soyabean extracts were compared using ELISA inhibition testing, two-dimensional sodium dodecyl sulfate (SDS) gel electrophoresis and IgE immunoblotting. To evaluate the effects of digestive enzymes and heat treatment, the soyabean extracts were heated or pre-incubated with or without simulated gastric and intestinal fluids. The IgE sensitisation rates to non-GM and GM soyabeans were identical (3.8% of allergic adults), and circulating IgE antibodies specific for the two extracts were comparable. The results of the ELISA inhibition test, SDS-polyacrylamide gel electrophoresis (PAGE), and IgE immunoblotting showed a similar composition of IgE-binding components within the non-GM and GM extracts. None of the subjects had a positive response to purified EPSPS protein in the skin prick test, ELISA, or IgE immunoblot analysis. These findings suggest that the IgE sensitisation rate to GM soyabean extracts is identical to that of non-GM soyabean extracts in adult allergy patients. In addition, based on both in vivo and in vitro methods, the allergenicity of non-GM and GM soyabean extracts was identical.

The aim of a study by Batista et al. (2007) was to characterise the proteins contained in soyabeans and, specifically, the IgE-reactive proteins, as well as to compare the IgE response in soyabean-allergic individuals to GM (glyphosate-tolerant) soyabeans and a non-GM. Two new potential soyabean allergens were identified in both types of soyabeans. One was a maturation-associated protein and apparently part of the abundant proteins found in late embryogenesis. The other was a cysteine proteinase inhibitor. None of the individuals tested reacted differentially to the GM versus non-GM soyabeans. It was concluded that the expression of endogenous soyabean allergen expression was not altered by genetic modification. As suggested by other researchers, Batista et al. (2007) suggested that proteomics (characterisation of the proteins contained in a food product) should be considered a powerful tool for the functional characterisation of plants and for food safety assessment.

Delaney et al. (2008b) published the results of a collaborative study conducted by scientists working at companies involved in the development of GM cultivars of soyabeans, the Mount Sinai School of Medicine (New York) and the Sungkyunkwan University School of Medicine, (Seoul, Korea). The report described the results of an allergenicity and toxicity assessment of the GAT4601 protein in 356043 soyabeans. Bioinformatic comparison of the amino acid sequence of GAT4601 did not identify any similarities to known allergenic or toxic proteins. In vitro studies demonstrated that the GAT4601 protein was rapidly degraded in simulated gastric fluid containing pepsin and in simulated intestinal fluid containing pancreatin, and was completely inactivated at temperatures above 56 °C. The GAT4601 protein expressed in the soyabeans is not glycosylated. Similar protein profiles were observed in flour extracts from the 356043 soyabeans and non-GM related conventional soyabeans (cultivar Jack). The authors concluded that the GAT4601 protein in GM soyabeans does not present any risk of an allergic effect in humans.

The above data indicate that the allergenicity of conventional and GM cultivars of soyabeans is similar.

Toxicity

One factor determining whether a protein food has toxic properties is its fate in the gastrointestinal tract. A possible result of the ingestion of a GM protein is the transfer of recombinant DNA in the food to bacteria or animal cells. This aspect has been investigated in several studies.

Harrison et al. (1996) measured the in vitro digestion of the CP4 EPSPS protein expressed in soyabeans resistant to glyphosate and reported a rapid breakdown by proteolytic enzymes. Aumaitre et al. (2002) reported data using a range of farm animals showing that the proteins expressed by glyphosate-tolerant soyabeans do not produce harmful effects in the digestive tract or in other tissues after ingestion.

Jennings et al. (2003a) conducted a study to investigate whether transgenic and endogenous plant DNA and transgenic protein are detectable in muscle from swine fed glyphosate-tolerant soyabean meal. Questions regarding the digestive fate of DNA and protein from transgenic feed have been raised with regard to human consumption and commercial trade of animal products (e.g. meat, milk, and eggs) from farm animals fed transgenic crops. Pork loin samples were analysed for the presence of fragments of transgenic and endogenous plant DNA and transgenic protein from animals fed meal prepared from conventional or glyphosate-tolerant soyabeans. Pigs were fed diets containing 240, 190 and 140 g/kg RR or conventional soyabean meal during grower, early-finisher and late-finisher phases of growth, respectively, and longissimus muscle samples were collected (12 per treatment) after slaughter. Total DNA was extracted from the samples and analysed by PCR. None of the extracted samples was positive for cp4 epsps or the endogenous soyabean gene le1 at the limit of detection. Neither the CP4 EPSPS protein nor the immunoreactive peptide fragments were detected in loin muscle homogenates. These results showed that neither small fragments of transgenic DNA nor immunoreactive fragments of transgenic protein are detectable in loin muscle samples from pigs fed a diet containing Roundup Ready soyabean meal.

Netherwood et al. (2004) studied the survival of transgene EPSPS from GM soyabeans in the small intestine of human ileostomic patients (individuals in which the terminal ileum is resected and digesta are diverted from the body via the stoma to a colostomy bag). The amount of transgene that survived passage through the small bowel varied among individuals, with a maximum of 3.7% recovered from the stoma of one individual. The transgene did not survive passage through the intact gastrointestinal tract of these human subjects fed GM soyabeans. Three of the seven ileostomic patients showed evidence of low-frequency gene transfer from GM soyabeans to the microflora of the small bowel before their involvement in these experiments. This low level of EPSPS in the intestinal microflora did not increase after consumption of a meal containing GM soyabeans. Therefore, it was concluded that gene transfer did not occur during the passage of food through the gastro-intestinal tract.

Rizzi et al. (2012) concluded that a small amount of fragmented dietary DNA may resist the digestive process, based on the results of animal feeding studies. Mammals have been shown to take up dietary DNA from the gut, but stable integration and expression of internalised DNA have not been demonstrated. They also concluded that despite the ability of several bacterial species to acquire external DNA by natural transformation, in vivo transfer of dietary DNA to bacteria in the intestine has not been detected in the few experimental studies conducted so far.

Hammond et al. (1996) conducted animal feeding studies with rats, broiler chickens, catfish and dairy cows as part of a safety assessment programme for a soyabean variety genetically modified to tolerate in-season application of glyphosate. These studies confirmed the results of other studies that demonstrated the safety of the introduced protein, a bacterial 5-enolpyruvyl-shikimate-3-phosphate synthase from *Agrobacterium* sp. strain CP4.

The effects of glyphosate-tolerant soyabeans was studied in mice, using testicular tissue as a sensitive biomonitor of potential toxic effects (Brake and Evenson, 2004). Pregnant mice were fed a diet containing glyphosate-tolerant soyabeans or non-GM (conventional) soyabeans through gestation and lactation. After weaning, the young male mice were maintained on the respective diets. At 8, 16, 26, 32, 63 and 87 days after birth, three male mice and an adult reference mouse were killed, the testes removed surgically, and the cell populations measured by flow cytometry. Multi-generational studies were conducted in the same manner. The results showed that the GM food had no effect on macromolecular synthesis or cell growth and differentiation. No differences were found in the proportions of testicular cell populations (haploid, diploid and tetraploid) in the transgenic soyabean-fed mice and those fed the conventional diet. Additionally, there were no differences in litter sizes and body weights of the two groups. It was concluded that the diet containing GM soyabeans had no negative effect on fetal, postnatal, pubertal or adult testicular development.

Zhu et al. (2004) conducted a nutritional assessment and a study of the fate of DNA of soyabean meal from glyphosate-tolerant or conventional soyabeans using rats. Eighty Sprague–Dawley rats (40 males and 40 females) were randomly allotted to one of four groups according to sex and body weight for a 13-week feeding experiment. The rats were fed maize-based diets containing

600 g/kg conventional soyabean meal, a mixture of 300 g conventional and 300 g/kg glyphosate-tolerant soyabean meal, or 600 or 900 g/kg glyphosate-tolerant soyabean meal. The two soyabean meals were similar in chemical analysis and amino acid composition. All diets were formulated to contain similar nutrient levels, except the diet containing 900 g/kg glyphosate-tolerant soyabean meal which could not be balanced exactly because of the high level of soyabean meal. Over the 13-week trial period, body weight and feed intake decreased only in rats fed the diet containing 900 g/kg GM soyabean meal during week 1. No treatment-related deaths occurred during the trial. Gross necropsy findings, hematological and urinalysis values, and clinical serum parameters showed no significant differences between rats fed the diets containing the conventional or GM soyabean meal. No adverse effects of glyphosate-tolerant soyabean meal on rats were observed, even at a dietary level as high as 900 g/kg.

Malatesta et al. (2008) conducted a long-term study on female mice fed a GM soyabean (glyphosate-tolerant) diet to test the effects on liver ageing. The rationale was that the liver represents a suitable model for monitoring the effects of a diet, due to its key role in controlling overall metabolism. The morpho-functional characteristics of the livers of 24-month-old mice, fed from weaning on control or GM soyabean diets, were investigated by combining a proteomic approach with ultrastructural, morphometrical and immunoelectron microscopical analyses. Several proteins related to hepatocyte metabolism, stress response, calcium signalling and mitochondria were differentially expressed in GM-fed mice, indicating a more marked expression of senescence (old age) markers in comparison to controls. Moreover, hepatocytes of GM-fed mice showed mitochondrial and nuclear modifications indicative of a reduced metabolic rate. The researchers concluded that the intake of GM soyabean intake may influence some liver features during ageing. In making this conclusion Malatesta et al. (2008) do not appear to have considered the possibility that the GM soyabeans used were contaminated with a mycotoxin.

As noted by Snell et al. (2012) the differences reported by Malatesta et al. (2008) are all quantitatively minor and are in the range of differences observed in other studies and could be due to genotypic differences or growth conditions of the plant material used. Snell et al. (2012) also pointed out that the paper by Malatesta et al. (2008) has been severely criticised by Williams and DeSesso (2010), in that the study suffers from six methodological errors. These are: (i) 'controlling for potential litter effect'; (ii) 'using an appropriate number of experimental animals per group and acquiring a sufficiently robust sample of independent observations'; (iii) 'establishing the representativeness of observations'; (iv) 'adhering to the principles for stereologic morphometry'; (v) 'using appropriate statistical methods (for study design as well as for data analysis)'; and (vi) 'controlling for potential confounding factors, including those related to differences in diet phytoestrogen contents'. Snell et al. (2012) also pointed out that none of the papers published by Malatesta and colleagues explicitly state the exact identity of the soyabean cultivars used. According to Snell et al. (2012) it seems likely that the plant material compared in these studies was not isogenic lines but rather GM material and non-GM material that the authors were able to purchase from different commercial sources. In addition, the authors made comments on the possible involvement of herbicide treatment to explain their results, which confirms that the plant material differs by additional parameters other than just the transgene insertion factor and the fact that the plant materials were not grown side by side in the same field. Therefore, these studies do not match the internationally recommended standards for a proper nutritional or toxicological assessment of a GM line.

Additional toxicological evidence has been published more recently by other researchers, none of which confirm the findings reported by Malatesta et al. (2008). Sakamoto et al. (2008) conducted a 104-week feeding study of GM (glyphosate-tolerant) soyabeans with F344 DuCrj rats to evaluate the safety of the GM soyabeans. The rats were fed a diet containing GM or conventional soyabeans at a dietary concentration of 300 g/kg. The conventional soyabeans were a closely related cultivar to the GM soyabeans. The two diets were formulated to provide an identical nutrient level. In addition, one group was fed a commercial control diet, described as containing some soyabean meal. The general condition of the animals was observed daily, and body weight and food consumption were recorded. At the end of the experiment (104 weeks) the animals were euthanised and subjected to hematology, serum biochemistry and pathological examinations. Body weight and food intake were similar for rats fed the diets containing GM and conventional soyabeans, as were gross necropsy findings, hematological and serum biochemical parameters, and organ weights. Pathological examination showed no increase in incidence or in any specific type of non-neoplastic or neoplastic lesions in either sex fed the diet containing GM soyabeans. These results indicated that the long-term intake of GM soyabeans at a dietary level of 300 g/kg has no apparent adverse effect in rats.

Delaney et al. (2008b) described the results of a toxicity assessment of the GAT4601 protein in GM soyabeans (356043). Acute toxicity of the GAT4601 protein was assessed by oral administration to mice of 2 g/kg of the purified protein via oral gavage. Control groups were given either water alone or bovine serum albumin at 2 g/kg. All mice survived the duration of the study and no clinical signs of systemic toxicity were observed in any of the treatment groups. No evidence of adverse effects was observed in mice following acute oral exposure to 2000 mg/kg of GAT4601 protein or doses of 10, 100 or 1000 mg/kg/day. Organ weight data are shown in Table 5.6. Based on these findings the authors concluded that the GAT4601 protein in GM soyabeans does not present any risk for adverse effects in humans.

Appenzeller et al. (2008) studied the effects of including Optimum GAT1 soyabeans (DP-356043-5, 356043) in the diet of rats. This cultivar contains proteins that confer tolerance to both glyphosate and sulfonylurea/imidazolinone

Table 5.6 Mean organ : body weight ratios of mice (±sd) fed diets containing GAT4601 soyabean protein at concentrations representing intakes of 0–1000 mg/kg/day for 27 days. Delaney et al., (2008b). Reproduced with permission of Oxford University Press.

Dietary GAT4601 protein, mg/kg/day	0	10	100	1000
MALES	*n*=5	*n*=5	*n*=5	*n*=5
Body wt. (g, final)	33.6±3.6	33.6±0.9	34.2±1.1	34.6±3.2
Adrenals	0.0328±0.0073	0.0270±0.0058	0.0258±0.0068	0.0306±0.0051
Brain	1.505±0.217	1.497±0.078	1.493±0.049	1.457±0.125
Testicular epididimides	0.354±0.066	0.385±0.129	0.394±0.100	0.346±0.044
Heart	0.669±0.154	0.591±0.054	0.637±0.131	0.625±0.072
Kidneys	1.844±0.226	1.768±0.162	1.858±0.235	1.682±0.186
Liver	5.245±0.523	5.429±0.438	5.365±0.317	5.465±0.658
Lungs	0.633±0.077	0.620±0.076	0.617±0.063	0.649±0.131
Spleen	0.330±0.077	0.301±0.066	0.284±0.025	0.281±0.031
Testes	0.752±0.115	0.728±0.080	0.728±0.126	0.711±0.051
Thymus	0.138±0.041	0.140±0.033	0.134±0.061	0.148±0.040
FEMALES	*n*=5	*n*=5	*n*=5	*n*=5
Body wt. (g, final)	28.4±1.1	27.8±1.6	27.0±2.4	28.4±1.5
Adrenals	0.055±0.015	0.055±0.014	0.052±0.014	0.048±0.006
Brain	1.803±0.131	1.839±0.160	1.806±0.238	1.872±0.169
Heart	0.566±0.037	0.642±0.108	0.602±0.057	0.565±0.037
Kidneys	1.416±0.132	1.577±0.146	1.473±0.221	1.587±0.117
Liver	5.298±0.332	5.215±0.467	5.229±0.366	5.780±0.504
Lungs	0.698±0.068	0.716±0.081	0.723±0.082	0.786±0.129
Ovary	0.0808±0.0362	0.0690±0.0081	0.0846±0.0238	0.0834±0.0198
Spleen	0.424±0.063	0.341±0.064	0.415±0.051	0.337±0.039
Thymus	0.182±0.054	0.201±0.049	0.219±0.028	0.184±0.047

herbicides (Tan et al., 2005). Dehulled/defatted toasted meal and toasted ground hulls were prepared from soyabeans from untreated plants (356043), herbicide-treated plants (356043+Gly/SU), non-transgenic isoline control (091) and three commercial non-transgenic reference varieties (93B86, 93B15 and 93M40). A series of diets containing 200 g/kg meal (w/w) and 15 g/kg hulls (w/w) was then prepared and fed to young adult Sprague–Dawley rats (12/sex/group) for 93 days. Compared with rats fed the isoline control or conventional reference diets, no biologically relevant adverse effects were observed in rats fed diets containing 356043 or 356043+Gly/SU soyabean with respect to body weight/gain, food consumption, food/weight gain, clinical signs, mortality, ophthalmology, neurobehavioural assessments (sensory response, grip strength, motor activity), clinical pathology (hematology, coagulation, serum chemistry, urinalysis), organ weights, and gross or histopathology. Based on these findings the authors concluded that GM soyabeans 356043 are as safe and nutritious as conventional non-GM soyabeans.

Chukwudebe et al. (2012) evaluated the nutritional and health status of rats fed diets containing GM soyabeans CV127 at incorporation levels of 110 and 330 g/kg. For control comparisons, rats were also exposed to similar incorporation levels of a nearly isogenic conventional soyabean variety (Conquista) and two other conventional soyabean varieties (Monsoy, Coodetec). In spite of phenotypic differences among these four soyabean cultivars, there were no quantitative differences in their respective proximate and nutrient contents, including proteins, amino acids, anti-nutrients and nutritional cofactors. All diets were prepared by incorporating the respective processed soyabean meal with ground Kliba maintenance meal ('GLP' meal supplied by Provimi Kliba SA, Switzerland) at high (330 g/kg) and low (110 g/kg) incorporation levels. The mixed diets were then fed to Wistar rats for about 91 days. The overall results demonstrated that CV127 soyabeans are similar to their nearly isogenic conventional counterpart, as well as other conventional soyabean cultivars, with respect to their nutritional value and health effects. Therefore, it was concluded that the introduction of the AHAS gene into CV127 soyabeans does not substantially alter its compositional properties, or adversely affect its nutritional or safety features as a food for mammals.

Qi et al. (2012) conducted a subchronic feeding study of stacked trait genetically modified soyabean (305423×40-3-2) with Sprague–Dawley rats. This GM soyabean cultivar has a higher concentration of oleic acid (18 : 1) relative to linoleic acid (18 : 2) compared with non-GM soyabeans. Also this cultivar is tolerant of glyphosate. In this study, diets containing three different concentrations (75, 150 and 300 g/kg) of 305423×40-3-2 or non-GM soyabeans were fed to Sprague–Dawley rats for 90 days. Nutritional and growth performance variables were evaluated followed by analysis of standard clinical chemistry, hematology and tissue variables. Compared to rats fed the non-GM control diet, some statistically significant differences were observed in rats fed the diet containing the GM soyabeans, but the differences were not considered treatment-related and commonly fell within the normal ranges for the control group consuming the commercial diet. Based on the findings the researchers concluded that GM soyabeans 305423×40-3-2 are as safe as non-GM soyabeans as a food source.

Studies have also been conducted using farm animals.

Ash et al. (2003) investigated the extent of deposition of genetically modified protein from glyphosate-tolerant soyabeans in the tissues and eggs of laying hens. Since a breakdown of the modified portion of the protein was expected during digestion, an immunoassay test was run. By using a double antibody sandwich format specific for the CP4EPSPS protein, a qualitative test of the new protein was done to determine the presence of modified proteins in various samples. Raw soyabeans, soyabean meal, a complete diet, whole egg, egg albumen, liver and droppings from laying hens were collected. Glyphosate-tolerant soyabeans, soyabean meal and complete diets were found to contain the GM proteins. However, whole egg, egg albumen, liver and droppings were all negative for GM protein. The results indicated that the digestive process of the laying hen effectively broke down the GM protein from the soyabean meal portion of the diet, explaining its absence in the liver, egg and droppings.

Phipps et al. (2002) investigated the possible presence of transgenic DNA in milk from cows receiving herbicide tolerant (CP4EPSPS) soyabean meal. Ten Holstein /Friesian cows with an average live weight of 630 kg and producing 25.3 kg milk/day received a total mixed ration in which the forage portion comprised 55% of the total dry matter (DM) and contained non-GM grass and maize silage in a 1 : 3 ratio (DM basis). During study weeks 1–3 the dietary dry matter also contained non-GM supplements of 18.5% cracked wheat and 26.1% rapeseed meal. During weeks 4–12 GM soyabean meal replaced rapeseed meal at 26.1% of the total diet in weeks 4–5 and 13.9% during weeks 6–12. Weekly milk samples were taken from all cows. Samples were spiked with DNA extracted from the soyabean meal to establish the limit of detection (LOD) of transgenic DNA using polymerase chain reaction (PCR) analyses designed to detect transgenic DNA fragments smaller than 200 base pairs (bp). Subsequent PCR analyses carried out in duplicate on diet and milk samples collected at weeks 3, 4 and 12 of the study used three DNA primer sets to establish the presence or absence of transgenic DNA, The LOD of transgenic DNA in milk was 7.5 mμg/l. Feed and milk samples analysed at week 3 were negative. The feed samples at weeks 4 and 12 were positive for DNA from GM soyabean meal, but all milk samples were negative. The results showed that transgenic DNA could not be detected in milk from cows receiving up to 26.1% of their diet DM as glyphosate-tolerant soyabean meal.

A study of the histopathology of internal organs of farm animals fed GM (MON 810) maize and GM (MON-40-3-2 glyphosate-tolerant) soyabean meal was conducted by Reichert et al. (2012). Samples of liver, kidney, spleen, pancreas, duodenum, jejunum, skeletal muscle and the bursa of Fabricius were collected from broiler chickens, laying hens, growing pigs and calves fed diets containing maize and soyabean meal. Histopathological examination of the tissues from these organs showed no significant differences between control animals fed diets containing no GM feeds and animals fed diets containing the GM feeds.

Hemre et al. (2005) conducted a three-month study in which post-smolt Atlantic salmon were fed one of three diets. The first was a standard diet based on fishmeal (control). The two other diets contained a full-fat soyabean meal (conventional or glyphosate-tolerant GM) in which the soyabean meal replaced fish meal at a level of 172 g/kg (130 g/kg of total protein). All diets were formulated to have the same proximate composition, starch and sugar levels, content of methionine and lysine, fatty acid profiles, and vitamin, mineral and pigment contents. There was, however, a slight difference in levels of anti-nutrients between the conventional and GM soyabean meal in that the levels of daidzein, genistein and phytate were slightly higher in the GM soyabean meal but still within the ranges found in the literature. According to these researchers the Monsanto Company had reported that an increase in anti-nutrients might be expected in Roundup Ready soyabeans.

The results showed that all fish tripled their weight, with no mortality. In addition no significant differences in feed utilisation, composition of whole body, liver and muscle or muscle fatty acid profiles were detected. The relative sizes of liver, kidney, head-kidney and brain were the same in all dietary groups, while the index of spleen size was significantly decreased during one growth period, from 0.13 in fish fed the GM soyabean diet to 0.09 in fish fed the conventional soyabean diet. The corresponding value for the fish fed the fishmeal control diet was intermediate at 0.11. The researchers were unable to explain the significance of this finding. The head-kidney serves as a major lymphoid organ in fish, in addition to the thymus and spleen (Press and Evensen, 1999, these organs being part of the immune system in fish). The detoxification system, measured as glutathione peroxidase (GPx) and lysozyme activities, showed equal values for all groups when measured in the plasma, liver and head-kidney, indicating similar responses to the soy proteins in GM and conventional soyabeans.

A similar study was conducted by Sanden et al. (2004), who investigated the survival of GM soyabean DNA fragments in Atlantic salmon. Fish were fed a series of test diets for 6 weeks, formulated to contain conventional or GM maize and conventional or GM (glyphosate-tolerant) full-fat soyabean flour. The conventional and GM soya flour replaced 17.2% of the fish meal in the diets. A control diet composed of fish meal as the only protein source was used for comparison purposes. The transgenic sequences (120 and 195 bp) and the lectin gene (180 bp) were detected in the diets containing GM soya. In the fish GI tract, however, only the smaller DNA fragment (120 bp) could be isolated from the contents of the stomach, pyloric region (approximately the fish's equivalent of the mammalian pancreas), mid-intestine and distal intestine. No transgenic or conventional soya DNA fragments could be detected in liver, muscle or brain tissues. These data indicated that transgenic GM soya sequences may survive passage through the GI tract but that they cannot be detected in fish tissues.

Sissener et al. (2009) conducted a seven-month feeding trial with GM glyphosate-tolerant soyabeans included in the diet of Atlantic salmon (initial weight 40 g) during the parr-smolt transformation. A near-isogenic parent soyabean cultivar was used in a control diet. The two test diets contained 250–262 g/kg full-fat soyabean meal and were formulated to be similar in nutrient content. The performance and health of the fish were assessed by growth, body composition, organ development, hematological parameters, clinical plasma chemistry and lysozyme levels, with samples collected both in the freshwater and seawater stages. The relative weights of sections of the intestinal were slightly different during some growth periods, with the mid-intestine being significantly

($p < 0.035$) smaller in the GM fed fish (index 0.134) than in the non-GM fed fish (index 0.145) and the distal intestine showing the reverse trend (index 0.565 vs. 0.505, the difference being significantly different at $p < 0.018$). There was no significant difference in intestinal weights at the end of the trial. Plasma triacylglycerol levels were higher in the GM group overall in the experiment, with the difference being greater around the time of seawater transfer. There were no differences in total growth during the trial. All other measured parameters showed no diet-related differences. The overall conclusion of these researchers was that the effects of feeding Atlantic salmon with diets containing GM soya at an inclusion level of 250–260 g/kg are minor.

All of the above studies can be interpreted as suggesting no problems of toxicity in humans or animals fed diets containing GM soyabeans. One report to the contrary was made by Ermakova (2005), a scientist at the Institute of Higher Nervous Activity and Neurophysiology of the Russian Academy of Sciences. The study involved a total of 15 female rats. One treatment consisted of adding GM (glyphosate-tolerant) soya flour (5–7 g) to the diet. Other females received a diet with non-GM soya or without soya. The experiment began two weeks before the rats conceived and continued through pregnancy and nursing. It was reported that 55.6% of the offspring of female rats fed GM soya flour died within three weeks, compared with 9% of the offspring of rats fed non-GM soya. In addition, offspring from the GM-fed group were significantly stunted: 36% weighed less than 20 g after 2 weeks, compared with 6.7% in the non-GM soya control group. The results were presented in 2005 to a symposium on genetic modification in Russia organised by the National Association for Genetic Security (NAGS). To date the experiment does not appear to have been repeated or the findings published in any peer-reviewed journal. The treatments were not described fully in the conference presentation and one possible explanation for the findings is that the GM soya used was unprocessed, whereas the conventional soya flour had been processed. Raw, unprocessed, soyabeans are known to contain several anti-nutritional factors that are known to interfere with growth and with digestion. Whatever the explanation, the findings are not of sufficient calibre for serious consideration, without more details of the study and without confirmation by other researchers.

Another report that questioned the safety of GM soyabeans was made by Carman et al. (2013). Pigs were fed commercial diets containing GM soyabean meal (glyphosate-tolerant) and GM maize ($n = 84$) or non-GM, non-isogenic soyabean meal and maize ($n = 84$) from weaning to slaughter (22.7 weeks). Each group contained equal numbers of males and females, in six replications. The GM maize was a mixture of double and triple-stacked varieties. Although the diets were formulated to contain similar levels of nutrients, no confirmatory analyses were conducted. Mycotoxin analyses indicated 2.08 µg/kg total aflatoxins and 3.0 mg/kg total fumonisins in a pooled sample of the GM feed, and no aflatoxins and 1.2 mg/kg total fumonisins in a pooled sample of the non-GM feed. No other mycotoxins were detected. No significant diet-related differences were recorded for feed intake, weight gain, mortality or routine blood biochemistry measurements. However, the researchers reported that the GM diet was associated with gastric and uterine changes, although no histopathological details were presented. GM-fed pigs had uteri that were 25% heavier than in non-GM fed pigs. Also, the GM-fed pigs had a higher rate of severe gastric 'inflammation' with a rate of 32% of GM-fed pigs compared to 12% of non-GM-fed pigs. The incidence of 'inflammation' was higher in GM-fed males compared to non-GM fed males by a factor of 4.0, and in GM-fed females compared to non-GM fed females by a factor of 2.2.

The findings have been criticised by several researchers and agencies, including Food Standards Australia New Zealand (FSANZ), based on the methodology, the type of statistical analysis used and the source of funding (partly provided by an anti-GM group). As was pointed out, the total incidence and rates of mild and moderate 'inflammation' were higher in the non-GM-fed pigs than in the GM-fed pigs. The difference in gastric 'inflammation' reported by the researchers was in the severe inflammation category only, making it difficult to suggest a reasonable hypothesis for the entire findings on 'inflammation'. Also, the incidence of gastric 'inflammation' was assessed visually and was not followed up by histopathological examination of gastric tissue. Feed intake was not affected by the reported gastric 'inflammation'.

It is possible that the differences reported in uterine weight could be explained by the higher content of fumonisins in the GM diet. These mycotoxins are known to cause increased levels of estrogen in pigs, resulting in earlier puberty.

Like the Séralini et al. (2012a) report on GM maize (Chapter 4), it is unfortunate that the study was not conducted in an impeccable fashion so that the findings could be received with more credibility by the scientific community. Studies such as these need to be repeated in well-conducted, large coordinated trials by competent scientists and the findings confirmed before alarmist reports are released to the general public.

Pesticide and chemical residues

A concern is the possibility of pesticide residues in oilseeds, either as a result of crop spraying and dusting, residues in the soil, or residues of pesticides used to protect harvested seed during storage. One particular concern is that GM cultivars of oilseeds that are resistant to herbicides, such as Roundup, may contain excessive amounts of herbicidal residue.

Arregui et al. (2004) monitored glyphosate and AMPA residues in soyabean plants and grains in field crops in Santa Fe Province, Argentina. Individual soyabean plants were sampled from emergence to harvest, dried and ground. Residues of glyphosate in the beans ranged from 0.01 to 1.6 mg/kg after two applications and up to 1.8 mg/kg after three applications. AMPA residues in the beans ranged from 0 to 0.9 mg/kg. In soyabean leaves and stems, glyphosate residues ranged from 1.9 to 4.4 mg/kg. These data indicated very low concentrations of residual glyphosate. The higher concentrations were detected when glyphosate was sprayed several times during the crop cycle, and when treatments occurred as the soyabeans approached the flowering stage.

A related concern is that growing crops might take up chemicals in the environment and deposit them in the harvested beans. These chemicals include some metals and some persistent pesticides remaining in the environment as a result of past use. It is not known whether GM soyabeans are more or less likely to be affected than conventional soyabeans since no comparative data appear to be available.

Countries that have approved the use of GM soyabeans for food and feed but do not cultivate them import these beans mainly from the USA. US data on pesticide contamination are therefore of particular interest. The USDA (United States Department of Agriculture) tested a total of 5512 samples of domestically produced food for pesticide residues in 2006. Soyabeans and soyabean products showed a violation rate of zero.

The most recent USDA dataset (year 2013) included domestically produced soyabeans (300 samples from the 2011 crop year). The soyabean samples were analysed for 93 parent pesticides, metabolites, degradates and/or isomers, plus five environmental contaminants. The test included determining the presence of glyphosate and its metabolite AMPA (aminomethylphosphonic acid) in the soyabean samples. Of the 300 samples tested, 271 (90.3%) of samples contained glyphosate at levels ranging from 0.26 to 18.5 mg/kg. The AMPA metabolite of glyphosate was detected in 287 (95.7%) of the samples at levels ranging from 0.26 to 20 mg/kg. The EPA tolerance for glyphosate in soyabeans is 20 mg/kg, which was not exceeded in any sample. Residues of 12 pesticides were detected in the soyabean samples, all lower than the tolerances established by the EPA. No distinction was made between GM and conventional cultivars in this test, but since most of the soyabeans grown in the USA are GM, and since a high percentage of samples contained traces of glyphosate, it may be inferred that the results relate to GM (mainly glyphosate-tolerant) soyabeans.

The available data indicate that soyabeans can contain residues, but that the residues are too small to be of concern in relation to human health. Some of the residues found were at the limit of detection, although some soyabeans had glyphosphate residues close to the EPA tolerance level. More studies need to be carried out to provide comparable data on residue levels in GM and conventionally grown oilseeds.

Mycotoxin contamination

Mycotoxin contamination of soyabeans appears to be much lower on a worldwide basis than that of cereal grains (Binder et al., 2007). A total of 122 soyabean samples from the Asia/Pacific region were tested by these researchers. The occurrence and concentration of mycotoxins were found to be relatively low.

Aflatoxin was found in only 0.02% of samples (maximum of 13 μg/kg, median 9 μg/kg), zearalenone in 0.17 (maximum 1078 μg/kg, median 57 μg/kg), ochratoxin in 0.13 (maximum 11 μg/kg, median 7 μg/kg), and DON and fumonisins each in 0.07% of samples analysed (DON: maximum 1347 μg/kg, median 264 μg/kg; fumonisins: maximum 331 μg/kg, median 154 μg/kg). A similar pattern was observed for other protein sources such as lupinseed and other pulses that originated mainly from Oceania and were sourced in the Asia-Pacific region. Deoxynivalenol was detected in 13 of 32 soyabean samples from Europe (median 370 μg/kg, maximum 840 μg/kg), while 1 out of 18 samples tested positive for zearalenone. The authors did not provide any information on the relationship between the mycotoxin levels found and the regulatory tolerances prescribed in various countries.

No comparable data on mycotoxin contamination in conventional and GM cultivars of soyabeans appear to be available.

Nutritional and organoleptic qualities

An important question relating to GM oilseeds is whether the GM cultivars are of the same nutritional quality as conventional cultivars. This aspect can be assessed from chemical analyses and from feeding studies with farm animals. Examples of published data on chemical analyses of soyabeans, oil and fat-extracted soyabean meal are shown in Table 5.7, 5.8 and 5.9. These data show that conventional and GM cultivars of soyabeans, the extracted oils and the meals obtained after fat-extraction are similar in nutritional composition. In addition they show that the contents of anti-nutritive factors are similar.

Feeding studies have also been conducted with the different cultivars of soyabeans. Such studies yield valuable data on nutritive value since the diets of farmed species of animals and poultry have to be formulated precisely to achieve efficient growth and feed utilisation.

Hammond et al. (1996) conducted feeding studies with rats, broiler chickens, catfish and dairy cows to compare glyphosate-tolerant soyabeans/meal and conventional soyabeans/meal as dietary protein sources. Both beans and processed meal were incorporated into the diets at the same concentrations as used commercially. Growth, feed conversion (rats, catfish, broilers), fillet composition (catfish) and breast muscle and fat pad weights (broilers) were compared for animals fed the conventional and GM cultivars. Milk production, milk composition, rumen fermentation and nitrogen digestibility were also compared for dairy cows. In all studies, measured variables were similar for animals fed both types of soyabeans/meal.

Soyabean meal from glyphosate-tolerant or conventional soyabeans were compared in diets for growing-finishing pigs (Cromwell et al., 2002). Both cultivars of soyabeans were processed at the same plant. The compositions of the two types of soyabeans and the processed soyabean meal were found to be similar (Table 5.9). A total of 100 cross-bred pigs from 24 to 111 kg body weight were

Table 5.7 Published data on the nutritional composition of conventional and GM soyabeans.

Reference	Padgette et al. (1996)		Sanden et al. (2006)		Flachowsky et al. (2007)		Sissener et al. (2009)		Qi et al. (2012)	
	Conventional (A5403, DM basis)	GM 40-3-2 (Glyphosate-tolerant, DM basis)	Conventional	Roundup Ready	Conventional (isogenic)	Roundup Ready	Conventional (near-isogenic)	Roundup Ready (40-3-2)	Conventional (JACK)	305423 (Hi oleic, glyphosate-tolerant)
Component, g/kg Unless stated										
Dry matter	918	918	910	893	918	915	926	929	917.3	908.5
Crude protein	416	414	344	366	432	413	367	385	318	320
Crude fibre	71.3	68.7	–	–	33	41	–	–	72.3	71.4
Ether extract	155.2	162.8	186	182	251	255	226	204	196	180
Ash	50.4	52.4	47	50	–	–	48	51	52.2	48.2
Carbohydrate	381	371	–	–	–	–	–	–	–	–
Digestible carbohydrate	–	–	4	7	–	–	–	–	–	–
Starch	–	–	–	–	33	37	18.0	16.0	–	–
Monosaccharides										
Glucose	–	–	0.6	0.9	–	–	–	–	–	–
Fructose	–	–	0.1	0.4	–	–	–	–	–	–
Sucrose	–	–	21.0	15.3	–	–	–	–	–	–
Raffinose	–	–	4.2	3.7	–	–	–	–	–	–
Stachyose	–	–	15.5	14.8	–	–	–	–	–	–
Minerals										
Calcium	–	–	–	–	2.0	1.4	–	–	1.15	0.92
Phosphorus (total)	–	–	–	–	6.5	7.1	–	–	3.9	3.3
Amino acids g/kg										
Histidine	10.6	10.3	–	–	–	–	–	–	–	–
Isoleucine	17.8	17.3	–	–	–	–	–	–	–	–

Continued

Table 5.7 Continued

Reference	Padgette et al. (1996)		Sanden et al. (2006)		Flachowsky et al. (2007)		Sissener et al. (2009)		Qi et al. (2012)	
	Conventional (A5403, DM basis)	**GM 40-3-2 (Glyphosate-tolerant, DM basis)**	**Conventional**	**Roundup Ready**	**Conventional (isogenic)**	**Roundup Ready**	**Conventional (near-isogenic)**	**Roundup Ready (40-3-2)**	**Conventional (JACK)**	**305423 (Hi oleic, glyphosate-tolerant)**
Leucine	30.5	29.7	–	–	–	–	–	–	–	–
Lysine	26.1	25.6	–	–	28.3	26.5	–	–	–	–
Methionine	5.5	5.5	–	–	6.1	6.1	–	–	–	–
Cystine	6.0	6.2	–	–	–	–	–	–	–	–
Phenylalanine	19.7	19.0	–	–	–	–	–	–	–	–
Tyrosine	14.5	14.0	–	–	–	–	–	–	–	–
Threonine	16.0	15.6	–	–	17.8	17.2	–	–	–	–
Valine	18.5	18.0	–	–	–	–	–	–	–	–
Fatty acids g/kg total FA										
C16:0	111.9	112.1	–	–	103	106	–	–	–	–
C18:0	40.4	41.4	–	–	42	43	–	–	–	–
C18:1	197.2	197.4	–	–	247	252	–	–	–	–
C18:2	525.2	523.1	–	–	498	488	–	–	–	–
C18:3	80.2	82.3	–	–	58	62	–	–	–	–
Other										
Trypsin inhibitor TIU mg	43	45	6.7 g/kg	35.7 g/kg	–	–	–	–	–	–
Saponins mM/kg	–	–	5.6	3.8	–	–	–	–	–	–
Phytoestrogens mg/kg										
Daidzein	734	721	250	100	–	–	–	–	–	–
Genistein	833	830	410	470	–	–	–	–	–	–

Table 5.8 Published data on the composition of conventional and GM soyabean oil.

Reference	Padgette et al. (1996)	
	Conventional (A5403)	GM 40-3-2 (Glyphosate-tolerant)
Fatty acids g/kg		
C6:0	1.6	2.0
C7:0	3.9	4.6
C16:0	104.6	105.0
C17:0	1.2	1.4
C18:0	40.9	41.9
C18:1	211.3	214.1
C18:2	522.0	517.1
C18:3	74.1	75.1
C20.0	1.3	2.7
C20.1	1.7	1.7
C22.0	5.5	5.2
C24.0	1.5	1.6

used. They were grouped by sex (male castrates and gilts), and then pigs within each group were allotted at random to two dietary treatments from outcome groups of initial weight. All pigs were scanned for backfat thickness at 107 kg mean body weight and all males were slaughtered at the end of the test for carcass measurements and tissue collection. Rate and efficiency of weight gain, scanned backfat and longissimus area, and calculated carcass lean percentage were not significantly different for pigs fed diets containing conventional or glyphosate-tolerant soyabean meal. In general the carcass traits of males were similar for the two types of soyabean meal. Longissimus muscle samples from males fed conventional soyabean meal tended to have less fat than from those fed glyphosate-tolerant soyabean meal, but water, protein, and ash contents were similar. Sensory scores of cooked longissimus muscles were not influenced by diet.

Flachowsky et al. (2007) reported a study in which pigs were fed a diet containing full-fat roasted soyabeans from glyphosate-tolerant or a conventional (isogenic) cultivar, from about 65 to 100 kg body weight. Feed intake, daily weight gain, feed conversion and slaughter data were not significantly different in the two groups.

Herman et al. (2011) studied the growth performance of broiler chickens fed diets containing DAS-68416-4 or conventional soyabean meal. The DAS-68416-4 cultivar is a GM soyabean that provides tolerance to the herbicides 2,4-dichlorophenoxyacetic acid (2,4-D) and glufosinate. A six-week study was conducted, with diets containing toasted DAS-68416-4 or conventional (non-GM) soyabean meal (400, 360 and 320 g/kg in starter, grower and finisher diets,

Table 5.9 Published data on the nutritional composition of conventional and GM soyabean meal (fat-extracted).

Reference	Padgette et al. (1996)		Cromwell et al. (2002)		Chukwudebe et al. (2012)	
	Conventional DM basis	Roundup Ready DM basis	Conventional dehulled	Roundup Ready dehulled	Conventional	Roundup Ready
Component, g/kg Unless stated						
Dry matter	906.3	944.2	903	910	906	929
Crude protein	540	522	515	512	478	487
Crude fibre	45.1	46.8	34.1	31.2	95	92
Ether extract	9.3	11.0	15.9	8.9	10	13
Ash	68.3	66.9	–	–	61	62
Carbohydrate	383	400	–	–	357	367
Starch	–	–	–	–	48	62
Acid detergent fibre	–	–	35.0	39.4	69.8	70.9
Neutral detergent fibre	–	–	49.5	48.5	139.9	131.8
In vitro digestibility	–	–	–	–	607.6	662
Minerals						
Calcium	–	–	2.4	2.0	3.25	3.08
Phosphorus (total)	–	–	7.3	8.3	7.12	7.15
Phytate P	19.1	18.2	–	–	–	–
Amino acids g/kg						
Histidine	–	–	14.4	14.0	15.4	17.8
Isoleucine	–	–	23.2	22.7	19.3	21.1
Leucine	–	–	39.6	38.9	32.4	39.5
Lysine	–	–	31.6	30.9	26.5	32.3
Methionine			7.0	7.1	5.0	5.0
Cystine	–	–	7.6	8.0	5.8	6.5

Phenylalanine	–	–	25.8	25.1	21.6	26.6
Tyrosine	–	–	18.0	17.6	14.1	16.6
Threonine	–	–	19.5	19.4	16.9	20.0
Tryptophan	–	–	7.7	7.5	6.6	6.7
Valine	–	–	24.4	24.3	21.1	24.4
Monosaccharides						
Raffinose	9.0	9.4	–	–	–	–
Stachyose	60.0	55.2	–	–	–	–
Other						
Lectin HU/mg	<0.01	<0.01	–	–	–	–
Urease ΔpH	0.17	0.09	–	–	1.1	0.05
Trypsin inhibitor TIU/mg	2.6	2.6	–	–	–	–

respectively). Toasting the meal was required to inactivate endogenous antinutrients that make raw soyabeans unsuitable for feeding to non-ruminant animals and poultry. No significant differences in growth performance parameters were noted.

Tan et al. (2012) reported similar findings with broiler chickens fed diets containing GM glyphosate-tolerant or conventional soyabean meal for 42 days. No treatment differences were observed in growth performance or carcass parameters. The study also showed no adverse effects of the GM soyabean meal on the ileal or cecal microflora.

Mejia et al. (2010) evaluated the nutritional equivalency of meal from soyabean meal from DP-305423-1 (Hi Oleic) soyabeans, using laying hens from 25 to 36 weeks of age. For comparison, a near-isogenic cultivar was used in addition to two conventional soyabean meals. Body weight, hen-day egg production, total weight of eggs produced, feed consumption, and feed efficiency for hens fed the transgenic soyabean meal were not significantly different from those of hens fed the diet containing the near-isogenic soyabean meal. In addition, egg composition, Haugh unit (a measure of albumen quality) and egg weight were similar with all of the diets.

Sissener et al. (2009, 2011) reviewed the available data on glyphosate-tolerant soyabeans as a protein source in fish feed. Among the species tested were salmon, channel catfish, rainbow trout and zebrafish. Results indicated that both types of cultivar had similar quality as feed ingredients for fish.

The overall results indicated that glyphosate-tolerant GM soyabeans and soyabean meal are essentially equivalent in composition and nutritional value to conventional soyabeans and soyabean meal as a protein source for humans and farmed species of animals and poultry.

No studies on the organoleptic quality of GM soyabeans appear to have been conducted.

Canola (rapeseed)

Canola now ranks second in world production of oilseed crops after soyabeans (Table 5.1), the seed containing about 44% oil. It is grown in several regions, including North America, Europe and Australia.

The transformation of rapeseed into canola was made by Canadian plant breeders, starting in the 1940s. The need for increased Canadian rapeseed production arose from a critical shortage of rapeseed oil, following the World War II blockade that prevented importation from European and Asian sources. The oil was urgently needed as a lubricant for marine engines in naval and merchant ships. Rapeseed was one of the very few oil sources that could be grown successfully in temperate climates.

Later efforts towards breeding rapeseed with improved agronomic characteristics and oil composition were conducted, principally by Keith Downey at the University of Saskatchewan and Baldur Stefansson at the University of Manitoba. Two features of the crop had to be changed to make it suitable as a food crop: the oil contained a high level of erucic acid and other long-chain fatty acids known to be detrimental to human health (e.g. Borg, 1975); and a high level of glucosinolates in the seed made the by-product remaining after oil extraction unsuitable for human or animal feeding. The glucosinolates are compounds present naturally in several members of the Brassicaceae family (cabbage, mustard, rapeseed, etc.) that are released by the enzyme myrosinase when the seed is damaged. Several of the released compounds (such as isothiocyanates) interfere with growth or are toxic to animals. The glucosinolate–myrosinase system is present in the plants as a natural defence system against damage from insects and so on.

The scientists were able to develop varieties low in erucic acid and meal low in glucosinolates using traditional breeding techniques. Studies indicated that the oil had a fatty acid profile that made it suitable as a salad and cooking oil. GRAS (Generally Recognized as Safe) status was granted by the FDA after a traditional FDA review of the data in the USA in 1985. Other studies indicated that the meal remaining after oil extraction was an excellent source of protein for poultry and livestock.

A change of name was then appropriate to differentiate the new cultivars from unimproved rapeseed meal. The name selected was Canola and the agreed international definition was as follows: 'Seeds of the genus Brassica (*Brassica napus, Brassica rapa* or *Brassica juncea*) from which the oil shall contain less than 2% erucic acid in its fatty acid profile and the solid component shall contain less than 30 micromoles of any one or any mixture of 3-butenyl glucosinolate, 4-pentenyl glucosinolate, 2-hydroxy-3 butenyl glucosinolate, and 2-hydroxy-4-pentenyl glucosinolate per gram of air-dry, oil-free solid.' In addition, the meal is required to have a minimum of 350 g per kg crude protein and a maximum of 120 g per kg crude fibre, moisture-free basis. The designation is licensed for use in at least 22 countries.

In the USA there is also a requirement that 'before the removal of dockage the seed shall contain not more than 10.0% of other grains for which standards have been established under the United States Grain Standards Act'.

More recent developments have involved the use of hybrid cultivars and the introduction of herbicide tolerance. Roundup Ready and Liberty Link cultivars were produced using genetic modification, and Clearfield cultivars were developed using mutagenesis. The transgenic gene inserted into the canola plant to produce Roundup Ready and Liberty Link InVigor herbicide resistance is a protein. All protein is removed from the canola oil during processing; therefore canola oil contains no GM material and is identical to canola oil from conventional canola plants.

GM canola cultivars have been approved for food and feed use in North America and several other countries (Appendix Table 5.1A).

The Australian Federal Office of the Gene Technology Regulator (OGTR) approved the commercial release of Bayer CropScience's InVigor® GM canola and Monsanto Australia's Roundup Ready® GM canola in 2003. The OGTR concluded that these varieties of canola pose no greater risk to human health or the environment than conventionally bred canola. GM canola was the first GM agricultural crop used primarily for food that was approved for general release in Australia. GM cottonseed had already been approved and was being grown widely in New South Wales and Queensland.

However, the approval process in Australia was affected by the same differences of opinion relating to GM food production existing in other countries. Approval of GM canola for cultivation in Australia led to some dissention in the state of Victoria and in 2004 the State Government concluded that the timing was not appropriate for the full commercial release of the two varieties of GM canola due to '... divisions and uncertainty within industry, the farming sector and regional communities about the impact of GM canola on markets'. Subsequently the Minister for Agriculture issued an Order declaring a four-year moratorium in Victoria on commercial-scale planting. In May 2007 the State Government established an Independent Panel, chaired by Sir Gustav Nossal, to review the moratorium on GM canola. The Panel examined information from public submissions, undertook discussions with a range of stakeholders and analysed a number of independent reports, including a detailed economic analysis prepared specifically for the Review. The Panel submitted its report in late October 2007. The Panel found that there were no valid trade or market grounds to justify a moratorium that prevented farmers from having a choice about the type of canola they wished to grow. On 27 November 2007 the State Premier announced that the government would accept the Panel's recommendation to allow the moratorium on GM canola to expire on 29 February 2008.

Another issue, that of the patent rights applying to a living organism, was highlighted when a Canadian farmer (Percy Schmeiser) was sued by Monsanto for growing glyphosate-tolerant canola without a license. The case sparked widespread interest worldwide as a case of 'David vs. Goliath'. Because the seeds of GM canola contain the gene for resistance to glyphosate and if planted will produce a new crop of GM canola, Monsanto requires that farmers growing the crop sign a Technology Use Agreement (TUA) when purchasing Roundup Ready seeds. The TUA is a license that allows farmers to use the seed, provided that no seed is kept for replanting or for inventory or sold to other farmers. Inspectors enforce the contract by inspecting fields.

In his defence Schmeiser claimed that it was not his intention to grow Roundup Ready canola and that the GM must have ended up in his field by accident, probably by falling from a truck. As a result he did not infringe Monsanto's patent. On the other hand, Monsanto argued that it was highly improbable that Schmeiser did not know about Roundup Ready canola, given its popularity and the publicity surrounding it. They claimed that Schmeiser willingly grew Roundup Ready without a license, and that by harvesting and selling the resulting grain he made a profit from which Monsanto should have benefitted. These claims, arguments and counter arguments continued for six years, involving hearings in three different courts. At one of the hearings when Mr Schmeiser's appeal was rejected, a judge made the point that independent tests had shown 1030 acres of Mr Schmeiser's canola to consist of plants that were 95–98% tolerant to Roundup herbicide. Accordingly the judge ruled that the seed could only have been of commercial quality and could not have occurred on Mr Schmeiser's land by accident. Finally, in May 2004, the Supreme Court of Canada ruled against Schmeiser, stating that he had violated the company's rights by growing the GM plants without a license. However, the court also overturned a previous judgement that ordered Schmeiser to pay back the sales profit to Monsanto and to pay for the company's legal fees.

During the legal wrangling, Schmeiser argued that since the Supreme Court of Canada in 2002 had ruled against patenting the Harvard transgenic mouse, the Supreme Court should also rule in his favour. His argument was that despite Monsanto's holding of a patent for the gene responsible for Roundup resistance, it could not claim to have property rights on the plants containing the patented genes, when the plants themselves are not being used to exploit the benefits provided by the gene. The Supreme Court did not accept that argument, ruling that since Monsanto holds a patent for a gene inside canola seeds, it therefore also has full control over the use of the plant.

The Harvard mouse case was somewhat different in that it involved a higher life form, namely a mammal. In 1985 the Canadian Supreme Court heard an appeal from Harvard University (Harvard College vs. Canada (Commissioner of Patents)), in an action seeking a Canadian patent for a transgenic mouse whose genome had been genetically altered to contain a cancer-promoting gene (oncogene). The claims to the mouse were rejected by the Commissioner of Patents and Patent Appeal Board as being outside the definition of an invention. That decision was upheld by the Federal Court Trial Division, but reversed on appeal. After hearing the case, the Supreme Court in 2002 ruled that the transgenic mouse was not patentable subject matter in Canada under the existing patent legislation. However, on 7 October 2003, Canadian patent 1,341,442 was granted to Harvard College. The patent was amended to omit the 'composition of matter' claims on the transgenic mice. The patent remains valid until 2020. In 1988 the US Patent and Trademark Office (USPTO) granted U.S. Patent 4,736,866 (filed Jun 22, 1984, issued Apr 12, 1988, expired April 12, 2005) to Harvard College claiming 'a transgenic non-human mammal whose germ cells and somatic cells contain a recombinant activated oncogene sequence introduced into said mammal ...'. The claim explicitly excluded humans, apparently reflecting moral and legal concerns about patents on human beings, and about modification of the human genome. Two separate patents were issued to Harvard College covering methods for providing a cell culture from a transgenic non-human animal (U.S. Patent 5,087,571; filed Mar 22, 1988, issued Feb 11, 1992, expired Feb 11, 2009) and testing methods using transgenic mice expressing an oncogene (U.S. Patent 5,925,803; filed Sep 19, 1991, issued Jul 20, 1999, expires July 20, 2016). A European patent was granted in 1992 (EP 0169672).

All of these legal decisions indicate clearly that companies have the legal right to patent genetically modified organisms, including food crops.

Safety aspects

Allergenicity

The structure of canola protein (conventional and GM) does not suggest that it is likely to be allergenic. This has been confirmed by the approval of several GM cultivars of canola by government agencies worldwide, assessing possible allergenicity on the basis of evidence submitted by applicant companies. EFSA, on the basis of available evidence, does not consider canola to be a common allergenic food (CEC, 2007). In addition, EFSA (2009b) reported that a search in the open scientific literature did not identify any data on the levels of any allergenic proteins in the pollen of the GM canola cultivar GT73 (glyphosate-tolerant).

A relative lack of peer-reviewed papers on allergenicity compared with publications on soyabeans also suggests that the possible allergenicity of GM canola is perceived by the scientific community as being much less of a problem than that in soyabeans. Another probable reason is that soyabeans and canola are used differently for food and feed use. Soyabeans (whole and processed) and the oil are used for food, the fat-extracted meal being used in animal feeding. Canola is processed into oil for food use, the fat-extracted meal being used in animal feeding. Therefore even if canola protein possessed allergenic properties, this would present a potential problem for animals not humans.

EFSA (2009a) issued grounds for its continued approval of InVigor canola (ACS-BN005-8 × ACS-BN003-6) for food (oil) and feed use. This cultivar has been modified using conventional crossing and biotechnology to possess traits for glufosinate tolerance, male sterility and fertility restoration. In making its assessment EFSA considered the information provided in the original application (2005) along with additional information submitted by the applicant. The new data included molecular and bioinformatic analyses and a 42-day feeding study on chickens. Data from the animal study showed that the tested material from this canola cultivar was nutritionally equivalent to its non-GM counterpart. In addition, bioinformatic studies using updated databases confirmed that no relevant similarities exist between the newly expressed proteins and known allergens or toxic proteins. Therefore EFSA confirmed the conclusions of its previous scientific opinion of 2005 that this canola cultivar (and its related cultivars) is unlikely to have any adverse effect on human and animal health or, in the context of its proposed uses, on the environment. This assessment also applies to the products of the cultivar.

EFSA (2009b) issued the grounds for its continued approval of MON-00073-7 (Westar Roundup Ready) canola for food (oil) and feed use. This cultivar has been modified genetically to possess traits for glyphosate tolerance and resistance. In making the assessment EFSA considered both the original evidence supporting the existing authorisation as well as additional information and information published in the scientific literature. The new data in the application included bioinformatic analyses using updated databases, which confirmed that no relevant similarities exist between the newly expressed proteins and known allergens and toxic proteins. EFSA reiterated, therefore, its previous conclusions that canola cultivar GT73 is unlikely to have an adverse effect on human and animal health and on the environment, in the context of its proposed uses. This also applies to products of the cultivar.

Governmental agencies in other countries (e.g. USA, Canada and Australia/New Zealand) have made similar assessments and have reached similar conclusions.

It is, therefore, unlikely that allergenicity is a problem with GM cultivars of canola. Furthermore, it has been noted that there is no documented proof that any approved GM commercial crop or crop product has caused allergic reactions (Goodman et al., 2008).

Toxicity

As with other GM foods, an important issue is the extent of breakdown of modified DNA in the gut during digestion and absorption. This issue has been studied in several animal species.

Alexander et al. (2002) studied the impact of feed processing and in vitro ruminal cultures on the persistence of recombinant and canola-specific endogenous DNA, using various canola substrates (whole seed, cracked seed, meal and diet). Despite this study being an analysis of plant DNA throughout in vitro incubations, the use of intact plant substrates instead of isolated DNA provides important information on the fate of endogenous and recombinant canola DNA in the ruminal environment. The finding that plant DNA was not detected in the supernatant but only in pellets containing plant debris, suggests that the presence of plant DNA in ruminant systems is directly related to intact plant cells. Essentially, disappearance of plant DNA fragments is analogous to the digestion of plant cells in the rumen. This appears to be the limiting step for transformation of rumen bacteria, because once DNA is released into the aqueous ruminal environment it is degraded almost immediately. Plant DNA fragments reaching the intestinal tract of ruminants and being available for absorption are therefore most likely the result of undigested plant material passing through the rumen to the intestines, where plant cells can then be degraded, releasing DNA.

Alexander et al. (2004) assessed the stability of the cp4 epsps transgene from glyphosate-tolerant canola during digestion in sheep. The concentration of cp4 epsps DNA in the diet declined more rapidly in duodenal fluid at pH 5 as compared to pH 7. The authors attributed this mainly to microbial activity at pH 7 and perhaps to plant endogenous enzymes at pH 5. This study showed that digestion of plant material and release of transgenic DNA can occur in the ovine small intestine. However, free DNA is rapidly degraded at neutral pH in duodenal fluid, thus reducing the likelihood that intact transgenic DNA would be available for absorption in the distal ileum.

All of the above data indicate that the genetic modification of canola does not result in its being toxic as a feed for animals.

Nutritional and organoleptic qualities

The US Food and Drug Administration, the Canadian Food Inspection Agency and the Australia/New Zealand Food Standards Authority are among agencies reporting that GM canola and conventional canola, and food/feed obtained from them, are of similar nutritional composition.

Data relating to canola seed from the crop years 1996–2001 were compiled by Daun (2004). According to these data GM canola seed generally had a higher oil content and a lower crude protein content than conventional non-GM canola seed, but not in all crop years. The oil from GM canola seeds tended to be slightly higher in unsaturated fatty acids than the oil from GM canola seed, and there was no difference in the contents of erucic acid, unsaturated fatty acids or linolenic acid between the two types of oil.

Results from the latest crop year (2001) are shown in Table 5.10, indicating close similarity in nutritional composition between conventional and GM canola seed.

The Australia/New Zealand Food Authority (FSANZ, 2000) published a full assessment report related to an application for the approval of glyphosate-tolerant

Table 5.10 Published data on the nutritional composition of conventional and GM canola seed (Daun, 2004).

	Conventional (Commercial, non-GM) 2001 crop year	Herbicide tolerant (Commercial) 2001 crop year
Component, g/kg Unless stated		
Crude protein 85 g/kg DM	223.2	224.9
Ether extract	426.8	426.1
Fatty acids mg/g oil		
Saturated	70.8	71.7
Linolenic acid	90.4	96.6
Unsaturated index[1]	1.286	1.314
Erucic acid	0.3	0.2
Glucosinolates μmoles/g		
Total	11.0	10.1

[1] Mean double bonds per fatty acid.

canola (cultivar GT73, Roundup Ready) as a food source. The evaluation included data on major constituents, nutrients, anti-nutritional factors and natural toxicants in the GM cultivar and in its non-transgenic parent cultivar Westar. All new varieties of canola oil must be analysed to ensure they meet CODEX (2001) specifications for canola. This includes 18 quality analyses that define canola oil and includes a fatty acid analysis and 17 other food chemical tests. The second year of data included analyses from GM canola which had been treated with glyphosate herbicide. An assessment of canola oil from the GM cultivar found it to be comparable in all respects to oil derived from conventionally produced canola. The results for all analyses of glyphosate-tolerant canola line GT73 were within CODEX specifications except for the values for four minor fatty acids: arachidic acid (C20:0), behenic acid (C22:0), lignoceric acid (C24:0) and nervonic acid (C24:1). However, the value for these four fatty acids exceeded the CODEX specifications in both the conventional cultivar and the GM (GT73) cultivar. This result is considered to reflect the natural variation within canola rather than any effect of the genetic modification on the canola line. Furthermore, there is no anti-nutritional or toxicological significance associated with these fatty acids. With the exception of four slightly elevated minor fatty acids (20:0, 22:0, 24:0 and 24:1) in GT73, the oil derived from glyphosate-tolerant canola line GT73 is comparable to oil derived from Westar canola.

Detailed data (Table 5.11) on the composition of the oil derived from glufosinate-ammonium tolerant and pollination controlled canola were published by Food Standards Australia New Zealand (FSANZ, 2003). The fatty acid composition of

Table 5.11 Fatty acid profiles of refined, bleached and deodorized oil from conventional and GM cultivars of canola (FSANZ, 2003).

Fatty acid, % of FA profile	1992		1993		
	Conventional (Westar)	GM (GT173, Roundup Ready)	Conventional (Westar)	GM (GT173, Roundup Ready)	GM (GT173, Roundup Ready) treated
16:0	3.9–4.2	3.98	3.8–4.3	4.1	4.1
16:1	0.3–0.4	0.32	0.2	0.2	0.2
18:0	1.4–2.0	1.72	1.4–1.9	1.7	1.8
18:1	58.8–62.5	61.4	60.1–62.8	62.9	62.8
18:2	18.9–20.2	18.9	18.8–20.6	18.7	18.7
18:3	8.1–12.1	10.8	8.6–10.13	9.65	9.73
20:0	0.6–0.8	0.72	0.6–0.7	0.65	0.68
20:1	1.7–2.0	1.58	1.57–2.0	1.49	1.51
20:2	0.1	0.17	0.1	0.09	0.1
22:0	0.3–0.4	0.4	0.4–0.5	0.4	0.43
22:1	0.3–0.6	0.12	0.15–0.57	0.04	0.0

the oil derived from a number of transgenic and conventional (isogenic) cultivars was analysed to provide data on 11 different key fatty acids, including erucic acid (C22:1). The samples included control varieties (e.g. Drakkar), different generations of male sterile lines (Ms1 and Ms8), different generations of fertility restorer lines (Rf1, Rf2 and Rf3), multiple backcrosses of Ms and Rf lines in different canola varieties and unrestored (Ms/control) and restored (Ms/Rf) hybrids. The seed samples were collected from plants grown at locations in Belgium, France, Sweden, Canada and the United Kingdom. The data showed that the contents of 11 key fatty acid components were comparable in all of the oils tested, both from a number of GM canola cultivars and a range of conventional (non-GM) cultivars. Variations due to environmental conditions were found to be greater than the variation between GM and conventional canola oil. The agency concluded, therefore, that oil derived from the genetically modified canola lines (T45, Topaz 19/2, Ms1, Ms8, Rf1, Rf2, Rf3 and their crosses), is equivalent to oil from non-GM canola in terms of safety and nutritional properties.

The meal from oil-extracted GM canola has been shown to be comparable in nutritional composition to the meal obtained from conventional canola cultivars (Table 5.12). In addition, the results of feeding tests on swine (Aalhus et al., 2003; Caine et al., 2007), trout (Brown et al., 2003), sheep (Stanford et al., 2003) and chickens (Stanisiewski et al., 2002; Taylor et al., 2004) confirm that the meals obtained from GM and conventional cultivars of canola are of similar nutritive value. Sissener et al. (2011) reviewed the available data on glyphosate-tolerant Ready canola meal as a protein source in fish feed. Among the species tested were salmon, channel catfish, rainbow trout and zebrafish. Results indicated that both types of cultivar had similar qualities as feed ingredients for fish.

Cottonseed (*Gossypium hirsutum* L.)

Cottonseed is important in world oilseed production as a source of oil and fibre, the major producing countries being China, the United States, Latin America, India and Pakistan. This crop ranks third in importance as a source of edible oil, after soyabeans and canola (Table 5.1). As with canola, only the oil obtained from cottonseed is used as human food. Similar to the situation with other oil crops, no transgenic material from GM cultivars is present in the oil; the transgenic material remains with the protein in the residue after oil extraction.

Whole, delinted, cottonseed is used commonly in feed for dairy cows. The meal remaining after oil extraction is also used in many countries as a protein source in a range of animal feeds.

GM cottonseed has been approved as a food source in several countries, including the EU, USA, Japan, China, India, Argentina and Brazil. The cultivars used have been modified genetically to provide resistance to insects and fungi, also herbicide tolerance.

Table 5.12 Published data on the nutritional composition of conventional and GM canola meal.

Reference	Brown et al. (2003)		Stanford et al. (2003)		Daun (2004)		Taylor et al. (2004)		Caine et al. (2007)	
	Conventional (Westar, DM basis)	Roundup Ready (GT73, DM basis)	Conventional (DM basis)	Roundup Ready (DM basis)	Conventional (Commercial) 2001 crop year	Herbicide tolerant (Commercial) 2001 crop year	Conventional (46A65)	Roundup Ready (RT73)	Conventional	Roundup Ready
Component, g/kg Unless stated										
Dry matter	892	890	945.4	948.9	–	–	954.5	951.9	–	–
Crude protein	368	373	443.2	446.8	417.8	420.1	428	427	–	–
Crude fibre	129	131	–	–	–	–	117	119	–	–
Ether extract	40	31	59.6	53.9	–	–	51.5	39.6	–	–
Ash	65	66	–	–	–	–	–	–	–	–
Organic matter	–	–	931.0	930.4	–	–	–	–	–	–
ADF	–	–	199.9	201.2	–	–	–	–	–	–
NDF	–	–	262.3	263.5	–	–	–	–	–	–
Amino acids mg/kg [1]										
Histidine	–	–	–	–	–	–	1.16	1.19	–	–
Isoleucine	–	–	–	–	–	–	1.79	1.81	–	–
Leucine	–	–	–	–	–	–	3.07	3.15	–	–
Lysine	–	–	–	–	–	–	2.45	2.39	–	–
Methionine	–	–	–	–	–	–	0.81	0.84	–	–
Cystine	–	–	–	–	–	–	1.01	1.04	–	–
Phenylalanine	–	–	–	–	–	–	1.72	1.75	–	–

Continued

Table 5.12 Continued

Reference	Brown et al. (2003)		Stanford et al. (2003)		Daun (2004)		Taylor et al. (2004)		Caine et al. (2007)	
	Conventional (Westar, DM basis)	**Roundup Ready (GT73, DM basis)**	**Conventional (DM basis)**	**Roundup Ready (DM basis)**	**Conventional (Commercial) 2001 crop year**	**Herbicide tolerant (Commercial) 2001 crop year**	**Conventional (46A65)**	**Roundup Ready (RT73)**	**Conventional**	**Roundup Ready**
Glucosinolates										
μmoles/g										
Allyl	–	–	–	–	–	–	–	–	<0.05	<0.05
3-butenyl	–	–	–	–	–	–	–	–	4.93	4.01
4-pentenyl	–	–	–	–	–	–	–	–	0.54	0.65
2-OH-3-butenyl	–	–	–	–	–	–	–	–	9.38	6.10
2-OH-4-butenyl	–	–	–	–	–	–	–	–	0.22	0.12
methylthiobutenyl	–	–	–	–	–	–	–	–	0.14	0.12
phenylethyl	–	–	–	–	–	–	–	–	0.15	0.27
methylthiopentenyl	–	–	–	–	–	–	–	–	0.08	0.10
4-OH-4-benzyl	–	–	–	–	–	–	–	–	0.18	<0.05
3-methylindolyl	–	–	–	–	–	–	–	–	0.71	0.40
40 OH-3-methylindolyl	–	–	–	–	–	–	–	–	3.87	3.27
Total	–	–	–	–	–	–	–	–	20.20	15.04

[1] The values for amino acids are very low and the unit used may be incorrect. However the comparison is valid.

Adoption of GM cultivars of cottonseed is highest in Australia, with almost 100% of the cotton crop being composed of transgenic cultivars. Australia was one of the first cotton producing nations (the other was the USA) to grow transgenic cultivars, starting in 1996. As with other transgenic crops, each genetic trait is individually assessed on a case-by-case basis by the Office of Gene Technology Regulator (OGTR), Food Standards Australia New Zealand (FSANZ) and the Australian Pesticides and Veterinary Medicines Authority (APVMA). Both FSANZ and the OGTR consider the allergenic potential of GM food and crops in their assessments. To date, all approved GM foods have been found to be as safe as comparable non-GM foods.

One of the admirable features of the Australia/New Zealand regulatory system is that the results of assessments relating to the safety of foods derived from GM crops are available to the public. For instance, the assessment of foods derived from glyphosate-tolerant cotton line GHB614 (FSANZ, 2008). In that assessment FSANZ stated that it had completed a comprehensive safety assessment of foods derived from glyphosate-tolerant cotton line GHB614, which included consideration of (i) the genetic modification introduced into the plant; (ii) the potential toxicity and allergenicity of the novel protein; and (iii) the composition of GHB614 cottonseed, compared with that from conventional cotton varieties. No public health and safety concerns were identified in the safety assessment. On the basis of the available evidence, including detailed studies provided by the Applicant, food derived from glyphosate-tolerant cotton line GHB614 was considered as safe and wholesome as food derived from other commercial cotton varieties.

Another admirable feature of the Australia/New Zealand system is that the assessment process allows for public comment. For instance, the initial assessment related to GHB614 was advertised for public comment between 12 December 2007 and 6 February 2008 and 13 submissions were received. The Draft Assessment Report was advertised for public comment between 6 August 2008 and 17 September 2008 and 82 submissions were received.

Another agency that publishes its assessments is EFSA (its opinions only since the final decisions on approval are made by the EU legislature). For instance, following the submission of an application (EFSA-GMO-NL-2005-16) under Regulation (EC) No 1829/2003 from Dow AgroSciences, the EFSA Scientific Panel on Genetically Modified Organisms was asked to deliver a scientific opinion on the safety of the insect resistant GM cotton 281-24-236 × 3006-210-23 (Unique identifier DAS-24236-5 × DAS-21023-5) for food and feed uses, import and processing. This cultivar (Widestrike) has insect resistance traits against certain lepidopteran pests, such as *Heliothis zea* (cotton bollworm), *Heliothis virescens* (tobacco budworm) and *Pectinophora gossypiella* (pink bollworm). Assessment of the available evidence (EFSA, 2010) indicated that cotton 281-24-236 × 3006-210-23 is compositionally, agronomically and phenotypically equivalent to its conventional counterpart, except for the expression of Cry1Ac,

Cry1F and PAT proteins. The safety assessment identified no concerns regarding toxicity and allergenicity of cotton 281-24-236 × 3006-210-23. A feeding study on broiler chickens confirmed the nutritional equivalence of this GM cotton to its conventional counterpart and two commercial non-GM cotton varieties. The intended uses of cotton 281-24-236 × 3006-210-23 excluded cultivation within the European Union. In addition the EFSA GMO Panel considered scientific comments raised by Member States in concluding that cotton 281-24-236 × 3006-210-23 is unlikely to have any adverse effect on human and animal health or the environment in the context of its intended uses.

A similar scientific opinion issued by EFSA (2011b) evaluated the risk assessment for the authorisation for continued marketing of genetically modified herbicide-tolerant cotton MON 1445 for food and feed produced from it. Stability of the inserted DNA was confirmed over several generations. Bioinformatic analyses and levels of recombinant proteins did not show any safety issues. No biologically relevant differences were identified in the compositional, phenotypic and agronomic characteristics of cotton MON 1445 in comparison to its conventional counterpart and its composition fell within the range of non-GM cotton varieties, except for the inserted CP4 EPSPS and NPTII proteins. No toxicity and allergenicity issues related to the newly expressed proteins were identified. Products from cotton MON 1445 did not contain viable plant parts. Considering the expected low frequency of gene transfer from cotton MON 1445 to bacteria compared to that between bacteria, and the very low exposure to DNA from cotton MON 1445, the EFSA GMO Panel concluded that the contribution of horizontal gene transfer to the environmental prevalence of NPTII genes was negligible. Potential interactions of cotton MON 1445 with non-target organisms and the abiotic environment were not considered to be an issue because of low exposure levels. The EFSA GMO Panel concluded that the information available for cotton MON 1445 addresses the questions raised by the member states and that MON 1445-derived products are as safe as products derived from the conventional counterpart in the context of their intended uses.

In addition, EFSA (2011c) issued a scientific opinion evaluating the risk assessment for the genetically modified insect-resistant cotton MON531. Stability of the inserted DNA was confirmed over several generations. Bioinformatic analyses and the levels of recombinant proteins did not show any safety concerns. Analysis of compositional, phenotypic and agronomic characteristics indicated that MON531 is not different from its conventional counterpart and is compositionally within the range observed among conventional cotton varieties, except for the inserted proteins Cry1Ac and NPTII. Safety assessment of Cry1Ac and NPTII proteins and cotton MON531 identified no concerns regarding potential toxicity and allergenicity. Products from MON531 did not contain viable plant parts. Considering the expected low frequency of gene transfer from MON531 to bacteria compared to that between bacteria, and the very low exposure to MON531 DNA, the GMO Panel concludes that gene transfer from MON531 to bacteria does not indicate a risk to human and animal health or to the environment in the context of its intended uses. The exposure of potentially sensitive non-target organisms to Cry1Ac protein is likely to be low and of no biological relevance. The EFSA GMO Panel considered that information available for cotton MON531 addresses the questions raised by the member states and that cotton MON531, as described in this application, is as safe as its conventional counterpart and is unlikely to have adverse effects on human and animal health or the environment in the context of its intended uses.

Safety aspects

Allergenicity

No documented studies indicating that cottonseed protein is allergenic and that genetic manipulation alters its allergenicity appear to have been published. Also, this feature of GM cottonseed has been considered by regulatory authorities in several countries and shown to be absent before GM cultivars have been approved.

As indicated in the section above, the Australia/New Zealand food authority and EFSA are among agencies that have examined the possibility of allergenicity in GM cottonseed and have found no evidence of allergenicity in GM cottonseed or its products.

Toxicity

Global cottonseed production has the potential to provide the protein requirements for half a billion people per year. However, it is underutilised because of the presence of a toxic compound, gossypol, within the seed glands. Gossypol is a natural toxin present in the cotton plant that protects it from insect damage. Therefore, elimination of gossypol from cottonseed has been a longstanding goal of geneticists. Attempts were made to meet this objective by developing so-called 'glandless cotton' in the 1950s by conventional breeding techniques, but the glandless varieties were not a commercial success because of the increased susceptibility of the plant to insect damage. The free gossypol content of cottonseed meal is similar in conventional and GM cultivars and the current approach is to control the content to acceptable levels by processing conditions (heating and crushing).

The inclusion of cottonseed meal in pig and poultry diets is limited because of the deleterious effects produced by residual-free gossypol found in the pigment glands of the seed. Cattle and sheep are less susceptible than pigs to gossypol poisoning because gossypol binds to proteins in the rumen. This problem does not occur with glandless cultivars of cottonseed.

Sunilkumar et al. (2006) described a biotech method of reducing the gossypol content of cottonseed by interfering with the expression of the δ-cadinene synthase gene during seed development. This resulted in a significant reduction in cottonseed-gossypol levels in a stable and heritable manner. The levels of gossypol and related terpenoids in the foliage and floral parts were not decreased, therefore their function in protecting the cotton plant against insects and diseases remained unaffected. These results suggest a valuable new approach to the gossypol problem associated with the use of cottonseed in human and animal diets.

Animal studies have been used to evaluate the safety of GM cottonseed as a food and feed source. For instance, Dryzga et al. (2007) compared diets containing 100 g/kg GM insect-protected cottonseed (WIDESTRIKE) or cottonseed meal prepared from four conventionally bred lines of cotton (three commercial non-transgenic lines (PHY72, PHY78 and 98M-2983), and a related non-GM control (PSC355) in a dietary study with rats. Following 90 days of feeding, no adverse effects were observed clinically or in any of the parameters measured in this study. The researchers concluded that rodent diets containing 100 g/ kg GM cottonseed meal did not produce any deleterious effects.

Singh et al. (2003) investigated the effect of feeding GM cottonseed (insect-resistant: B.t.) and non-GM cottonseed on blood biochemical constituents in lactating Murrah buffaloes. Animals received a diet with 395 g/kg non-GM or GM cottonseed in the concentrate mixture. The experiment lasted for 35 days. There was no significant difference in the dry matter intake between the two groups of buffaloes, or in weight gain, total erythrocyte count, hemoglobin content or packed cell volume. The concentration of plasma glucose, serum total proteins, albumin, globulin, triglycerides and high density lipoprotein did not differ significantly between the two groups. It was concluded that transgenic and non-transgenic cottonseeds have similar nutritional values and have similar effects on the health status of buffaloes as assessed by blood chemistry.

In a study related to potential toxicity, Jennings et al. (2003b) determined whether transgenic or endogenous plant DNA was detectable in dairy milk or beef organs. The animals were fed a diet containing conventional or GM cottonseed or maize. Genomic DNA was extracted from milk and analysed for fragments of the cry1Ac transgene and an endogenous cotton gene, acp1, from cows fed a diet containing whole cottonseed. Additionally, milk, liver, kidney and spleen were assessed for fragments of the cry1Ab transgene and an endogenous maize gene (sh2), from animals fed a diet containing maize. These results indicated that DNA, whether derived from conventional or transgenic cottonseed or maize, is not present at detectable levels in bovine milk or organs.

These research findings confirm the conclusions of regulatory authorities that GM cottonseed is as safe as conventional non-GM cottonseed as a foodstuff and is unlikely to have adverse effects on human and animal health.

Nutritional and organoleptic qualities

Composition of the oil from GM cultivars of cottonseed has been studied by regulatory authorities as part of the approval process. For instance, the UK Advisory Committee on Novel Foods and Processes (2002) issued a favourable opinion on the equivalence of cottonseed oil and oil derived from glyphosate-tolerant cottonseed. This committee also noted that the host plant used for the genetic modification has a history of use as a source of food products.

The Committee obtained analytical information from the applicant company regarding the possible presence of protein and DNA in the refined oil and was satisfied that processing eliminated these products in the refined oil. The refined cottonseed oil from RRC line 1445 and the isoline parent line Coker 312 were found to be comparable in quality to commercially processed cottonseed oil from non-GM sources. These results indicate that the levels of the important fatty acids, as well as the toxic cyclopropenoid fatty acids, are comparable in refined cottonseed oil fractions produced from RRC line 1445 and the control line. There was no detectable gossypol in refined oil, and levels of alpha-tocopherols were also comparable in both lines and were within the range published for other commercial varieties. The insertion of the genes to provide herbicide tolerance did not alter the processing characteristics of the cottonseed or the quality of refined oil.

Extensive compositional analyses were performed by the Committee on cottonseed from Roundup® Ready cotton line 1445 and the parental line obtained from field trials in the USA. Although there were clear fluctuations between the control and test lines over a two-year period, the differences fell within the limits of previously reported ranges found in the literature for refined cottonseed oil. It was also demonstrated that the introduction of the EPSPS protein, which catalyses a step in the aromatic amino acid biosynthetic pathway, and the application of Roundup® herbicide on the modified line had no effect on the amino acid profile of the seeds. In addition the Committee reported that the values obtained for the major toxicants were shown to be statistically equivalent between the two lines throughout the field trials.

Based on the available data the Committee concluded that line RRC 1445 is comparable in terms of protein, oil, carbohydrate, moisture, ash and calories in composition to the parental line and to commercial varieties. There was no novel genetic material present in the refined oil. The Committee is therefore of the opinion that oil derived from herbicide-tolerant cottonseed line RRC 1445 is substantially equivalent to oil from conventional cottonseed lines, in terms of composition, nutritional value, metabolism, intended use and level of undesirable substances. These conclusions confirm the findings of researchers (e.g. Berberich et al., 1996; Nida et al., 1996) (Table 5.13).

Berberich et al. (1996) studied the composition of cottonseed that had been modified genetically by the introduction of a gene encoding an insecticidal protein from *Bacillus thuringiensis* ssp. *kurstaki* (B.t.k.) to confer protection to the growing plants against major lepidopteran insect pests. The insect-protected lines and the parental control were shown to contain levels of nutrients comparable to those of other commercial varieties. Nutrients included protein, fat, carbohydrate, moisture, ash, amino acids and fatty acids. The levels of the anti-nutrients gossypol, cyclopropenoid fatty acids and aflatoxin in the seed from the insect-protected lines were similar to or lower than the levels present in the parental variety and reported for other commercial varieties. Sissener et al. (2009, 2011) reviewed the available data on Roundup Ready cottonseed meal as a protein source in fish feed. Among the species tested were salmon, channel catfish, rainbow trout and zebrafish. Results indicated that both types of cultivar had similar quality as feed ingredients for fish.

These results demonstrate that seed from the insect-protected cotton lines is compositionally equivalent to, and as nutritious as, seed from the parental and other commercial cotton varieties.

Nida et al. (1996) reported that the composition of seeds from glyphosate-tolerant cotton was equivalent to that from conventional cottonseed (Table 5.13). The compositions of the cottonseed and oil (Table 5.14) from two glyphosate-tolerant cultivars (1445 and 1698) were compared to that of the parental variety Coker 312 and to published values for other commercial cotton varieties. The nutrients measured in the cottonseed were protein, fat, fibre, carbohydrate, calories, moisture, ash, amino acids and fatty acids. The anti-nutrients measured in the cottonseed included gossypol, cyclopropenoid fatty acids and aflatoxin. In addition, the fatty acid profile and α-tocopherol levels were measured in the refined oil. These analyses demonstrated that the glyphosate-tolerant cotton lines are compositionally equivalent and as safe and nutritious as the parental and conventional cotton varieties commercially available.

Hamilton et al. (2004) evaluated the composition and nutrient content of Bollgard II (insect-protected) cottonseed and conventional cottonseed cultivars from a total of 14 US field sites over a two-year period. Results showed that the cottonseed (Table 5.13) and cottonseed oil (Table 5.14) from Bollgard II cotton were comparable in their composition to those of the conventional control cotton line and other commercial varieties. The compositional data were supported by nutritional safety studies conducted with dairy cows, catfish and quail. Results from these studies showed that Bollgard II gave results similar to those obtained with the conventional cottonseed cultivars. The researchers concluded that Bollgard II cottonseed is compositionally and nutritionally equivalent to conventional cottonseed.

The composition of oil and fatty acids in B.t. and non-B.t. cotton (*G. hirsutum*) cultivars (JKCH-2245, K-5038, RCH-2, NCEN-3R, JK-Ishwar and JKCH-266) was studied by Nagappa and Khadi (2011). On average, the oil content of B.t. cultivars (174 g/kg) was similar to that of conventional non-GM cultivars (174 g/kg). The contents of unsaturated fatty acids were also similar. Based on their findings the researchers concluded that there is no significant difference in oil content or fatty acid composition of oil in B.t. cotton cultivars in comparison with conventional non-GM cultivars.

Whole cottonseed is used commonly in some countries as a dietary component for dairy cows. Castillo et al. (2004) conducted two experiments to evaluate dry matter intake, milk yield and milk composition in Argentinean Holstein dairy cows fed diets containing different sources of GM whole cottonseed. Twenty-four lactating multiparous Argentinean Holstein dairy cows were used in two experiments, with cows averaging 565 kg body weight and 53 days in milk at the beginning of the experiments. Treatments in Experiment 1 were:

Table 5.13 Published data on the nutritional composition of conventional and GM cottonseed.

Reference	Berberich et al. (1996)				Nida et al. (1996)			Hamilton et al. (2004)	
Cultivar	**Conventional (Non-GM)**	**Insect protected (GM)**			**Conventional (Non-GM)**	**Glyphosate-tolerant (GM)**		**Conventional (Non-GM)**	**Insect protected (GM)**
	Croker 312	531	757	1076	Croker 312	1445	1698	DP50	Bollgard II
Component, g/kg DM unless stated[1]									
Dry matter	876.4	865.7	868.2	894	933	925	931	951.2	951.4
Crude protein	270	275.6	276	265.7	288	306	294	261.2	262.6
Crude fibre	–	–	–	–	–	–	–	162.2	160.7
Ether extract	229.6	232.3	229.5	208	244	253	249	208.5	213.3
Ash	46.3	45.3	44.5	44.5	44	45	43	43.2	42.8
Carbohydrate	454	446.8	449.9	481.7	424	396	414	485.5	479.5
ADF	–	–	–	–	–	–	–	252.6	256.8
NDF	–	–	–	–	–	–	–	389.7	387.5
kcal/kg (Atwater)	4.96	4.98	4.97	4.86	5.05	5.08	5.08	4.86	4.80
Minerals									
Calcium	–	–	–	–	–	–	–	1.4	1.4
Copper mg/kg	–	–	–	–	–	–	–	7.42	7.08
Iron mg/kg	–	–	–	–	–	–	–	52.0	48.88
Magnesium	–	–	–	–	–	–	–	0.41	0.40
Manganese mg/kg	–	–	–	–	–	–	–	14.2	14.03
Phosphorus (total)	–	–	–	–	–	–	–	0.72	0.68
Potassium	–	–	–	–	–	–	–	1.12	1.11
Sodium	–	–	–	–	–	–	–	0.18	0.18
Zinc mg/kg	–	–	–	–	–	–	–	40.19	39.69

Continued

Table 5.13 Continued

Reference	Berberich et al. (1996)				Nida et al. (1996)			Hamilton et al. (2004)	
Cultivar	**Conventional (Non-GM)**		**Insect protected (GM)**		**Conventional (Non-GM)**	**Glyphosate-tolerant (GM)**		**Conventional (Non-GM)**	**Insect protected (GM)**
Amino acids mg/kg protein, DM basis unless stated [4]								**Amino acids g/kg total AA**	
Histidine	2.72	2.94	2.99	2.74	3.0	3.0	3.0	30.9	30.6
Isoleucine	3.26	3.02	3.14	3.22	2.9	2.8	2.9	35.6	35.6
Leucine	5.98	5.93	6.09	5.98	5.9	5.8	5.9	65.3	65.3
Lysine	4.50	4.73	4.80	4.57	4.6	4.5	4.5	51.8	51.7
Methionine	1.48	1.94	1.92	1.49	1.5	1.5	1.50	17.1	17.1
Cystine	1.60	1.68	1.73	1.60	1.6	1.6	1.6	18.3	18.0
Phenylalanine	5.32	5.28	5.35	5.26	5.2	5.1	5.1	57.0	56.5
Tryptophan	–	–	–	–	1.1	1.1	1.1	10.6	10.6
Tyrosine	2.92	3.08	3.05	2.96	2.7	2.7	2.7	28.2	28.0
Threonine	3.40	3.42	3.45	3.50	3.4	3.4	3.4	35.1	34.8
Valine	4.50	4.09	4.26	4.39	4.0	4.0	4.0	49.9	50.1
Vitamins mg/kg									
α-tocopherol	638	568	597	689	670	588	624	–	–
Gossypol g/kg DW									
Total [2]	11.6	11.0	10.8	10.4	9.0	10.2	8.8	9.8	9.6
Free [3]	ND	ND	ND	ND	7.7	9.0	7.5	8.5	8.7

[1]Protein fat, ash, carbohydrate and calories reported on a dry-weight basis.

[2]Not typically measured in raw cottonseed.

[3]Not detected.

[4]The unit used by Berberich and Nida may be incorrect since the values are very small. The comparison is, however, valid.

Table 5.14 Published data on the composition of refined oil from conventional and GM cottonseed.

Reference	Berberich et al. (1996)				Nida et al. (1996)			Hamilton et al. (2004)	
Cultivar	Conventional (Non-GM)	Insect protected (GM)			Conventional (Non-GM)	Glyphosate-tolerant (GM)		Conventional (Non-GM)	Insect protected (GM)
	Croker 312	531	757	1076	Croker 312	1445	1698	DP50 mg/g total FA	Bollgard II mg/g total FA
Fatty acids mg/g refined oil									
C14:0	9.8	7.7	8.6	8.8	9.5	8.4	9.3	10.6	13.2
C16:0	254.2	250.8	249.6	259.4	255.4	251.4	254.2	253	239
C16:1	6.4	5.8	6.0	6.3	6.4	6.1	6.3	7.8	8.32
C17:0	1.9	1.0	1.1	1.1	ND[1]	ND[1]	ND[1]	<1.0	<1.0
C18:0	25.3	26.7	26.2	23.8	24.6	24.1	25.3	20.4	20.4
C18:1	149.2	158.9	154.9	136.4	150.3	145.3	145.1	147	151
C18:2	502.7	508.8	500.9	508	501	512.7	504.4	549	556
C18:3	1.6	1.7	1.5	1.5	1.4	1.6	1.4	1.45	1.71
C20:0	2.1	3.0	2.6	2.7	2.6	2.7	2.4	1.78	1.76
C22:0	1.2	1.3	1.2	1.3	1.2	0.8	1.1	<1.0	<1.0
Malvalic C17	3.6	4.6	4.2	4.1	3.5	5.6	4.6	3.77	3.78
Sterculic C18	4.8	4.3	5.8	6.0	4.4	5.0	5.3	2.17	2.05
Dihydrosterculic C19	2.2	1.6	3.5	2.6	2.3	2.3	3.6	1.46	1.65

[1] Not detected.

Bollgard cotton containing the cry1Ac gene; Bollgard II cotton containing cry1Ac and cry2Ab genes; glyphosate-tolerant cotton containing the cp4 epsps gene; and a control non-GM but genetically related cottonseed. In Experiment 2, two commercial sources, a parental control line, and the GM cotton containing both cry1Ac and cp4 epsps genes were used as treatments. All cows received the same total mixed ration but with different whole cottonseed sources. Cottonseed was included to provide 2.50 kg per cow daily (dry matter basis) or about 100 g per kg total diet DM. Feed intake, milk yield, milk composition, body weight and body condition scores did not differ among treatments, and no sample of milk was found to contain GM or plant DNA fragments. It was concluded that cottonseed from the GM cultivars used in this study resulted in similar performance in lactating dairy cows when compared to non-GM cultivars and reference cottonseed.

Few detailed data have been published on the nutritional composition of fat-extracted GM cottonseed meal (Table 5.15), but since GM and conventional cultivars of cottonseed have been shown to be similar in nutritional composition it may be inferred that the meals are of similar composition when processed in a similar manner. This may also be inferred from the results of feeding studies using poultry and other farm animals.

Elangovan et al. (2006) compared the growth of broiler chickens fed diets containing 100 g/kg meal from GM (insect-resistant) cottonseed or non-GM low-gossypol cottonseed meal (Table 5.15). Day-old broiler chicks were used. At the end of 6 weeks 10 birds were removed from each treatment, euthanised and the carcass traits, yield of organs and histopathological changes in organs were measured. Feed intake, feed conversion efficiency, nitrogen retention, dressing percentage, weight of liver weight and heart, and blood biochemical constituents did not differ significantly between the dietary treatments.

Li et al. (2008) conducted three experiments with channel catfish to assess the relative nutritional quality of cottonseed meals (Table 5.15) from genetically modified (GM) cotton crops (Roundup Ready Flex, Bollgard Roundup Ready, Bollgard II Roundup Ready, and Bollgard II Roundup Ready Flex). Diets containing 200 g/kg cottonseed meal were used. Although there was considerable variation (Table 5.15) in proximate composition, amino acids and gossypol concentrations, the researchers concluded that the GM cottonseed meals were not significantly different from a conventional control and reference cottonseed meals. Juvenile catfish were fed the test diets for 8 weeks. Rate of weight gain, efficiency of feed conversion, survival rate and fillet composition of catfish fed the diets containing GM cottonseed meals were found to be similar to those obtained with a control diet or a diet containing commercial non-GM cottonseed meal. Based on the findings, the researchers concluded that the GM cottonseed meals were nutritionally equivalent to conventional non-GM cottonseed meals.

Reported effects of dietary inclusion of conventional or GM oilseeds on the productivity of animals, poultry, fish and lab animals are summarised in Table 5.16. These data show comparable productivity in these animals when fed diets containing GM and conventional non-GM oilseed products.

Table 5.15 Published data on the composition of processed meal from conventional and GM cottonseed.

Reference	Elangovan et al. (2006)			Li et al. (2008)						
Cultivar	Conventional	Insect protected	Commercial	Conventional (isogenic to Mon 88913)	Mon 88913 (Herbicide-tolerant)	Conventional (DP5415)	Mon 531 × Mon 1445 (Insect-resistant, herbicide-tolerant)	Mon 15985 × Mon1445 (Insect-resistant, herbicide-tolerant)	Conventional (DP5415)	Mon 531 × Mon 1445 (Insect-resistant, herbicide-tolerant)
Component, g/kg DM unless stated							As-fed basis			
Dry matter	969.1	927.8	925.7	940	910	896	881	911	941	895
Crude protein	331.8	344	351.7	454	384	364	423	388	281	330
Crude fibre	126.5	111.2	166.5	92	91	123	109	112	–	–
Ether extract	5.0	9.1	7.7	41.1	34.8	57.3	13.7	26.6	9.3	8.6
Ash	50.6	53.2	50.7	75.0	71.9	73.8	67.1	73.4	65.2	62.9
NFE	486.1	482.5	423.4	–	–	–	–	–	–	–
Amino acids g/kg protein as-fed basis										
Lysine	–	–	–	20.2	19.1	10.5	11.4	11.5	8.7	10.3
Methionine + Cystine	–	–	–	14.2	12.6	7.8	8.5	8.2	6.0	7.3
Gossypol g/kg										
Total	5.8	7.4	7.2	11.9	11.2	10.2	6.4	12.5	8.9	14.6
Free	0.2	0.2	0.3	0.7	0.6	0.5	0.2	0.9	0.2	1.2

Table 5.16 Reported effects of dietary inclusion of conventional or GM oilseeds on the productivity of animals, poultry and fish.

Animal species	Reference	GM feedstuff	Reported effects
		Meat/lab animals	
Buffaloes	Singh et al., 2003	Bt cottonseed	No significant differences in milking performance.
Sheep	Stanford et al., 2003	Herbicide-tolerant canola	No significant differences in feed intake or growth performance (lambs).
Pigs	Aalhus et al., 2003	Meal from glyphosate-tolerant canola	No significant difference in growth performance.
Pigs	Cromwell et al., 2002	Herbicide-tolerant soyabean meal	No significant difference in growth performance.
Pigs	Świątkiewicz et al., 2013	Glyphosate-tolerant soyabean meal and Bt maize	No significant effects on growth or reproductive performance.
Broilers	Elangovan et al., 2006	Bt-cottonseed meal	No significant difference in growth performance.
Broilers	Hammond et al., 1996	Glyphosate-tolerant soyabean meal	No significant differences in growth performance or livability.
Broilers	Kan et al., 2001	Meal from dehulled Bt soyabeans	No significant difference in growth performance.
Broilers	Kan and Hartnell, 2004	Bt soyabean meal	No significant difference in growth performance.
Broilers	Stanisiewski et al., 2002	Glyphosate-tolerant canola	No significant difference in growth performance.
Broilers	Taylor et al., 2004	Glyphosate-tolerant canola meal	No significant difference in growth performance.
Rabbits	Maertens et al., 1996	Glyphosate-tolerant canola	No significant difference in digestibility between canola types.
Catfish	Hammond et al., 1996	Glyphosate-tolerant soyabean meal	No significant differences in feed: gain ratio or livability. Improved growth with genetically modified soya.
Channel catfish	Li et al. 2008.	Range of transgenic cottonseed meals.	Rate of weight gain, efficiency of feed conversion, survival rate and fillet composition similar.
Rainbow trout	Chainark et al., 2006	Meal from glyphosate-tolerant soybeans	No significant effect on growth performance.
Rainbow trout	Brown et al., 2003	Meal from glyphosate-tolerant canola	No significant effect on growth performance.
Rats	Zhu et al., 2004	Meal from glyphosate-tolerant soybeans	No significant effect on growth performance.
Milk			
Dairy cows	Hammond et al., 1996	Meal from glyphosate-tolerant soyabeans	Production of fat-corrected milk greater with transgenic soya.
Dairy cows	Castillo et al., 2001	Herbicide-tolerant cottonseed, Bt cottonseed	No significant differences in body condition or milk yield.
Dairy cows	Castillo et al., 2004	Glyphosate-tolerant cottonseed, Bt cottonseed	No significant differences in body condition or milk yield.
Dairy cows	Singhal et al., 2006	Cry1ac+ cr2ab cottonseed.	No significant differences in dry matter intake or milk production.

No tests on the organoleptic quality of cottonseed oil or other cottonseed products for humans appear to have been conducted.

Summary

Approved GM oilseed crops being grown worldwide include soyabeans, canola and cotton, which are major sources of oil and protein for human and animal feeding. Soyabeans are the most important of these crops since the beans and extracted oil are used as human food. Only the oil from canola and cottonseed is used as human food, the fat-extracted meals being used as animal feed. A striking feature is the high proportion of GM cultivars of these crops being grown, in some countries approaching 100%. Several GM cultivars of these crops have been approved as food/feed sources by regulatory authorities, based on evidence of safety and nutritional composition. Several countries have also approved their cultivation domestically, but this has not been the case in Europe. Results of scientific investigations confirm the approval granted to GM cultivars by regulatory authorities, and contrary findings published by some investigators have not been accepted as credible by the scientific community. Claims of higher contents of pesticide residues in GM oilseed crops are not supported by the scientific evidence. Based on the available scientific evidence it is clear that the approved cultivars of GM oilseeds produce food and feed that is as safe and nutritious as non-GM cultivars.

References

Aalhus, J. L., Dugan, M. E. R., Lien, K. A., et al. (2003). Effects of feeding glyphosate-tolerant canola meal on swine growth, carcass composition and meat quality. *Journal of Animal Science* 81: 3267 (Abstract).

Advisory Committee on Novel Foods and Processes, UK, (ACNFP) (2002). Request for an Article 5 opinion on the substantial equivalence of cottonseed oil and food ingredients derived from Roundup® Ready cotton. http://www.food.gov.uk/multimedia/pdfs/cottonseedoil RRfinalopinion.pdf (accessed 22 August 2013).

Alexander, T. W., Sharma, R., Okine, E. K., et al. (2002). Impact of feed processing and mixed ruminal culture on the fate of recombinant EPSP synthase and endogenous canola plant DNA. *FEMS Microbiology Letters* 214: 263–269.

Alexander, T. W., Sharma, R., Deng, M. Y., et al. (2004). Use of quantitative real-time and conventional PCR to assess the stability of the cp4 epsps transgene from Roundup Ready ®canola in the intestinal, ruminal, and fecal contents of sheep. *Journal of Biotechnology* 112: 255–266.

Appenzeller, L. M., Munley, S. M., Hoban, D., et al. (2008). Subchronic feeding study of herbicide-tolerant soybean DP-356043-5 in Sprague-Dawley rats. *Food and Chemical Toxicology* 46: 2201–2213.

Arregui, M. C., Lenardón, A., Sanchez, D., et al. (2004). Monitoring glyphosate residues in transgenic glyphosate-resistant soybean. *Pest Management Science* 60: 163–166.

Ash, J., Novak, C. and Scheideler, S. E. (2003). The fate of genetically modified protein from Roundup Ready soybeans in laying hens. *Journal of Applied Poultry Research* 12: 242–245.

Aumaitre, A., Aulrich, K., Chesson, A., et al. (2002). New feeds from genetically modified plants: substantial equivalence, nutritional equivalence, digestibility, and safety for animals and the food chain. *Livestock Production Science* 74: 223–238.

Batista, R., Martins, I., Jeno, P., et al. (2007). A proteomic study to identify soya allergens – the human response to transgenic versus non-transgenic soya samples. *International Archives of Allergy and Immunology* 144: 29–38.

Batista, R., Nunes, B., Carmo, M., et al. (2005). Lack of detectable allergenicity of transgenic maize and soya samples. *Journal of Allergy and Clinical Immunology* 116: 403–410.

Berberich, S. A., Ream, J. E., Jackson, T. L., et al. (1996). The composition of insect-protected cottonseed is equivalent to that of conventional cottonseed. *Journal of Agricultural and Food Chemistry* 44: 365–371.

Binder, E. M., Tan, L. M., Chin, L. J., et al. (2007). World-wide occurrence of mycotoxins in commodities, feeds and feed ingredients. *Animal Feed Science and Technology* 137: 265–262.

Borg, K. (1975). Physiopathological effects of rapeseed oil: a review. *Acta Medica Scandinavica Supplement* 585: 5–13.

Brake, D. G. and Evenson, D. P. (2004). A generational study of glyphosate-tolerant soybeans on mouse fetal, postnatal, pubertal and adult testicular development. *Food and Chemical Toxicology* 42: 29–36.

Brookes, G. and Barfoot, P. (2012) GM crops: global socio-economic and environmental impacts 1996–2010. PG Economics Ltd, UK. http://www.pgeconomics.co.uk/page/33/global-impact-2012 (accessed February 2015).

Brown, P. B., Wilson, K. A., Jonker, Y. and Nickson, T. E. (2003). Glyphosate tolerant canola meal is equivalent to the parental line in diets fed to rainbow trout. *Journal of Agricultural and Food Chemistry* 51: 4268–4272.

Caine, W. R., Aalhus, J. L., Dugan, M. E. R., et al. (2007). Growth performance, carcass characteristics and pork quality of pigs fed diets containing meal from conventional or glyphosate-tolerant canola. *Canadian Journal of Animal Science* 87: 517–526.

Carman, J. A., Vlieger, H. R., Ver Steeg, L. J., et al. (2013). A long-term toxicology study on pigs fed a combined genetically modified (GM) soy and GM maize diet. *Journal of Organic Systems* 8: 38–54.

Castillo, A. R., Gallardo, M. R., Maciel, M., et al. (2001). Effect of feeding dairy cows with either BollGard®, BollGard II®, Roundup Ready® or control cottonseeds on feed intake, milk yield and milk composition. *Journal of Dairy Science* 84 (Suppl. 1): 413 Abstract 1712.

Castillo, A. R., Gallardo, M. R., Maciel, M., et al. (2004). Effects of feeding rations with genetically modified whole cottonseed to lactating Holstein cows. *Journal of Dairy Science* 87: 1778–1785.

Chainark, P., Satoh, S., Hino, T., et al. (2006). Availability of genetically modified soybean meal in rainbow trout Oncorhynchus mykiss diets. *Fisheries Science* 72: 1072–1078.

Chow, C. K. (2007). *Fatty Acids in Foods and their Health Implications*, third edition. *CRC Press: p.* 241.

Chukwudebe, A., Privalle, L., Reed, A., et al. (2012). Health and nutritional status of Wistar rats following subchronic exposure to CV127 soybeans. *Food and Chemical Toxicology* 50: 956–971.

CODEX (2001). Standards, Section 2. Codex Standards for Fats and Oils from Vegetable Sources 210-1999. CODEX Alimentarius Commission, Food and Agriculture Organisation, World Health Organisation. Rome.

Commission of the European Communities (CEC) (2007). Commission directive 2007/68/ec. Official Journal of the European Union. 28.11.2007. L 310/14.

Cromwell, G. L., Lindemann, M., Randolph, J., et al. (2002). Soybean meal from Roundup Ready or conventional soybeans in diets for growing-finishing swine. *Journal of Animal Science* 80: 708–715.

Daun, J. K. (2004). Quality of genetically modified (GM) and conventional varieties of canola (spring oilseed rape) grown in western Canada, 1996–2001. *Journal of Agricultural Science* 142: 273–280.

Delaney, B., Astwood, J. D., Cunny, H., et al. (2008a). *Evaluation of protein safety in the context of agricultural biotechnology. ILSI International Food Biotechnology Committee Task Force on Protein Safety.* Food and Chemical Toxicology 46 Supplement 2: S71–97.

Delaney, B., Zhang, J., Carlson, G., et al. (2008b). A gene-shuffled glyphosate acetyltransferase protein from Bacillus licheniformis (GAT4601) shows no evidence of allergenicity or toxicity. *Toxicological Sciences* 102: 425–432.

Domingo, J. L. (2007). Toxicity studies of genetically modified plants: a review of the published literature. *Critical Reviews in Food Science and Nutrition* 47: 721–733.

Dryzga, M. D., Yano, B. L., Andrus, A. K. and Mattsson, J. L. (2007). Evaluation of the safety and nutritional equivalence of a genetically modified cottonseed meal in a 90-day dietary toxicity study in rats. *Food and Chemical Toxicology* 45: 1994–2004.

Dubois, V., Breton, S., Linder, M., et al. (2007). Fatty acid profiles of 80 vegetable oils with regard to their nutritional potential. *European Journal of Lipid Science and Technology* 109: 710–732.

EFSA (European Food Safety Authority) (2009a). Scientific opinion on an application (EFSA-GMO-RX-MS8-RF3) for renewal of the authorisation for continued marketing of existing (1) food and food ingredients produced from genetically modified glufosinate-tolerant oilseed rape Ms8, Rf3 and Ms8 × Rf3, and (2) feed materials produced from genetically modified glufosinate-tolerant oilseed rape Ms8, Rf3 and Ms8 × Rf3, under Regulation (EC) No 1829/2003 from Bayer CropScience. *EFSA Journal* 7 No. 9 Article 1318.

EFSA (European Food Safety Authority) (2009b). Scientific opinion on applications (EFSA-GMO-RX-GT73[8.1.a] and EFSA-GMO-RX-GT73[8.1.b/20.1.b]) for renewal of the authorisation for continued marketing of existing (1) food and food ingredients produced from oilseed rape GT73; and of (2) feed materials, feed additives and food additives produced from oilseed rape GT73, all under Regulation (EC) No 1829/2003 from Monsanto. *EFSA Journal* 7 No. 12 Article 1417.

EFSA (2010). EFSA Panel on Genetically Modified Organisms (GMO); Scientific Opinion on application (EFSA-GMO-NL-2005-16) for the placing on the market of insect resistant genetically modified cotton (*Gossypium hirsutum* L.) 281-24-236/3006-210-23 for food and feed uses, import and processing under Regulation (EC) No 1829/2003 from Dow AgroSciences. EFSA Journal 8:1644. [32 pp.]. doi:10.2903/j.efsa.2010.1644. (1644 – Scientific Output number). Available online: www.efsa.europa.eu (accessed February 2015).

EFSA (2011a). European Food Safety Authority (EFSA), Parma, Italy. Scientific opinion on application (EFSA-GMO-UK-2007-43) for the placing on the market of herbicide tolerant genetically modified soyabean 356043 for food and feed uses, import and processing under Regulation (EC) No 1829/2003 from Pioneer. *EFSA Journal* 9: 2310.

EFSA (2011b). Scientific opinion on application EFSA-GMO-RX-MON1445 for renewal of the authorisation for continued marketing of cottonseed oil, food additives, feed materials and feed additives produced from cotton MON 1445 that were notified as existing products under Articles 8(1)(a), 8(1)(b) and 20(1)(b) of Regulation (EC) No 1829/2003 from Monsanto. EFSA Panel on Genetically Modified Organisms. European Food Safety Authority (EFSA), Parma, Italy. *EFSA Journal* 9: 2479–2507.

EFSA (2011c). Scientific Opinion on application EFSA-GMO-RX-MON531 for renewal of the authorisation for continued marketing of existing cottonseed oil, food additives, feed materials and feed additives produced from MON 531 cotton that were notified under Articles 8(1)(a), 8(1)(b) and 20(1)(b) of Regulation (EC) No 1829/2003 from Monsanto. European Food Safety Authority Panel on Genetically Modified Organisms. European Food Safety Authority (EFSA), Parma, Italy. *EFSA Journal* 9: 2373–2403.

EFSA (2012a). European Food Safety Authority (EFSA), Parma, Italy. Scientific opinion on an application (EFSA-GMO-NL-2005-24) for the placing on the market of the herbicide tolerant genetically modified soyabean 40-3-2 for cultivation under Regulation (EC) No 1829/2003 from Monsanto. EFSA Panel on Genetically Modified Organisms. *EFSA Journal* 10: 2753.

EFSA (2012b). European Food Safety Authority (EFSA), Parma, Italy. Scientific opinion on application (EFSA-GMO-NL-2010-78) for the placing on the market of herbicide-tolerant, increased oleic acid genetically modified soybean MON 87705 for food and feed uses, import and processing under Regulation (EC) No 1829/2003 from Monsanto. EFSA Panel on Genetically Modified Organisms. *EFSA Journal* 10: 2909.

Elangovan, A. V., Tyagi, P. K., Shrivastav, A. K., et al. (2006). GMO (Bt-Cry1Ac gene) cottonseed meal is similar to non-GMO low free gossypol cottonseed meal for growth performance of broiler chickens. *Animal Feed Science and Technology* 129: 252–263.

Engvall, E. and Perlmann P. (1971). Enzyme-linked immunosorbent assay (ELISA). Quantitative assay of immunoglobulin G. *Immunochemistry* 8: 871–874.

Ermakova, I. (2005). Influence of GM-soya on the posterity of rats. http://www.iatp.org/files/421_2_77176.pdf (accessed 1 July 2013).

Flachowsky G., Aulrich, K, Bohme, H. and Halle, I. (2007). Studies on feeds from genetically modified plants (GMP) – Contributions to nutritional and safety assessment. *Animal Feed Science and Technology* 133: 2–30.

FSANZ (2000). *Draft Risk Analysis Report. Application A363. Food Produced from Glyphosate-Tolerant Canola Line GT73*. Canberra, Australia and Wellington, New Zealand: Australia/New Zealand Food Authority.

FSANZ (2003). Oil Derived From Glufosinate-Ammonium Tolerant and Pollination Controlled Canola: a Safety Assessment. Technical Report Series No. 16. Canberra, Australia and Wellington, New Zealand: Australia/New Zealand Food Authority.

FSANZ (2008). Final Assessment Report, Application A614. Food Derived From Glyphosate-Tolerant Cotton Line Ghb614. Canberra, Australia and Wellington, New Zealand: Australia/New Zealand Food Authority.

Goodman, R.E., Vieths, S., Sampson, H. A., et al. (2008) Allergenicity assessment of genetically modified crops-what makes sense. *Nature Biotechnology* 26: 73–81.

Hamilton, K. A., Pyla, P. D., Breeze, M., et al. (2004). Bollgard II cotton: Compositional analysis and feeding studies of cottonseed from insect-protected cotton (*Gossypium hirsutum* L.) producing the Cry1Ac and Cry2Ab2 proteins. *Journal of Agricultural and Food Chemistry* 52: 6969–6976.

Hammond, B., Vicini, J., Hartnell, G., et al. (1996). The feeding value of soybeans fed to rats, chickens, catfish and dairy cattle is not altered by genetic incorporation of glyphosate tolerance. *Journal of Nutrition* 126: 717–727.

Harrison, L.A., Bailey, M. R., Naylor, M. W., et al. (1996). The expressed protein in glyphosate-tolerant soybean, 5-enolpyruvylshikimate-3-phosphate synthase from Agrobacterium sp. strain CP4, is rapidly digested in vitro and is not toxic to acutely gavaged mice. *Journal of Nutrition* 126: 728–740.

Hemre, G. I., Sanden, M., Bakke-Mckellep, et al. (2005) Growth, feed utilization and health of Atlantic salmon *Salmo salar L.* fed genetically modified compared to non-modified commercial hybrid soybeans. *Aquaculture Nutrition* 11:157–167.

Herman, R. A., Dunville, C. M., Juberg, D. R., et al. (2011). Performance of broiler chickens fed diets containing DAS-68416-4 soybean meal. *GM Crops* 2: 169–175.

ISAAA (2014). GM approval database (2014). ISAAA: Ithaca, New York, http://www.isaaa.org/gmapprovaldatabase/default.asp (accessed 27 October 2014).

James, C. (2013). Global Status of Commercialized Biotech/GM Crops: 2013. ISAAA Brief No. 46. ISAAA: Ithaca, NY. saaa.org/resources/publications/briefs/46/ (accessed 11 August 2014).

Jennings, J. C., Kolwyck, D. C., Kays, S. B., et al. (2003a). Determining whether transgenic and endogenous plant DNA and transgenic protein are detectable in muscle from swine fed Roundup Ready soybean meal. *Journal of Animal Science* 81: 1447–1455.

Jennings, J. C., Whetsell, A. J., Nicholas, N. R., et al. (2003b) Determining whether transgenic or endogenous plant DNA is detectable in dairy milk or beef organs. *Bulletin of the International Dairy Federation* 383: 41–46.

Kan, C.A. and Hartnell, G. F. (2004). Evaluation of broiler performance when fed insect-protected, control, or commercial varieties of dehulled soybean meal. *Poultry Science* 83: 2029–2038.

Kan, C. A., Versteegh, H. A. J., Uijttengoogaart, T. G., e al. (2001). Comparison of broiler performance and carcass characteristics when fed B.t., parental control or commercial varieties of dehulled soybean meal. *Poultry Science* 80 (Suppl. 1):203 Abstract 841.

Kim, S. H., Kim, H. M., Ye, Y. M., et al. (2006). Evaluating the allergic risk of genetically modified soyabean. *Yonsei Medical Journal* 47: 505–512.

Lack, G. (2002). Clinical risk assessment of GM foods. *Toxicology Letters* 127: 337–340.

Li, M. H., Hartnell, G. F., Robinson, E. H., et al. (2008). Evaluation of cottonseed meal derived from genetically modified cotton as feed ingredients for channel catfish, *Ictalurus punctatus*. *Aquaculture Nutrition* 14: 490–498.

Maertens, L., Luzi, F.and Huybechts, I. (1996). Digestibility of non-transgenic and transgenic oilseed rape in rabbits. *Proceedings 6th World Rabbit Congress*, Toulouse, France: Vol. I: 231–235.

Malatesta, M., Boraldi, F., Annovi, G., et al. (2008). A long-term study on female mice fed on a genetically modified soybean: effects on liver ageing. *Histochemistry and Cell Biology* 130: 967–977.

Mejia, L., Jacobs, C. M., Utterback, P. L., et al. (2010). Evaluation of the nutritional equivalency of soybean meal with the genetically modified trait DP-3O5423-1 when fed to laying hens. *Poultry Science* 89: 2634–2639.

Nagappa, H. and Khadi, B. M. (2011). Studies on composition of oil and fatty acid in Bt and non Bt cotton (G. hirsutum). *Proceedings World Cotton Research Conference* 5, 7–11 November 2011, Mumbai, India: pp. 568–571.

Netherwood, T., Martín-Orúe, S. M., O'Donnell, A. G., et al. (2004). Assessing the survival of transgenic plant DNA in the human gastrointestinal tract. *Nature Biotechnology* 22: 204–209.

Nida, D. L., Patzer, S., Harvey, P., et al. (1996). Glyphosate-tolerant cotton: the composition of the cottonseed is equivalent to that of conventional cottonseed. *Journal of Agricultural and Food Chemistry* 44: 1967–1974.

Padgette, S. R., Taylor, N. B., Nida, D. L., et al. (1996). The composition of glyphosate-tolerant soybean seeds is equivalent to that of conventional soybeans. *Journal of Nutrition* 126: 702–716.

Phipps, R. H., Beever, D. E. and Humphries, D. J. (2002). Detection of transgenic DNA in milk from cows receiving herbicide tolerant (CP4EPSPS) soyabean meal. *Livestock Production Science* 74: 269–273.

Press, C. M. and Evensen, 0. (1999). The morphology of the immune system in teleost fishes. *Fish and Shellfish Immunology* 9: 309–318.

Qi, X., He, X., Luo, Y., et al. (2012). Subchronic feeding study of stacked trait genetically-modified soybean (305423×40-3-2) in Sprague-Dawley rats. *Food and Chemical Toxicology* 50: 3256–3263.

Reichert, M., Kozaczyński, W., Karpińska, T. A., et al. (2012). Histopathology of internal organs of farm animals fed genetically modified corn and soybean meal. *Bulletin Veterinary Institute Pulawy* 56: 617–622.

Rizzi, A., Raddadi, N., Sorlini, C., et al. (2012). The stability and degradation of dietary DNA in the gastrointestinal tract of mammals: implications for horizontal gene transfer and the biosafety of GMOs. *Critical Reviews in Food Science and Nutrition* 52: 142–161.

Sakamoto, Y., Tada, Y., Fukumori, N., et al. (2008). A 104-week feeding study of genetically modified soybeans in F344 rats. *Shokuhin Eiseigaku Zasshi (Journal of the Food Hygienic Society of Japan)* 49: 272–282.

Sanden, M., Bruce, I. J., Rahman, M. A. and Hemre, G. I. (2004) The fate of transgenic sequences present in genetically modified plant products in fish feed, investigating the survival of GM soybean DNA fragments during feeding trials in Atlantic salmon, *Salmo salar* L. *Aquaculture* 237: 391–405.

Sanden, M., Krogdahl, A., Bakke-Mckellep, A. M., et al. (2006). Growth performance and organ development in Atlantic salmon, *Salmo salar* L. parr fed genetically modified (GM) soybean and maize. *Aquaculture Nutrition* 12: 1–14.

Singh, M., Tiwari, D. P., Kumar, A. and Kumar, M. R. (2003). Effect of feeding transgenic cottonseed vis-a-vis non-transgenic cottonseed on haematobiochemical constituents in lactating Murrah buffaloes. *Asian Australasian Journal of Animal Sciences* 16: 1732–1737.

Singhal, K. K., Kumar, R. S., Tyagi, A. K. and Rajput, Y. S. (2006). Evaluation of Bt cottonseed as protein supplement in the ration of lactating dairy cows. *Indian Journal of Animal Sciences* 76: 532–37.

Sissener, N. H., Sanden, M., Bakke, A. M., et al. (2009). A long term trial with Atlantic salmon (*Salmo salar* L.) fed genetically modified soy; focusing general health and performance before, during and after the parr-smolt transformation. *Aquaculture* 294: 108–117.

Sissener, N. H., Sanden, M., Krogdahl, Å., et al. (2011). Genetically modified plants as fish feed ingredients. *Canadian Journal of Fisheries and Aquatic Sciences* 68: 563–574.

Snell, C., Bernheim, A., Bergé, J., et al. (2012). Assessment of the health impact of GM plant diets in long-term and multigenerational animal feeding trials: A literature review. *Food and Chemical Toxicology* 50: 1134–1148.

Stanford, K., Aalhus, J. L., Dugan, M. E. R., et al. (2003). Effects of feeding transgenic canola on apparent digestibility, growth performance and carcass characteristics of lambs. *Canadian Journal of Animal Science* 83: 299–305.

Stanisiewski, E. P., Taylor, M. L., Hartnell, G. F., et al. (2002). Broiler performance when fed Roundup Ready® (event RT73) or conventional canola meal. *Poultry Science* 81 (Suppl. 1): 95 abstract 408.

Sunilkumar, G., Campbell, L. M., Puckhaber, L., et al. (2006). Engineering cottonseed for use in human nutrition by tissue-specific reduction of toxic gossypol. *Proceedings of the National Academy of Sciences USA* 28: 18054–18059.

Świątkiewicz, M., Bednarek, D., Markowski, J., et al. (2013). Effect of feeding genetically modified maize and soybean meal to sows on their reproductive traits, haematological indices and offspring performance. *Bulletin Veterinary Institute, Pulawy, Poland* 57: 413–418.

Tan, S., Evans, R. R., Dahmer, M. L., et al. (2005). Imidazolinone-tolerant crops: History, current status and future. *Pest Management Science* 61: 246–257.

Tan, J., Liu, S., Sun, Z., et al. (2012). Comparison of broiler performance, carcass yields and intestinal microflora when fed diets containing transgenic (Mon-40-3-2) and conventional soybean meal. *African Journal of Biotechnology* 11: 12371–12378.

Taylor, M. L., Stanisiewski, E. P., Riordan, S. G., et al. (2004). Comparison of broiler performance when fed diets containing roundup ready (Event RT73), nontransgenic control, or commercial canola meal. *Poultry Science* 83: 456–461.

US Environmental Protection Agency (1992). Pesticide tolerance for glyphosate. *Federal Registry* 57: 8739 40, 10–98.

USDA (2007). US Department of Agriculture Pesticide Data Program. Annual Summary, Calendar Year 2006. http://www.ams.usda.gov/pdp (accessed 28 April 2013).

USDA (2010). US Department of Agriculture Pesticide Data Program. Annual Summary, Calendar Year 2008. http://www.ams.usda.gov/pdp> (accessed 28 April 2013).

USDA (2013). US Department of Agriculture Pesticide Data Program. Annual Summary, Calendar Year 2011. http://www.ams.usda.gov/pdp (accessed 28 April 2013).

USDA (2014). United States Department of Agriculture, Foreign Agricultural Service. Oilseeds: World Markets and Trade. http://apps.fas.usda.gov/psdonline/circulars/oilseeds.pdf (accessed 23 October 2014).

Williams, A. L. and DeSesso, J. M. (2010). Genetically-modified soybeans. A critical evaluation of studies addressing potential changes associated with ingestion. *The Toxicologist* 114: 1154. Annual Meeting of the Society of Toxicology Salt Lake City, Utah March 7–11, 2010.

Zhu, Y., Li, D., Wang, F., Yin, J. and Hong, J. (2004). Nutritional assessment and fate of DNA of soybean meal from roundup ready or conventional soybeans using rats. *Archives of Animal Nutrition* 58: 295–310.

Appendix Table 5.1A Approved cultivars of GM soyabeans (total 30), canola (34), cotton (total 48) and flaxseed (total 1) (ISAAA, 2014). Reproduced with permission of ISAAA.

Soyabeans (*Glycine max* L.)		
Name	**Code**	**Trade name**
260-05 (G94-1, G94-19, G168)	DD-026005-3	not available
A2704-12	ACS-GM005-3	Liberty Link soybean
A2704-21	ACS-GM004-2	Liberty Link soybean
A5547-127	ACS-GM006-4	Liberty Link soybean
A5547-35	ACS-GM008-6	Liberty Link soybean
CV127	BPS-CV127-9	not available
DAS44406-6	Code: DAS-44406-6	not available
DAS68416-4	DAS68416-4	Enlist Soybean
DAS68416-4 x MON89788	DAS-68416-4 x MON-89788-1	not available
DAS81419	DAS-81419-2	not available
DP305423	DP-305423-1	not available
DP305423 x GTS 40-3-2	DP-305423-1 x MON-04032-6	not available
DP356043	DP-356043-5	Optimum GAT
FG72 (FG072-2, FG072-3)	MST-FG072-3	not available
GTS 40-3-2 (40-3-2)	MON-04032-6	Roundup Ready soybean
GU262	ACS-GM003-1	Liberty Link soybean
MON87712	MON-87712-4	not available
MON87701	MON-87701-2	not available
MON87701 x MON89788	MON-87701-2 x MON-89788-1	Intacta Roundup Ready 2 Pro
MON87705	MON-87705-6	Vistive Gold
MON87705 x MON89788	MON-87705-6 x MON-89788-1	not available
MON87708	MON-87708-9	Genuity Roundup Ready2 Xtend
MON87708 x MON89788	MON-87708-9 x MON-89788-1	not available
MON87751	MON-87751-7	not available
MON87769	MON87769-7	not available
MON87769 x MON89788	MON-87769-7 x MON-89788-1	not available
MON89788	MON-89788-1	Genuity Roundup Ready 2 Yield
SYHTOH2	SYN-000H2-5	Herbicide-tolerant Soybean line
W62	ACS-GM002-9	Liberty Link soybean
W98	ACS-GM001-8	Liberty Link soybean
Canola (*B. napus*)		
Name	**Code**	**Trade name**
23-18-17 (Event 18)	CGN-89111-8	Laurical Canola
23-198 (Event 23)	CGN-89465-2	Laurical Canola
61061	DP-061061-7	not available
73496	DP-073496-4	Optimum Gly canola
GT200 (RT200)	MON-89249-2	Roundup Ready Canola
GT73 (RT73)	MON-00073-7	Roundup Ready Canola
HCN10 (Topas 19/2)	not available	Liberty Link Independence
HCN28 (T45)	ACS-BN008-2	InVigor Canola

Appendix Table 5.1A Continued

Name	Code	Trade name
HCN92 (Topas 19/2)	ACS-BN007-1	Liberty Link Innovator
MON88302	MON-88302-9	TruFlex Roundup Ready Canola
MPS961	not available	Phytaseed Canola
MPS962	not available	Phytaseed Canola
MPS963	not available	Phytaseed Canola
MPS964	not available	Phytaseed Canola
MPS965	not available	Phytaseed Canola
MS1 (B91-4)	ACS-BN004-7	InVigor Canola
MS1 x RF1 (PGS1)	ACS-BN004-7 x ACS-BN001-4	InVigor Canola
MS1 x RF2 (PGS2)	ACS-BN004-7 x ACS-BN002-5	InVigor Canola
MS1 x RF3	ACS-BN004-7 x ACS-BN003-6	InVigor Canola
MS8	ACS-BN005-8	InVigor Canola
MS8 x RF3	ACS-BN005-8 x ACS-BN003-6	InVigor Canola
MS8 x RF3 x GT73 (RT73)	ACS-BN005-8 x ACS-BN003-6 x MON-00073-7	not available
OXY-235	ACS-BN011-5	Navigator Canola
PHY14	not available	not available
PHY23	not available	not available
PHY35	not available	not available
PHY36	not available	not available
RF1 (B93-101)	ACS-BN001-4	InVigor Canola
RF2 (B94-2)	ACS-BN002-5	InVigor Canola
RF3	ACS-BN003-6	InVigor Canola
Canola (*B. rapa*)		
HCR-1	not available	not available
ZSR500	not available	Hysyn 101 RR Roundup-Ready
ZSR502	not available	Hysyn 101 RR Roundup-Ready
ZSR503	not available	Hysyn 101 RR Roundup-Ready
Cottonseed (*Gossypium hirsutum* L.)		
19-51a	DD-01951A-7	not available
281-24-236	DAS-24236-5	not available
281-24-236 x 3006-210-23 (MXB-13)	DAS-24236-5 x DAS-21023-5	WideStrike Cotton
3006-210-23	DAS-21023-5	not available
3006-210-23 x 281-24-236 x MON1445	DAS-21023-5 x DAS-24236-5 x MON-01445-2	WideStrike Roundup Ready Cotton
3006-210-23 x 281-24-236 x MON88913	DAS-21023-5 x DAS-24236-5 x MON-88913-8	Widestrike Roundup Ready Flex Cotton
31707	not available	BXN Plus Bollgard Cotton
31803	not available	BXN Plus Bollgard Cotton

Continued

Appendix Table 5.1A Continued

Name	Code	Trade name
31807 x 31808	not available	not available
31807	not available	BXN Plus Bollgard Cotton
31808	not available	BXN Plus Bollgard Cotton
42317	not available	BXN Plus Bollgard Cotton
BNLA-601	not available	not available
BXN10211 (10211)	BXN-10211-9	BXN Cotton
BXN10215 (10215)	BXN-10215-4	BXN Cotton
BXN10222 (10222)	BXN-10222-2	BXN Cotton
BXN10224 (10224)	BXN-10224-4	BXN Cotton
COT102 (IR102)	SYN-IR102-7	VIPCOT Cotton
COT102 x COT67B	SYN-IR102-7 x SYN-IR67B-1	VIPCOT Cotton
COT102 x COT67B x MON88913	SYN-IR102-7 x SYN-IR67B-1 x MON-88913-8	VIPCOT™ Roundup Ready Flex Cotton
COT67B (IR67B)	SYN-IR67B-1	not available
Event1	not available	JK 1
GFM Cry1A	GTL-GFM311-7	not available
GHB119	BCS-GH005-8	not available
GHB614	BCS-GH002-5	GlyTol
GHB614 x LLCotton25	BCS-GH002-5 x ACS-GH001-3	GlyTol Liberty Link
GHB614 x LLCotton25 x MON15985	BCS-GH002-5 x ACS-GH001-3 x MON-15985-7	not available
GHB614 x MON15985	BCS-GH002-5 × MON-15985-7	not available
GHB614 x T304-40 x GHB119	BCS-GH002-5 x BCS-GH004-7 x BCS-GH005-8	Glytol x Twinlink
GK12	not available	not available
LLCotton25	ACS-GH001-3	Fibermax Liberty Link
LLCotton25 x MON15985	ACS-GH001-3 x MON-15985-7	Fibermax Liberty Link™ Bollgard II
MLS 9124	not available	not available
MON 88701-3	MON88701	not available
MON1076	MON-89924-2	Bollgard Cotton
MON1445	MON-01445-2	Roundup Ready Cotton
MON15985	MON-15985-7	Bollgard II Cotton
MON15985 x MON1445	MON-15985-7 x MON-01445-2	Roundup Ready Bollgard II Cotton
MON1698	MON-89383-1	Roundup Ready Cotton
MON531	MON-00531-6	Bollgard Cotton
MON531 x MON1445	MON-0531-6 x MON-01445-2	Roundup Ready Bollgard Cotton
MON757	MON-00757-7	Bollgard Cotton
MON88913	MON-88913-8	Roundup Ready Flex Cotton
MON88913 x MON15985	MON-88913-8 x MON-15985-7	Bollgard II™ Cotton
Ngwe Chi 6 B	not available	Ngwe Chi 6 Bt
SGK321	not available	not available
T303-3	BCS-GH003-6	not available
T304-40	BCS-GH004-7	not available
T304-40 x GHB119	BCS-GH004-7 x BCS-GH005-8	TwinLink™ Cotton
	Flax (*Linum usitatissumum* L.)	
FP967 (CDC Triffid)	CDC-FL001-2	CDC Triffid Flax

CHAPTER 6
Fruits and vegetables

Several GM cultivars of fruits and vegetables have been approved for food and feed use, although not all are currently marketed commercially.

Fruit

The list of approved GM cultivars includes apple, cantaloupe, plums and papaya, but only one type of GM fruit, the Hawaiian papaya with resistance to ringspot virus, is currently available to consumers.

Cantaloupe (*Cucumis melo*)

In 1999 the US Food and Drug Administration (FDA) approved two GM cultivars of cantaloupe with a modified fruit ripening phenotype that had been developed by Agritope. The modification resulted in a reduced production of ethylene in the fruit, a delayed ripening process and a reduction in losses due to over-ripe fruit in the field, packaging, handling and distribution systems. Approval was based on safety and nutritional assessment data submitted by the company showing that the new cultivars were not materially different in composition, safety and other relevant parameters from cantaloupes currently being marketed.

These GM cultivars of cantaloupe do not appear to be marketed currently.

Papaya (*Carica papaya* L.)

Papaya is a tropical fruit that has edible flesh and seeds. The fruit is grown commercially in many tropical areas including Brazil, Chile, Indonesia, India and Mexico. In the USA, papayas are cultivated in Hawaii and Florida. Hawaiian varieties are the papayas commonly found in supermarkets.

Genetic Modification and Food Quality: A Down to Earth Analysis, First Edition.
Robert Blair and Joe M. Regenstein.

Unfortunately, the fruit is susceptible to papaya ringspot virus (PRV), a pathogen belonging to the potyvirus group of plant viruses and considered to be the main factor limiting commercial papaya production throughout the world. According to Mochida (2007) the Hawaiian papaya industry was devastated in the early 1990s due to the disease, with production decreasing from over 25,000 tonnes in 1993 to less than 16,000 tonnes in 1998 – a decline of 40%. As a result, efforts were made to develop a GM papaya resistant to the virus since there was no alternative method of controlling infection from the pathogen.

A GM papaya line 55-1 with resistance to PRV was approved by the FDA in 1997, based on a safety and nutritional assessment. It was concluded that the new cultivar was not materially different in composition, safety, or other relevant parameters from papaya varieties currently on the market. Health Canada also approved the food use of papaya cultivars 'Rainbow' and 'SunUp', derivatives of the GM 55-1. This agency conducted a comprehensive assessment of this virus-resistant papaya according to the International Guidelines for the Safety Assessment of Novel Foods and concluded that virus-resistant papaya is as safe and nutritious as currently available commercial papaya cultivars. The data included analytical results for total soluble solids, carotenoids, vitamin C and minerals (potassium, phosphorus, calcium, magnesium, sodium, iron, copper, zinc and boron). The results indicated that the compositional details of the GM papaya were within the range of levels found in non-GM cultivars grown in Hawaii.

Approval was later granted for the GM papaya line X17-2, which also had resistance to PRV. Again, approval was granted based on a safety and nutritional assessment of the GM papaya.

Most of the current production of Hawaiian papaya is of PRV-resistant cultivars, often on small farms. Other countries have adopted a similar approach to dealing with infection from this virus. For instance, a GM papaya line (TPY10-4) that is resistant to both PRV and papaya leaf distortion mosaic virus (PLDMV) has been developed and is being grown in Taiwan. Consequently, papaya is regarded as the first commercialised GM fruit crop.

Safety

Fermín et al. (2011) conducted an assessment of the potential allergenicity of the PRV coat protein that is expressed in GM Rainbow papaya. The Rainbow variety was derived from GM papaya line 55-1 and carries the coat protein gene of the PRV. The researchers found that the sequence of amino acids in the coat protein was less than 35% homologous with that of known allergens, and that it did not have a sequence of eight amino acids found in known allergens (common bioinformatic methods are used for assessing similarity to allergen proteins). The coat protein was also tested for stability in simulated gastric and intestinal fluids and under various heat treatments. The results showed that the coat protein was degraded under conditions for which allergenic proteins are considered to be stable relative to non-allergens. The researchers also estimated the relative intake of coat protein by consumers purchasing Rainbow, SunUp (another GM PRV-resistant variety) and naturally infected non-GM papaya. On the basis of consuming 1 fruit per month, the estimated annual consumption of coat protein was 42.9, 1.7 and 331 mg, respectively. On the basis of consuming 1 fruit per day the corresponding intakes were estimated to be 1310, 51.8 and 10100 mg, respectively. Based on these findings the researchers concluded that the transgene-derived PRV coat protein does not pose a risk of food allergy.

Powell et al. (2008) conducted a 90-day feeding study with rats in which male and female Wistar rats (n = 56) received diets formulated to contain GM (PSV-resistant line 52.3) or non-GM (parent line) papayas at twice (100 g/kg) the equivalent of the average daily human consumption of fresh papayas. A control group received standard rat chow alone. Serum chemistry parameters were measured, including total protein, cholesterol, triglycerides, alanine transaminase (ALT), aspartate transaminase (AST), acid phosphatase (AP) and alkaline phosphatase (ALP). Livers and kidneys were collected from the euthanised animals and weighed; then tissue from these organs was examined microscopically. Weights of the spleen and intestines were determined. Hepatic and kidney enzymes (ALT, AST, AP and ALP) and liver lipids were also measured. Similar levels of plasma protein were found across all treatment groups and no significant adverse effects on animal behaviour or differences in body and organ weights were observed. No significant difference in the activities of ALT, AST, AP and ALP in plasma or in liver and kidney were recorded. The cholesterol levels in the plasma and liver of all animals were comparable, as were plasma triglyceride levels. However, the plasma cholesterol values across all three groups were higher (35–65%) than the average reported for rats. All triglyceride values were, however, within the normal range for rats (26–145 mg/dl). The livers of all rats showed normal morphological structure with preservation of the lobular architecture. Minor non-specific changes included fat within hepatocytes and centilobular venous congestion. Normal glomerular and tubular structures were found in the kidneys in rats from all groups. Microvesicular or macrovesicular fat was observed in all groups except those fed diets consisting of non-GM papaya pulp and GM pulp and peel. The results suggested that consumption of GM papayas had no adverse effects on the test animals.

Researchers in Taiwan studied the effect of a GM papaya line (TPY10-4) that is resistant to both PRV and papaya leaf distortion mosaic virus (PLDMV) on immune responses in ovalbumin-sensitised mice (Chen et al., 2011). A related native non-GM cultivar (TCK) was used for comparison. Since no allergenic proteins have been reported in papaya, both green and ripe papaya fruits were used. The fresh papaya fruit samples were harvested. These were peeled and the seeds removed. The fruit flesh was then lyophilised and ground into powder. Doses of both green and ripe papaya fruit were administered at low (0.2 g powder per kg body weight) and high (1.6 g powder per kg bodyweight) doses to mice by intragastric gavage for five weeks. The results showed that treatment with either the GM TPY10-4 or native TCK papaya fruit did not significantly affect body, visceral organs and relative tissue weights, total IgE antibody levels, OVA-specific IgE and IgG1 antibody titres or OVA-stimulated interferon- (IFN-γ), and interleukin-2 (IL-2), IL-4, IL-5 and IL-10 secretions using splenocytes. However, treatment with the GM papaya fruit markedly increased serum total IgM levels. It was concluded that intake of GM TPY10-4 papaya fruits did not result in any allergic response in ovalbumin-sensitised mice but might have a protective effect on immunity by increasing the serum total IgM level.

Lin et al. (2013) conducted a series of detailed studies to test the toxicity of 3 GM PRV-resistant papaya cultivars (823, 2210 and 823-2210) and the parent non-GM (TN-2) cultivar, using rodents. Fruit was obtained from each cultivar, lyophilised and ground into powder. Results showed that the coat protein was rapidly hydrolysed in simulated gastric fluid and that no fragments were detected in organs (brain, heart, liver, spleen, lung, kidneys and testes) or in the stomach contents, cecal contents, colon contents, urine, whole blood and serum of rats fed the ground fruit at a rate of 2 g/kg per day for seven consecutive days. Results of an Ames test (in-vitro bacterial reverse-mutation test) and a chromosomal aberration test showed that, except for the positive control mutagens, all samples of GM papaya fruits were negative when treated with or without S9 metabolic activation. A micronucleus test in mice showed that both the percentage of reticulocytes and frequency of micronuclei in the peripheral blood erythrocytes showed no significant difference between groups given non-GM and GM papaya. In a final test rats were fed lyophilised fruit powder at the rate of 2 g/kg body weight for 28 days or distilled water, in addition to the regular diet. No deaths or abnormalities were observed in any of the groups, and no biological or toxicological significances between groups fed the non-GM and GM papaya were recorded. No adverse effects were observed in the clinical hematology, biochemistry and blood coagulation parameter response variables of treatment groups. Also, no significant biological differences were found in organ weights (brain, heart, thymus, liver, kidney, adrenal, spleen, ovary and testis) and urinary parameters. Based on their findings the researchers concluded that the GM papaya fruit studied can be regarded as equivalent to traditional (non-GM) papaya in terms of food safety.

In 2011 Japan approved the importation of GM Rainbow papaya from Hawaii for food use following an assessment that was completed in 2011. Approval by the Japanese regulatory body confirmed the safety of the Hawaiian product (Gain Report, 2011).

Nutritional and organoleptic qualities

The nutritional composition of GM Rainbow papaya at three ripening stages was reported by Tripathi et al. (2011). According to these researchers this cultivar accounts for about 70% of Hawaii's papaya acreage and GM papaya has been widely consumed since 1998 without any reported adverse health effects. Fruit from a conventional non-GM papaya of related parentage was used as a control. Overall, the conventional and GM fruits were similar in nutritional composition when compared at three ripeness stages: (i) mature green fruit, (ii) fruit harvested at colour break then ripened off the tree, or (iii) fruit ripened on the tree. These data confirm earlier findings on the composition of GM papaya (Roberts et al., 2008; Wall, 2006) (Table 6.1).

The data published by Tripathi et al. (2011) relate to ripeness at stage (ii), the most common commercial process of ripening. At this stage the GM fruit was slightly smaller and had a significantly higher content of vitamin A. The data published by Roberts et al. (2008) relate to fruit ripened for six days. No significant differences in nutritional composition were reported.

The data published by Wall (2006) relate to fruit typically harvested at the colour break to the 1/8th ripe stage for the commercial market. Although considerable variation was recorded in this study no significant differences in nutritional composition due to genetic type was observed. The mean vitamin C content was 512 mg/kg, with no difference among cultivars. The mean content of provitamin A carotenoids was 2320 mg β-carotene and 5940 mg β b-cryptoxanthin per kg, and vitamin A ranged from 187 to 740 mg RAE per kg. Lycopene was not detected in the yellow-fleshed cultivars, Kapoho, Laie Gold and Rainbow, but the red-fleshed Sunrise and SunUp fruit contained 13,500–36,700 mg lycopene per kg. Papayas (100 g) contained 9% of the dietary reference intake (DRI) for copper, 6–8% of the DRI for magnesium, but less than 3% of the DRI for other minerals.

An interesting consumer study was conducted to test Hawaiian consumer responses to GM fruit (Shehata, 2011). At each of six grocery stores in Oahu, papayas labelled 'Hawaiian Grown GMO Papayas' or 'Hawaiian Grown Papayas – Genetically Modified Organism' were displayed next to non-labelled papayas in the produce department, and the sales were monitored. The results showed that labelling papaya fruits increased their sales, which was possibly attributed to a consumer perception that a labelled product implies a better quality item, a desire by the consumer to sample GM fruits, or to a general lack of knowledge regarding GMOs. Given the importance of papaya to the economy of this region and the publicity related to the introduction of GM cultivars, it is difficult to believe that Hawaiian consumers are unaware of GMOs.

Some other fruits have been approved in several countries but are not available in commercial quantity to consumers. Others are at various stages of being introduced to the commercial market.

Apples (*Malus domestica*)

Currently no GM apples are being marketed. However, as of 2015 an apple that has been genetically modified to resist browning (a non-browning Arctic apple produced by Okanagan Specialty Fruits in Canada) has received regulatory approval in the USA and Canada. Browning in apples and potatoes results from polyphenol

Table 6.1 Nutritional composition of non-GM and GM Rainbow papaya fruit.

Component fresh-weight basis	Tripathi et al. (2011)		Roberts et al. (2008)				Wall (2006)				
	Conventional (non-GM)	Rainbow (GM)	Conventional (non-GM)	52.22 (GM)	52.24 (GM)	52.3 (GM)	Kapoho (non-GM)	Laie Gold (GM)	Rainbow (GM)	Sunrise (GM)	Sunup (GM)
Weight, g	533.8	516.8	–	–	–	–	–	–	–	–	–
Length, cm	14.3	14.3	–	–	–	–	–	–	–	–	–
Girth, cm	30.2	30.2	–	–	–	–	–	–	–	–	–
Moisture g/kg	866	861	830	819	821	810	860	858	861	862	877
Energy kJ/kg	2198	2290	–	–	–	–	–	–	–	–	–
Crude fibre g/kg	5.76	5.6	25.3	27.0	24.3	27.0	–	–	–	–	–
Ash g/kg	3.64	3.88	8.5	7.0	8.8	8.0	–	–	–	–	–
Carbohydrate[1]	–	–	150	162.5	157.5	172.4	–	–	–	–	–
Crude fat g/kg	1.69	1.71	2.9	5.2	5.2	2.2	–	–	–	–	–
Soluble solids (°Brix)	–	–	–	–	–	–	12.1	12.5	12.4	12.1	10.9
Crude protein g/kg	8.29	8.43	8.6	6.3	7.5	7.4	–	–	–	–	–
Major minerals mg/kg											–
Calcium	213	119	–	–	–	–	98	204	160	124	139
Phosphorus (total)	55.7	50.7	–	–	–	–	80	62	68	50	49
Magnesium	198	191	–	–	–	–	192	248	249	327	211
Potassium	1220	1380	–	–	–	–	897	1529	1730	2214	1701
Sodium	28.1	23.4	–	–	–	–	56	89	81	64	69
Trace minerals mg/kg											
Copper	0.3	0.2	–	–	–	–	–	–	–	–	–
Iron	0.7	0.5	–	–	–	–	–	–	–	–	–
Manganese	0.1	0.1	–	–	–	–	–	–	–	–	–
Zinc	0.4	0.3	–	–	–	–	–	–	–	–	–
Vitamins											–
Vitamin A IU/kg	876	1560	–	–	–	–	–	–	–	–	–
Vitamin A µg RAE/100g	–	–	–	–	–	–	29.9	48.2	50.3	45.6	20.4
Vitamin C, mg/kg	658	683	–	–	–	–	454	513	518	556	453
Papain mg/kg	55.5	58.6	–	–	–	–	–	–	–	–	–
BITC mg/kg	0.42	0.41	–	–	–	–	–	–	–	–	–

[1]Carbohydrate calculated as 1000–(protein + fat+ moisture +ash).

oxidase (PPO), an enzyme that produces melanin, a compound that contains iron and gives cells a brown tint. The Arctic apples produce less than 10% of the PPO produced by conventional apples and therefore do not turn brown when sliced.

Plums (*Prunus domestica* L.)

A GM plum cultivar ARS-PLMC5-6 (C5), developed by the USDA Agricultural Research Service in collaboration with the French Institute National de la Recherche Agronomique (Paris) and Cornell University (Ithaca, New York), was approved by the US FDA in 2009. The new cultivar (HoneySweet) was developed to be resistant to plum pox virus by the introduction of the plum pox virus coat protein gene (PPV-CP). Approval was based on data relating to the safety and nutritional assessment of the GM plum that had been submitted by USDA-ARS to ensure that this product complied with the legal and regulatory requirements, and that plums derived from the new cultivar were not materially different in composition, safety and other relevant parameters from plums currently being marketed.

Commercial production of the GM cultivar has not yet taken place in the USA, although small-scale production may be occurring in other countries in which plum pox virus is a problem.

Vegetables

The list of approved GM cultivars includes lucerne (alfalfa), potatoes, squash, sugar beet, sweet corn (maize) and tomatoes.

Lucerne (alfalfa) (*Medicago sativa* L.)

This perennial herbaceous legume has a high nutritional quality, making it one of the most important legume forages in the world. A major source of protein for livestock, it is a basic component in diets for dairy cattle, beef cattle, horses, sheep and goats. Ground dehydrated lucerne is also used in diets for pigs and poultry.

Regulatory authorities in the USA, Canada, Australia/New Zealand, Mexico and the Philippines have approved lucerne cultivars J101 and J163 for feed and food use. These herbicide-tolerant cultivars are being grown mainly for use in animal feeding, such that any lucerne sprouts purchased in food stores by consumers are unlikely to be from GM cultivars. However, should they be from these approved GM cultivars, consumers can be assured that the products have been approved as being safe for food use.

Planting of GM cultivars started in the USA in 2005 but was halted in 2007 when a District Court found that the risks of cross-contamination with non-GM lucerne had not been sufficiently assessed by the USDA. The ban was reversed in

June 2010 by the US Supreme Court and planting resumed in January 2011. A low-lignin cultivar developed by Monsanto and Forage Genetics International received non-regulated status (approval for cultivation in the USA) by the FDA in 2014.

Safety

Approval of these cultivars by the regulatory authorities was based on evidence related to safety and nutritional composition. The conclusion of their assessment was that food and feed derived from the new alfalfa cultivars are not materially different in composition, safety and other relevant parameters from food and feed derived from alfalfa varieties currently being marketed.

Combs and Hartnell (2008) found that inclusion of J101 × J163 (herbicide-tolerant) lucerne hay at a level of 39.7% of the diet had no effect on feed intake, milk production, milk composition, body condition score or somatic cell count in the milk (a measure of the incidence of mastitis) in Holstein dairy cows, in comparison with diets containing similar levels of three conventional cultivars. Cows fed the test diet lost less weight than the cows fed the other diets. The trial lasted for four 28-day periods.

Nutritional and organoleptic qualities

Approval of the GM cultivars J101 and J163 for feed and food use by regulatory authorities was based partly on an assessment indicating substantial equivalence with conventional cultivars of lucerne. Additional information was provided by McCann et al. (2006) who showed that the forage from glyphosate-tolerant lucerne J101 × J163 was compositionally equivalent to forage from conventional alfalfa. The test forage was harvested at the late bud to early bloom stage from five field sites in the establishment crop year and from four field sites in the third year of the crop. Among the nutrients measured were proximate constituents, fibre types, amino acids, coumestrol and minerals. Some partial data are shown in Table 6.2. While some differences were found, the data overall were within the range observed for this crop and the differences were therefore not considered to be of biological significance.

Potato (*Solanum tuberosum* L.)

Potatoes are grown in many countries but only about 25% are grown for table use. The bulk of the potato crop is used for processing into starch and other products and for feeding to animals.

During the period from 1995–98 the regulatory authorities in the USA approved four GM cultivars (New Leaf Plus) of potato that had been developed by Monsanto with resistance to attack by Colorado beetles. These cultivars contain two genes isolated from bacteria, and the resulting two proteins these new genes produce. One gene produces a protein derived from *Bacillus thuringiensis (B.t.)* ssp. *tenebrionis* which provides the protection from insect attack. A second gene

Table 6.2 Composition of lucerne forage from GM (GTA J101 × J163) and conventional lucerne (McCann et al., 2006).

Component g/kg dry-weight basis unless stated	**2003 field season**	
	Conventional non-GM	**GTA J101 × J163 GM**
Moisture, g/kg fresh basis	786.8[a]	761[a]
Crude protein g/kg	210.7	214.2
Crude fat g/kg	34.1[a]	28.1[a]
Ash g/kg	92.0	90.2
Carbohydrate	663.3	670.8
Acid-detergent fibre	301	303.8
Neutral-detergent fibre	378	377
Lignin	63.5	68.1
Coumestrol	376.6	474.2
Amino acids % of total AA		
Alanine	6.07	6.07
Arginine	5.51	5.38
Aspartic acid	14.76	15.46
Cystine	1.56	1.5
Glutamic acid	10.81	10.78
Glycine	5.28	5.24
Histidine	2.61	2.64
Isoleucine	4.68	4.51
Leucine	8.33	8.20
Lysine	7.32	7.46
Methionine	1.86	1.76
Phenylalanine	5.53	5.37
Proline	4.83	4.86
Serine	5.34[a]	5.52[a]
Threonine	4.75	4.77
Tryptophan	1.24	1.27
Tyrosine	3.76	3.62
Valine	5.75	5.58
Major minerals g/kg dry-weight		
Calcium	13.9[a]	12.2[a]
Magnesium	2.9	2.6
Phosphorus	3.5	3.4
Potassium	29.4	27.8
Sodium	1.2	2.0
Trace minerals mg/kg dry-weight		
Copper	7.51[a]	6.47[a]
Iron	195.68	250.27
Manganese	33.36	33.17
Zinc	29.15	29.69

[a]Values on the same row with different superscripts are significantly different at $p < 0.05$.

produces an enzyme which functions as a biological marker to allow researchers to identify the modified plants.

The regulatory agencies in Australia/New Zealand (FSANZ, 2001) approved the New Leaf Plus potatoes for food use, following an assessment that indicated the cultivar was as safe for human consumption as other commercial potato cultivars. Based on the available data, the agencies concluded that New Leaf potatoes could be regarded as substantially equivalent to food derived from non-GM potatoes and that no mandatory labelling was required.

In 2010 the EU approved the introduction of the GM cultivar BPS-25271-9 (EH92-527-1) (Amflora), which had been developed by BASF to have an increased ratio of amylopectin to amylose in the starch fraction. This cultivar was intended for use in processing, the starch for use in technical non-food products such as paper, and the pulp for use as animal feed or other conventional, non-food purposes.

A GM potato with traits similar to those of the Arctic apple, such that it contained lower levels of acrylamide during cooking and resisted bruising, received regulatory approval from the FDA in 2014. Three cultivars that had been developed by the J. R. Simplot Co. were approved for cultivation in the USA.

Currently no GM cultivars appear to be being grown commercially in North America, Europe or Australia/New Zealand due to low market share and consumer resistance (see below).

Safety

Dr Arpad Pusztai of the Rowett Research Institute in Aberdeen, Scotland, startled the international scientific community in 1998 when he reported that diets containing GM potatoes expressing a lectin *Galanthus nivalisagglutinin* (GNA) from the snowdrop (*Galanthus nivalis* L.) had a deleterious effect on the gut of rats. Initial presentation of his findings was made in a very unusual way. Commonly scientists outline their findings at conferences of their peers, receive criticisms and comments, and subsequently publish the findings in peer-reviewed scientific journals. Later publication in magazines aimed at the general public may follow. According to the account of Fedoroff (2011), Dr Pusztai announced his findings when he appeared on the British TV show 'World in Action' on 10 August 1998. In the course of that interview, he is reported to have stated that his experiments demonstrated that rats fed a diet of GM potatoes expressing the gene coding for GNA showed stunted growth and reduced immune function. He is further quoted as saying that he would not eat GM food and that he found it 'very, very unfair to use our fellow citizens as guinea pigs'. The announcement made headlines around the world. The most important aspect of the pronouncement by Pusztai was that not only was the GM potato that he was studying unsafe to eat, but that GM food in general was unsafe. This was the first major criticism of GM foods to be made by a respected scientist.

The pronouncement had several repercussions. An audit committee appointed at the Rowett Institute later that year studied Pusztai's laboratory reports and data, and concluded that the findings did not support the conclusion that the GM potatoes had a deleterious effect on growth, organ development or immune function in rats. On the other hand, more than 20 scientists who had studied Dr Pusztai's findings issued a memorandum supporting him since he was regarded as an authority in the field of lectin chemistry.

The Royal Society UK (1999) reviewed all the relevant material and issued a summary statement in June of that year. The consensus of the experts appointed by the Society was that the experiments were poorly designed, the statistical analysis inappropriate and the results inconsistent. Their recommendation was that the experiments be repeated and the results published.

Pusztai published some results in 1999 as a member of the Department of Pathology, University of Aberdeen (his employment contract at the Rowett Research Institute had been cancelled). The paper appeared in a medical journal *The Lancet* (Ewen and Pusztai, 1999) as a Research Letter and provided summarised data on the effects of diets containing parent-line potatoes (Desiree), GM (GNA) potatoes or parent-line potatoes supplemented with 25.4 μg/g GNA on the gut histology of rats. GNA was selected for insertion into potatoes with the intention of making the crop resistant to insects and nematodes, as had been done with other crop plants. Also, it had been shown that the initial effect of this mannose-specific lectin (GNA) on the rat small intestine was minimal. ELISA analysis confirmed that the expression level of GNA in raw GM potatoes was 25.4 μg/g dry matter and was 4.9 μg/g dry matter after boiling for 1 hour. Six rats were randomly allocated to each group, and were fed the diets for 10 days. All the diets were isocaloric and contained an average of 6% protein. Mean crypt length (μm) in the jejunum was found to be 75, 78 and 78, respectively, with boiled potatoes and was 57, 64 and 90, respectively, with raw potatoes. Statistical analysis found the effect of the genetic transformation to be significant at $p < 0.041$. Therefore the authors suggested that 'the promotion of jejunal growth was the result of the transformation of the potato with the GNA gene, since the jejunum of rats was shown to be stimulated only by GM potatoes but not by dietary GNA'. That conclusion is debatable since the stimulatory effect was seen only with raw potatoes, the practical significance being that potatoes are never eaten raw by humans. Overall mean values for the caecum were 95, 98 and 70, respectively, and for the colon were 146, 177 and 139, respectively. Mean intra-epithelial lymphocyte counts per 48 villi were stated to be 7.6 in rats fed boiled parent-line potatoes, compared with 10.3 in rats fed diets containing boiled GM potatoes ($p < 0.01$). With raw potato diets, the intra-epithelial lymphocyte counts were found to be significantly different at 5.3 and 9.3 with the parent and GM potatoes, respectively ($p < 0.01$). Peyer's patches were stated to appear normal in all rats and GNA binding in the jejunum and ileum was found to be about the same, irrespective of whether the potato diets contained a supplement of GNA or contained GM potatoes. No data supporting the earlier claim of stunted growth were presented.

This was a relatively small trial and it seems clear that no definitive conclusions could be made from the data presented, either as to the safety for the human consumer of the specific GM potatoes in question or GM foods in general. A great deal of follow-up work would be necessary before any firm conclusions could be drawn.

A fact that appears to have been omitted in the discussions related to this whole matter is that consumers were never exposed to any possible dangers from eating the GM GNA potatoes in question. These potatoes were at the testing stage and had not been approved for release to the public by any regulatory authority. Any alarm at the possible exposure of consumers to the GM GNA potatoes was, therefore, somewhat premature.

More recent research on GM potatoes has not identified any problems similar to those reported above. Korean researchers (Rhee et al., 2005) conducted a detailed study with rats fed diets containing GM Norland potatoes modified by insertion of the herbicide-resistant bar gene into the genome. A first-generation of 25 male and 25 female rats (150 animals total) was fed one of three diets for 10 weeks prior to mating; standard rat pelleted diet (control), a pelleted diet containing 95% standard rat chow and 5% non-GM potato powder, or a pelleted diet containing 95% standard rat chow and 5% GM potato powder. The dietary treatments were then repeated on progeny selected randomly from the litters produced and the treatments continued for a total of five generations (F0-F4). The non-GM and GM potatoes were found to be similar in composition.

No statistically significant difference in body weight gain or feed consumption between the treatment groups was recorded, including the pregnancy and lactation periods, nor were there any treatment-related clinical signs in any of the generations. The mating indices and fertility of the groups ranged from 85 to 100% and were similar in all groups. However, in the F0 generation, the fertility indexes of the group fed the diet containing GM potato appeared to be lower (male 72.0%, female 78.3%) than those of the control and non-GM groups. The estrous cycle patterns were not altered by dietary treatment and the gestation length ranged from 21 to 22 days in all groups. The mean litter size of the group fed GM potatoes ranged from 10 to 13 and was similar to that of the control and non-GM potato groups. Relative organ weights of rats fed the GM-potato diet were not significantly different from those in the non-GM and control groups. Sperm motility ranged from 80 to 100% in all groups and there were no significant differences. Histopathological examination of the reproductive tissues (testis, ventral prostate, epididymis, seminal vesicle, ovary, uterus and vagina) was performed on five males and five females from each group. No significant treatment differences were found. In addition, the DNA extracted from the testis, epididymis, seminal vesicle, ventral prostate, ovary, uterus, liver, kidney and spleen of all parental animals that had been fed the GM potato diet was found to be devoid of any bar gene. Similarly, in each generation the bar gene was not found in any of the reproductive organs of male and female rats fed the diet containing GM potato, and the litter-related indexes did not show any treatment-related changes. These findings indicated no adverse effects in five generations of rats fed a diet containing GM potato powder.

El-Sanhoty et al. (2004) conducted a safety evaluation of GM Spunta potatoes containing the Cry V gene (which confers resistance to the potato tuber moth). Three cultivars were tested, a non-GM parent-line Spunta (control) and two GM cultivars (G2 and G3 which contained the Cry V gene plus other slightly different genetic elements necessary for the gene to be expressed). The Spunta G2 and G3 cultivars were obtained from local fresh markets in Egypt.

Forty-eight weanling male Albino rats of about 50 g were fed a basal diet for one week. They were then divided randomly into 4 groups (n = 12) and fed one of the following diets for 30 days: Group 1 (control) basal diet; Group 2 basal diet with 30% of the corn starch replaced with 30% freeze-dried non-GM potato Spunta; Group 3 basal diet with 30% corn starch replaced with 30% freeze-dried GM potato Spunta G2; and Group 4 basal diet with 30% corn starch replaced with 30% freeze-dried GM potato Spunta G3. Water and diet were provided ad libitum. Results showed that animals in each group grew well and without any marked differences in appearance. No statistical differences were found in feed intake, daily body weight or feed efficiency. However, final body weight was slightly and significantly higher in the groups fed the diets containing the GM potato cultivars G2 and G3 than in the other two groups. No significant differences were found in serum biochemical values (total protein, albumin, glucose, total cholesterol, urea, GOT and GPT) between groups, or in relative organ weight (liver, spleen, heart, kidney and testes). Based on these findings the researchers concluded that the GM potato cultivars Spunta G2 and G3 produce the same biochemical results as conventional potatoes when fed to rats.

In a follow-up trial the same group (El-Sanhoty et al., 2006) expanded their safety assessment of GM Spunta potatoes G2 and G3 by studying the degradation of GM DNA in the gastrointestinal tract of rats and its deposition in organs. The dietary regime, sex of animals and number were similar to those used in their previous study (El-Sanhoty et al., 2004). After feeding ad libitum for 30 days, no statistical differences were found in feed intake, daily body weight gain or feed efficiency. DNA ingested in the feed was found to be partially resistant to the mechanical, chemical and enzymatic activities of the rat gastrointestinal tract, and was absorbed in the intestine or excreted via the fecal matter. Survival of non-GM and GM potato DNA was comparable during feed passage in the gastrointestinal tract. Fragmentary DNA from the GM constructs in the G2 and G3 cultivars was not detected in any tissue samples.

Other work on Spunta G2 was done by Quemada et al. (2010) who carried out a safety evaluation of the Cry1Ia1 protein contained in this GM cultivar. Included were toxicity and allergenicity evaluations of Cry1Ia1 protein and protein equivalency tests for the Cry1Ia1 protein from Spunta G2 and bacterially produced Cry1Ia1 protein. Comparison of the Cry1Ia1 amino acid sequence to known amino acid sequences showed no significant homology to known toxins or known allergens. Digestibility and thermo-stability assays determined that Cry1Ia1 was fully digested within 30 seconds of exposure to pepsin and was inactive after 3 to 4 minutes at 100 °C, indicating that it would not be a potential allergen. Acute toxicity studies using rodents were used to calculate an acceptable daily intake (ADI). One treatment group of 10 male and 10 female mice was administered the test Cry1IA1 protein at a dose level of 2000 mg/kg and at a dose volume of 12.5 ml/kg while a control group of 20 animals (10 males and 10 females) received control buffer at a dose volume of 12.5 ml/kg. The acute toxicity test established a no adverse effect level (NOAEL) of 2000 mg/kg body weight per day, from which an ADI of 20 mg/kg body weight per day was calculated. Based on the ADI value and an estimated human daily consumption of 7.9 g Spunta G2 potatoes/kg body weight in South Africa (where the study was conducted), it was calculated that the margin of exposure (MOE) was 2,222,222, that is more than 22,000 times greater than the commonly used target MOE of 100.

As noted above, the Cry1Ia1 protein was found to be digested within 30 seconds in simulated gastric fluid, being degraded specifically by pepsin. The speed of digestion was considered to be consistent with other non-allergenic proteins and in particular other Cry proteins that have obtained regulatory approval in several countries worldwide. One other criterion used to assess the allergenic potential of the Cry1Ia1 was its stability after being exposed to heat, using the survival of tobacco hornworm larvae as an index. Results showed that survival of tobacco hornworms fed Cry1Ia1 was at 50% of the controls by 2 min, and at the same level as the controls at 3 min. These results indicate that the Cry1Ia1 protein was inactivated by heating to 100 °C for 3 to 4 min.

In a further study by these researchers, young Sprague–Dawley rats (female $n = 30$, male $n = 30$), weighing between 49 and 78 g were divided randomly into six groups after an adaptation period. Each group of male and female rats was fed one of the following diets: (control) standard pellets only, pellets containing 30% (w/v) freeze-dried Spunta potatoes, or pellets containing 30% (w/v) freeze-dried Spunta G2 potatoes. No significant differences were found in body weight, organ weight (liver, heart, kidney, spleen, lungs, ovaries and testes), feed consumption, clinical sign changes, clinical pathological alterations, clinical chemistry or hematology. It was therefore concluded that the test animals responded in the same way to GM Spunta G2 potatoes as to conventional non-transgenic Spunta potatoes.

Langkilde et al. (2012) conducted a toxicological analysis of a GM potato cultivar SGT 9-2 which has a reduced α-solanine content. Potatoes contain two toxic glycoalkaloids (GAs), α-chaconine and α-solanine, which are believed to act in synergy. Their degree of toxicity may depend on their ratio in the potato.

A 90-day feeding study with Syrian Golden hamsters was conducted, comparing GM SGT 9-2 potatoes and the parental control cultivar (Desirée wild-type) which has a traditional α-solanine : α-chaconine ratio. The control cultivar was found to contain 84 mg/kg α-solanine and 260 mg/kg α-chaconine and the GM cultivar 8 and 207 mg/kg, respectively, dry matter basis. Diets containing up to 60% freeze-dried potato powder from each cultivar were used. Results showed a small number of significant differences in animal responses to the diets with different α-solanine : α-chaconine ratio but none were considered to raise safety concerns with respect to human (or animal) consumption.

Nutritional and organoleptic qualities

Rogan et al. (2000) presented detailed information on the composition of potatoes from insect and virus-resistant potato cultivars. The potatoes were GM cultivars of Russet Burbank and Shepody potatoes and had been approved by Canadian and US regulatory agencies for commerce in North America. The contents of nutritional, quality and anti-nutritional components were measured. The results (Table 6.3) showed that tubers produced by insect- and virus-protected varieties were substantially equivalent to tubers produced by conventional potato varieties. The contents of key nutrients, anti-nutrients, macronutrients and minor nutrients produced by tubers harvested from NewLeaf Y and NewLeaf Plus cultivars were consistent with those found in conventional potatoes.

El-Sanhoty et al. (2004) investigated the composition of GM Spunta potatoes with the Cry V gene (G2 and G3) relative to that of the Spunta control line. The analyses included proximate constituents plus carbohydrate, starch, reducing sugars, non-reducing sugars, sodium, calcium, potassium, phosphorus and ascorbic acid. Some toxins and anti-nutrient compounds were also determined. There were no significant differences between GM potatoes G2, G3 and Spunta control potato line in terms of the proximate chemical composition (Table 6.4). The levels of glycoalkaloids were also similar in GM and conventional potato cultivars, the levels being in agreement with a safety level recommended by FAO/WHO (200 mg/kg) for acute toxicity. Levels of protease inhibitor activity and total phenol were estimated and no significant differences were found between the GM and conventional Spunta potato cultivars.

Langkilde et al. (2012) studied the relative composition of a GM potato cultivar SGT 9-2, which has a reduced α-solanine content (Table 6.5).

Although the data indicated some differences in composition between the GM cultivar and its conventional parent, the recorded values were within the range reported by other researchers and were not considered substantive.

Squash (*Curcurbita pepo* L.)

Two GM cultivars of squash are currently being grown and marketed commercially in North America.

Table 6.3 Composition of potatoes from insect and virus-resistant cultivars (Rogan et al., 2000).

Component fresh-weight basis unless stated	Russet Burbank	NewLeaf Plus Russet Burbank			Russet Burbank	NewLeaf Y Russet Burbank	Shepody	NewLeaf Y Shepody	
	non-GM	RBMT21-129 (GM)	RBMT21-350 (GM)	RBMT22-082 (GM)	non-GM	RBMT15-101 (GM)	non-GM	SEMT15-02 (GM)	SEMT15-15 (GM)
Dry matter g/kg	215	216	219	210	207	207	227	223	226
Crude protein g/kg DM	99.0	98.6	99.4	99.3	123	117.5	110.3	114.3	107.6
Soluble protein g/kg DM	50	50	51	50	54	51	63	66	64
Crude fat g/kg DM	1.6	2.0	1.9	2.0	2.1	1.9	1.4	1.7	1.9
Ash g/kg DM	47.5	46.8	47.0	47.8	60.4	58.1	46.9	46.3	46.4
Crude fibre g/kg DM	16.8	16.4	15.5	16.1	16.6	16.9	15.3	13.3	14.2
Carbohydrate g/kg DM	851.8	852.4	851.7	850.9	814.4	822.5	841.4	837.7	844.1
Calories/kg	3818	3823	3821	3819	3777	3769	3819	3823	3824
Glucose g/kg	0.99	0.87	0.94	1.13	2.1	2.4	2.3	2.2	2.1
Sucrose g/kg	1.99	1.82	2.01	1.77	1.8	1.8	2.9	2.8	3.1
Vitamins mg/kg									
Vitamin C	100	101	99	104	134	145	238	227	239
Niacin	20.3	20.15	20.0	21.4	20.3	20.55	22.15	22.75	23.9
Pyridoxine	2.6	2.55	2.65	2.6	2.6	2.6	2.6	2.8	2.5
Minerals mg/kg									
Copper	1.60	1.95	1.50	1.65	1.60	1.50	1.95	2.05	2.42
Magnesium	257.7	260.9	228.65	234.9	257.7	248.85	271.1	265.7	284.75
Potassium	5403.7	5363	5133	5191	5403.7	4983	5810	5486	5676
Amino acids mg/kg									
Aspartic acid	6151	6400	7095	6481	6151	5969	5010	4596	4971
Threonine	736	734	808	762	736	690	915	918	1011
Serine	773	756	875	793	773	723	939	956	1008
Glutamic acid	3913	3854	4290	3893	3913	3753	4830	4327	4884
Proline	596	618	704	643	596	584	822	857	907

Glycine	606	610	632	627	606	571	809	828	896
Alanine	582	588	663	619	582	536	732	747	816
Cystine	310	307	314	420	310	296	382	394	419
Valine	1088	1083	1225	1187	1088	1042	1129	1095	1245
Methionine	282	293	311	312	282	272	361	376	417
Isoleucine	697	699	753	738	697	655	820	798	919
Leucine	1102	1122	1202	1147	1102	1035	1459	1520	1658
Tyrosine	718	638	751	745	718	605	755	737	855
Phenyalanine	834	834	921	899	840	788	1008	1002	1130
Histidine	411	391	413	419	411	392	435	423	471
Lysine	1167	1174	1205	1223	1167	1135	1374	1384	1520
Arginine	998	978	1095	1075	998	934	1209	1101	1252
Tryptophan	212	215	212	226	212	213	214	212	230
Glycoalkaloids mg/kg	43	54	48	51	117	106	46	55	53

Table 6.4 Composition of potatoes from GM Spunta cultivars (El-Sanhoty et al., 2004).

	Spunta	**Spunta G2**	**Spunta G3**
Component **DM basis unless stated**	**non-GM**	**(GM)**	**(GM)**
Dry matter (DM) g/kg fresh wt. basis	181.7	197.5	187.6
Crude protein g/kg	112.5	111.5	112.0
Crude fat g/kg	12	10	11
Ash g/kg	23.1	21.5	22.6
Crude fibre g/kg	17.4	17.3	17.2
Carbohydrate g/kg	653.3	642.2	649.6
Starch g/kg	71.5	71.2	70.3
Reducing sugars g/kg	10.6	12.0	10.7
Non-reducing sugars g/kg	27.6	27.4	27.7
Vitamins mg/kg			
Vitamin C mg/kg	318.7	328.6	336.6
Minerals mg/kg			
Calcium	82	79	84
Phosphorus	341.1	334.9	338.2
Potassium	28.8	28.2	27
Sodium	13.5	12.9	13.7
Amino acids mg/g protein			
Alanine	226	238	241
Arginine	301	310	294
Aspartic acid	92.5	93.43	94.15
Cystine	77.69	82.15	83.09
Glutamic acid	73.38	77.9	76.19
Glycine	68.11	71.14	69.74
Histidine	121	125.03	119.76
Methionine	97.16	88.74	104.27
Tryptophan	94.17	91.36	96.04
Tyrosine	23.6	24.3	22.9
Serine	25.4	23.9	24.4
Proline	24	23.3	24.2
Fatty acids mg/kg			
Palmitic	87.3	81.4	80.6
Stearic	22.1	20.5	21.3
Palmitoleic	13.9	14.1	13.1
Oleic	547	551	547
Linoleic	237	230	232
Linolenic	917	103	106
Glycoalkaloids mg/kg			
α-chaconine	14.7	14.1	15.4
α-solanine	4.8	4.4	4.1
Protease inhibitor activity mg/kg	614.4	655.8	705.3
Total phenols mg/kg	547.7	556.3	562.7

Table 6.5 Composition (dry-weight basis) of GM potato cultivar SGT 9-2 and non-GM control line Desirée (Langkilde et al., 2012).

	Control Desirée (non-GM)	SGT 9-2 (GM)
Component (g/kg DM basis) unless stated		
α-Solanine mg/kg	84	8
α-Chaconine (mg/kg)	260	207
Crude protein	77[a]	87[b]
Crude fat	6	6
Ash	29[b]	27[a]
Crude fibre	26	22
Starch	856	865
Sugars	10	9
Minerals		
Calcium	0.2	0.2
Chromium (mg/kg)	<0.25	<0.25
Cobalt (mg/kg)	<0.05	<0.05
Copper (mg/kg)	4.0	4.0
Iron (mg/kg)	27.9	23.4
Magnesium	0.8	0.8
Manganese (mg/kg)	7.3[a]	7.8[b]
Phosphorus (g/kg)	2.2	2.4
Potassium	13.7[b]	12.4[a]
Selenium (mg/kg)	<0.02	<0.02
Zinc (mg/kg)	12.1	12.8
Amino acids		
Alanine	1.6[a]	1.7[b]
Arginine	2.8[a]	3.3[b]
Aspartic acid	20.3[a]	23.3[c]
Cystine	1.0	1.0
Glutamic acid	15.2[a]	17.8[b]
Glycine	1.7[a]	1.9[b]
Histidine	1.0	1.1
Isoleucine	2.2[a]	2.5[b]
Leucine	3.0[a]	3.3[b]
Lysine	3.3[a]	3.8[b]
Methionine	0.9[a]	1.3[b]
Phenylalanine	2.1	2.4
Proline	1.7	1.7
Serine	2.1	2.2
Threonine	2.2[a]	2.4[b]
Tryptophan	0.6[a]	0.7[b]
Tyrosine	2.0[a]	2.5[b]
Valine	3.3[a]	4.0[b]
Fatty acid distribution%		
Lauric acid	0.4	0.4
Myristic acid	0.6	0.7

Continued

Table 6.5 Continued

	Control Desirée (non-GM)	SGT 9-2 (GM)
Pentadecanoic acid	0.4	0.4
Palmitic acid	30.8	30.1
Heptadecanoic acid	0.6	0.7
Stearic acid	11.2[a]	11.7[b]
Oleic acid	5.1[a]	4.0[b]
cis-Vaccenic acid	1.2	1.4
Linoleic acid	33.4[b]	31.8[a]
Linolenic acid	4.3[a]	5.6[b]
Arachidic acid	2.8[a]	3.1[b]
Behenic acid	2.4	2.5
Tricosanoic acid	1.0	1.1
Lignoceric acid	2.1[a]	2.8[b]
Others (not identified)	3.6	3.6
Vitamins mg/kg unless stated		
Thiamin	2.0	2.5
K_1, μg/kg	16.0	20.9
Nicotinic acid	26.5	27.7
Pantothenic acid	11.4	12.3
Folic acid (μg/kg)	573[a]	778[c]
Vitamin C	573[a]	605[c]

[a,b,c]Values on the same line with different superscripts are significantly different at $p<0.05$. Solanine and chaconine values were not compared statistically.

Safety

The FDA in 1994 approved the GM squash cultivar ZW20 (SEM-0ZW20-7) for food use. This cultivar of yellow crookneck squash has resistance to zuchini yellow mosaic virus and watermelon mosaic virus 2. Another virus-resistant GM cultivar was approved by the FDA in 1997, CZW3 (SEM-0CZW3-2), which has resistance to cucumber mosaic virus, zucchini yellow mosaic virus and watermelon mosaic virus 2. Approval of these cultivars was based on an assessment of evidence related to the safety and nutritional analysis that had been submitted by the applicant company (Seminis). It was concluded that the new cultivars were not materially different in composition, safety, or other relevant parameters from squash currently being marketed.

Health Canada approved the introduction of these GM squash cultivars for food use in 1998, following a comprehensive assessment of the data according to its Guidelines for the Safety Assessment of Novel Foods. It was concluded that the WMV2 and ZYMV coat proteins inserted into the GM cultivars do not possess characteristics typical of known protein allergens. Also there were no regions of homology when the sequences of these introduced proteins were compared to the amino acid sequences of known protein allergens. In addition the WMV2 and ZYMV coat protein sequences were compared to databases of known protein toxins and did not show homology with known mammalian protein toxins. The history of known safe consumption of these proteins from virus-infected plant products provided additional evidence of a lack of toxicity.

Nutritional and organoleptic qualities

Approval of these cultivars by the FDA was based in part on an assessment of evidence related to the nutritional analysis that had been submitted by the applicant company. It was concluded that the GM cultivars were substantially equivalent to conventional squash in terms of nutritional composition.

Assessment of the analysis of nutrients in the GM and conventional cultivars by Health Canada showed only small differences that were within the range of variability normally reported for squash. Nutrients measured included protein, moisture, fat, ash, total dietary fibre, carbohydrates, calories, fructose, glucose, sucrose, lactose, maltose, vitamin C, β-carotene, vitamin A, calcium, iron and sodium. Therefore it was concluded that consumption of GM squash would have no significant impact on the nutritional quality of the Canadian food supply. One nutritional component studied by Health Canada was the content of cucurbitacins, bitter compounds (alkaloids) produced by cucurbit plants. The assessment, which included a taste test, found that the GM squash and the parental yellow crookneck squash contained very low levels of cucurbitacins and were not bitter.

Sugar beet (*Beta vulgaris* L.)

Sugar beet is grown in several countries for processing into sugar for human use, the extracted beet residue being used as animal feed. About 50–60% of the US refined sugar is produced from sugar beet. Three GM cultivars of sugar beet have been approved for food and feed use by the regulatory authorities in the USA (T120-7, glufosinate-tolerant, 1998; GTS B77, glyphosate-tolerant, 1998; and H71, glyphosate-tolerant, 2004). The T120-7 cultivar has also been approved for production in Canada. Approval by these regulatory authorities was based on safety and nutritional assessments in which it was concluded that the new cultivars were not materially different in composition, safety, or other relevant parameters from conventional cultivars of sugar beet. Following approval, the GM cultivars achieved a rapid and high rate of adoption by farmers.

In 2010 a District Court in California placed an injunction on the growing of glyphosate-tolerant GM sugar beet on the grounds that a complete environmental assessment had not been conducted. The injunction was overturned by the Ninth Court of Appeals in 2011. In 2012 the US Department of Agriculture announced its decision to deregulate glyphosate-tolerant GM sugar beet following a thorough environmental impact statement (EIS) and plant pest risk assessment (PPRA), the holding of three public meetings, and consideration and analysis of thousands of comments. Much of the current production in North America is now from GM cultivars. Other countries may allow the cultivation of GM cultivars while others, such as Australia/New Zealand and the European Union, have approved the importation of sugar and sugar-containing processed foods from GM sugar beet but have not yet approved the domestic cultivation of GM cultivars.

Safety

Allergenicity is one of the issues addressed by regulatory authorities. For instance, FSANZ (2003) concluded that none of the source organisms used in the development of sugar beet line 77 are known to be allergenic to humans. Based on this assessment and an assessment of other safety aspects, approval was granted for the use of sugar derived from this GM cultivar as human food. A subsequent review of glyphosate-tolerant sugar beet was conducted in 2005 under the Australia New Zealand Food Standards Code (FSANZ, 2005). In their review FSANZ carried out a full safety assessment and determined that the GM line H7-1 was safe for human consumption and did not pose public health or safety concerns. Further they found no novel DNA or proteins were present in refined sugars derived from this line and thus approved the use and sale of foods derived from it.

The FDA in 2004 concluded that glyphosate-tolerant sugar beet cultivar H7-1 is not materially different in safety or other relevant parameters from sugar beet now grown, marketed and consumed. That decision was based on data and information supplied by the development company. In addition the FDA considered the consultation on glyphosate-tolerant GM sugar beet event H7-1 to be complete. In its assessment the FDA noted that there was no immunologically significant amino acid sequence homology between the CP4 EPSPS protein in GM cultivar H7-1 and the amino acid sequences of allergens in the ALLERGEN3 database. Also, no biologically relevant sequence similarities between CP4 EPSPS protein in cultivar H7-1 and known toxins were observed. Included in the assessment were the results of an acute toxicity study conducted by the development company in which mice were dosed by gavage with up to 572 mg/kg of CP4 EPSPS. No adverse results events were observed at any dose level. The half-life of the CP4 EPSPS protein was reported to be less than 15 s in gastric fluid and less than 10 min in intestinal fluid. The agency further noted that the processing of sugar beet to refined sugar would remove all protein.

Health Canada in 2004 concluded that the potential for toxicity for the CP4 EPSPS protein expressed in GM sugar beet line H7-1 could be considered remote. This conclusion was based on the low amount of CP4 EPSPS protein found in the edible part of the plant, the absence of demonstrated acute toxicity to CP4 EPSPS in mice at doses which were orders of magnitude greater than the range associated with proteins, the lack of sequence homology between CP4 EPSPS protein and known anti-nutrients, and the likelihood that the protein would be degraded under conditions similar to those in the human gastrointestinal tract. There were no additional health concerns regarding endogenous anti-nutrients from the product derived from sugar beet H7-1 since it had been demonstrated that these were present at identical concentrations in GM and non-GM sugar beet. In both cases, the concentrations of these natural anti-nutrients in sugar derived from sugar beet sugar are not a human health concern. The possibility that the CP4 EPSPS protein would be an allergen in sugar beet lines containing event H7-1 was also considered unlikely. This was based on the negligible amount of protein in the edible parts of the sugar beet plant, the lack of sequence homology between CP4 EPSPS protein and known allergens, and the fact the EPSPS protein was degraded rapidly by gastric and intestinal enzymes. These results combined with the absence of CP4 EPSPS protein in refined sugar suggests that sugar derived from sugar beet lines containing event H7-1 would not pose any greater allergenic risk than non-GM sugar beet sugar.

Most of the studies on food safety issues relating to GM sugar beet have been conducted on farm animals, which is understandable since only the extracted sugar is used as human food. Beet pulp and molasses are the major feed by-products of sugar beets used in livestock feeding. Oguchi et al. (2009) tested sugar products for the presence of residual DNA. Polymerase chain reaction (PCR) analyses with taxon-specific primers indicated that sugar beet DNA was degraded at an early stage of sugar processing. The calculated concentrations of extracted DNA was 2.2 ng/ml in white superior sugar and 4.3 ng/ml in brown sugar #1, both very low based on the Japanese standard method for quantitative PCR analysis. The results showed that sugar beet DNA is degraded rapidly in the raw juice and during subsequent processing so that no GM products are found in the refined sugar. These findings suggest that any residual GM material remains in the residue after sugar extraction, a situation analogous to that of extracted oil from GM oilseeds. Based on the results of this investigation the Japanese government decided that mandatory GM labelling should not be applied to sugar products.

Nutritional and organoleptic qualities

Compositional analysis is another factor used by regulatory authorities in their assessment of products from GM plants as human food. The FDA in 2004, as part of an assessment on whether glyphosate tolerant sugar beet cultivar H7-1 is as nutritious as conventional sugar beet varieties currently being produced, compared the data on composition of H7-1 sugar beet and corresponding non-GM, related control lines derived from the breeding process and eight commercial sugar beet varieties. Samples were collected at five locations in Europe in traditional sugar beet production areas (France, Germany, Italy, Spain and the United Kingdom). The H7-1 plants were treated with Roundup herbicide during the growing season. Tops (leaves) and brei (pulp from processed roots) were analysed chemically for contents of ash, crude fibre, fat, crude protein, dry matter, carbohydrates (by calculation), amino acids, acid detergent fibre, neutral detergent fibre and (for pulp only) percentage sucrose (by optical polarisation), invert sugar, sodium, potassium and α-amino nitrogen. No statistically significant differences were found in the mean levels of nutrients between pulp from sugar beet H7-1 and non-GM controls. Statistically significant differences were found between cultivar H7-1 and non-GM controls in the levels of 2 out of 18 amino acids analysed, with alanine higher and glutamic acid (actually glutamic acid plus glutamine) lower in cultivar H7-1. However, these differences were not deemed to be biologically significant since the mean values of these two amino acids were within the ranges reported for both the non-GM control and conventional varieties. It was therefore concluded that the pulp from GM sugar beet H7-1 is compositionally equivalent to that of the non-GM control and other commercial sugar beet varieties. A similar conclusion was reached for the tops. Root and top samples were analysed for saponin content. Saponins are triterpenoid glycosides that occur naturally in many plant foods such as legumes, potatoes, tea and sugar beet. They have a bitter taste and can act as a deterrent to foraging. Statistical analysis of the chemical data from sugar beet tissue showed that the saponin levels in H7-1 sugar beet were not significantly different from the levels in non-GM controls and were comparable to the levels in the tissues of the commercial reference varieties.

Canadian regulators in 2001 approved the introduction of glufosinate-tolerant sugar beet products derived from T120-7 as animal feed (Canadian Food Inspection Agency, 2001). Beet pulp and molasses are the major feed by-products of sugar beets used in livestock feeding. Approval was based on proximate, amino acid, mineral and sugar analyses of leaf and root samples from each of 32 cultivars of conventional sugar beets and two hybrids from T120-7 grown in Europe. These data showed that for the leaves, there was no difference in nutrient composition between the control and hybrid lines. For the root samples, there were no differences in proximate analysis between the control and the T120-7 hybrids. Of a total of 63 variables measured, three (calcium, manganese and arginine) were observed to be different in the T120-7 hybrids compared with the control lines.

Another set of data related to sugar beets grown in the USA. Sugar beet roots and tops from T120-7 and a control non-GM variety from seven sites in three locations were analysed. The data indicated that there were no overall differences between the T120-7 lines and the control. Analysis by location showed small but statistically significant differences in calcium, fibre in the roots and sodium in the leaves in one location, but these differences were not consistent with other locations or with the differences shown in the first study. Further analysis of processed fractions, that is, pulp, molasses and refined sugar were also conducted, and confirmed that there were no nutritionally significant differences between the T120-7 lines and the control line. Beet pulp, beet molasses and other sugar beet by-products are approved for use in livestock feeds, and these ingredients are listed in Schedule IV, Part 1 of the Feeds Regulations. Based on the above data, this approval extends also to sugar beet T120-7.

Böhme et al. (2001) studied the chemical composition and digestion of GM (glufosinate-tolerant) sugar beet in pigs and cattle. No significant differences were found in the nutrient composition of beets or beet tops of conventional, GM or non-GM parent line sugar beets (Table 6.6).

Table 6.6 Chemical analysis and nutritive value of sugar beet for pigs and cattle (Böhme et al., 2001).

Component g per kg DM unless stated	Conventional	GM (glufosinate-tolerant) Basta	Non-GM parent line
BEETS			
Dry matter (as received)	233	249	255
Organic matter	970	974	970
Crude protein	72	63	60
Ether extract	3	4	4
Crude fibre	56	47	46
NFE (nitrogen-free extractives)	839	860	860
Sugar	736	738	744
Digestibility % Pigs			
Organic matter	89.4[a]	92.5[a]	93.8[a]
Crude protein	50.4	50.2	55.9
NFE	95.7	96.4	96.8
Energy value MJ/kg DM			
Digestible energy	14.9	15.0	15.3
Metabolisable energy	13.7	14.0	14.2
Digestibility % Cattle			
Organic matter	91.9	91.6	92.1
Crude protein	45.3	47.6	49.9
Crude fibre	59.9	64.0	62.3
NFE	98.2	96.7	96.9
Energy value MJ/kg DM			
Digestible energy	15.1	15.4	15.4
Metabolisable energy	13.2	13.2	13.3
Net energy lactation	8.5	8.5	8.6
SILAGE (tops)			
Dry matter (as received)	150	157	152
Organic matter	759	766	737
Crude protein	167	154	160
Ether extract	18	22	24
Crude fibre	154	153	157
NFE (nitrogen-free extractives)	420	437	396
Digestibility % Cattle			
Organic matter	74.7	77.0	74.5
Crude protein	66.2	63.8	64.1
Crude fibre	72.4[b]	68.5[a,b]	66.2[a]
NFE	80.6[a]	86.1[b]	83.2[a,b]
Energy value MJ/kg DM			
Digestible energy	10.4	10.7	9.9
Metabolisable energy	8.7[a,b]	9.1[a]	8.6[b]
Net energy lactation	5.2[a,b]	5.5[a]	5.1[b]

[a]Values on the same line not followed by the same letter are significantly different at $p < 0.05$.

Digestibility of GM and non-GM sugar beets differed slightly in pigs but the differences were not considered to be meaningful biologically. Digestibility of sugar beet roots and sugar beet top silage for cattle was also found to be within the biological range reported by other researchers.

Based on the above findings it is clear that approved cultivars of GM sugar beet are similar to conventional non-GM sugar beet in chemical composition and nutritive value for farm animals.

Sweet corn (*Zea mays* L.)

Several GM cultivars of sweet corn have been developed and marketed, major companies involved being Monsanto and Syngenta. These cultivars are resistant to herbicide and/or to insect attack and have been approved by regulatory authorities in the USA and Canada for cultivation and human use.

Safety

The safety of these cultivars has been accepted by regulatory authorities, based on evidence submitted by the development companies. No independent publications relating to food safety appear to be available in peer-reviewed scientific journals.

Nutritional and organoleptic qualities

The assessment of GM sweet corn by regulatory authorities includes compositional and nutritional analyses. Approval of the GM cultivars signifies that the regulatory authorities are satisfied that the GM cultivars are substantially equivalent in composition and nutrient content to the related non-GM cultivars. No independent publications relating to composition and nutrient content appear to be available in peer-reviewed scientific journals.

An interesting consumer study on sweet corn was conducted by Canadian researchers (Powell et al., 2003). In this farm-to-fork trial, GM *B. thuringiensis*, (B.t.) sweet corn and B.t. potatoes were grown side-by-side with conventional cultivars in the 2000 growing season at a farm and market in Hillsburgh, Ontario, Canada. The B.t. sweet corn required no insecticides. From an economic perspective, only the first planting had pest pressure high enough to warrant the higher seed cost of the GM cultivar cultivar. The sweet corn harvested throughout the trial was segregated and labelled, and direct consumer evaluation of purchasing preferences was conducted. Overall, the B.t. sweet corn outsold the conventional sweet corn by a margin of 680 dozen (or 8160 cobs) to 452.5 dozen (or 5430 cobs). A limited number of intercept interviews were conducted after consumers made their purchasing decision. The majority of consumers interviewed said they were more concerned about pesticides than genetic modification; however, taste and quality also had a strong influence on purchasing decisions.

Tomatoes (*Lycopersicon esculentum* L.)

The first commercially grown GM food crop was a tomato introduced by Calgene (later Monsanto) in 1994 and named the FlavrSavr (Bruening and Lyons, 2000). Although technically a fruit, tomatoes are usually marketed as vegetables.

The GM tomatoes were engineered to suppress the activity of the enzyme polygalacturonase (PG) and hence delay the process of softening after ripening due to reduced pectin degradation. The rationale was that these tomatoes could be left to ripen on the vine and still have a long shelf life, thus allowing them to develop their full flavour. Conventionally, tomatoes are picked before they are ripe and then ripened artificially.

Several other tomato cultivars were introduced later in North America, but at present no GM cultivars are being grown commercially in this region, Europe or Australia/New Zealand.

In the UK, Zeneca developed a tomato using technology similar to that used with the FlavrSavr (www.wikipedia.org/Flavr_Savr) for tomato paste production. The paste was cheaper to produce than conventional tomato paste, resulting in the cost being 20% lower to the consumer. Between 1996 and 1999, 1.8 million cans that were labelled as containing a genetically modified product were reported to have been sold in Sainsbury's and Safeway supermarkets in the UK. For a time the paste outsold conventional tomato paste, but sales fell in the autumn of 1998 (www.wikipedia.org/Flavr_Savr). The UK Advisory Committee on Novel Foods and Processes (1995) later published a report stating that the decline in sales during this period was linked to changing consumer perceptions of genetically modified crops. The report identified several possible factors, including product labelling and perception of choice, lobbying campaigns and media attention. Another factor was a perceived increase in negative opinion among consumers with regard to GM. This was attributed to media reports following the statements of Dr Arpad Pusztai, who made a claim on television about detrimental health effects found in laboratory rats fed a diet of GM potatoes during the testing stage (see Potato section above). Although his claims were later shown to be without foundation and the potatoes in question were never approved for sale, the Sainsbury's and Safeway supermarkets responded to the negative publicity by pledging that none of their house brand products would contain GM ingredients.

Another factor that may also have been partly responsible for the decline in sales of the FlavrSavr tomatoes was that an antibiotic resistance gene (neo) had also been inserted into the genome as a selectable marker to identify transformed plants during tissue culture regeneration and multiplication. The gene encodes the enzyme neomycin phosphotransferase II (NPTII), which inactivates aminoglycoside antibiotics such as kanamycin and neomycin. Concern was expressed in some quarters that dietary exposure to the GM tomato might result in the development of resistance to the antibiotics kanamycin and neomycin, which are used in human and veterinary medicine. However it should be noted that the report issued by the UK Advisory Committee on Novel Foods and Processes (1995) found that the NPTII enzyme was inactivated in simulated gastric fluid, so that there was no significant reduction in kanamycin activity after 4 h. Consequently the Committee concluded that the use of an antibiotic marker gene in GM tomatoes did not raise any specific safety concerns.

GM cultivars are being developed in North America and in some other countries. For instance, India is developing tomatoes with a variety of improved traits (Rai et al., 2013) but to date no GM cultivars have been approved for food use in that country. B.t. cotton is the only GM crop allowed at present in India.

Safety

The UK Advisory Committee on Novel Foods and Processes (1995) approved the use of paste from GM tomatoes developed by Zeneca (see above), following an assessment of data relating to food safety and nutritional composition.

Over the period 1994–1998, the FDA in the USA approved five GM cultivars of tomato (8538; 135-4; 35-1-N; B, Da, F; and 5345). These cultivars had been modified genetically to possess traits of delayed ripening/softening due to reduced ethylene synthesis, delayed softening due to reduced pectin degradation, and/or resistance to certain lepidopteran insect pests. As with other applications, approvals were based on assessments that indicated that the new cultivars were not materially different in safety, composition, or any other relevant parameter from tomato cultivars currently being marketed commercially.

Tomatoes, and other members of the genus *Solanaceae*, have the potential to accumulate deleterious secondary plant constituents known as glycoalkaloids, mainly α-tomatine. The level of α-tomatine decreases as the fruit matures so that the amounts in ripe vine-ripened red tomatoes are negligible. If a genetic modification results in shorter ripening time, the levels of tomatine, which decrease through maturation, could be influenced. Therefore an assessment of this inherent plant toxin/anti-nutrient is considered to be very important. Solanine and chaconine are the principal alkaloids of potato, but have been found in tomatoes in lower amounts. Also, considerable amounts of oxalate can be found in tomatoes.

GM tomatoes intended for processing (especially for tomato paste) were analysed for α-tomatine, solanine and chaconine in both GM fresh fruit and paste samples (UK Advisory Committee on Novel Foods and Processes (1995). The results show that the glycoalkaloid levels in the GM tomato paste were within the range of glycoalkaloid levels of commercially available pastes, and that genetic modification did not alter the levels of glycoalkaloids in the paste. Analysis of tomato paste samples indicated a level of 58 mg α-tomatine/kg for the GM cultivar and 74 mg/kg for the conventional cultivar. All the other cultivars and also the fresh fruit samples showed less than 15 mg/kg tomatine. The amounts of α-chaconine and α-solanine were below the limit of detection of 5 mg/kg.

Health Canada approved the FlavrSavr tomato for food use in 1995. This followed a comprehensive assessment of the available evidence, according to its

Guidelines for the Safety Assessment of Novel Foods. The agency noted that the GM tomatoes ripened normally but experienced reduced pectin breakdown that benefitted all stages of harvesting and processing. One component investigated was α-tomatine, the main naturally occurring glycoalkaloid in tomatoes. Health Canada noted that a comparison of the levels of α-tomatine in GM and non-GM tomatoes did not show any statistically significant difference and the levels were within the normal range reported for tomatoes. Other than reduced polygalacturonase (PG) activity, the disease, pest and other agronomic characteristics of the FlavrSavr tomatoes were found to be comparable with conventional cultivars. The reduced synthesis of native PG in the GM cultivar was judged not to have any potential for additional human toxicity or allergenicity.

Health Canada in 1995 also approved the introduction of the GM tomato cultivar 1345-4, which was developed to exhibit reduced ethylene accumulation and thus delayed ripening. Other GM tomato cultivars approved for food use in Canada in 1998 were 1401F, H282F, 11013F and 7913F, which had been developed to exhibit reduced pectin degradation through a suppression of PG activity. As in other cases, approval was based on a comprehensive assessment of evidence relating to the safety and nutritional composition of these cultivars. The level of α-tomatine in tomato paste prepared from the GM F cultivars was 58 mg/kg and was 74 mg/kg for the unmodified TGT7 parental line. The additional glycoalkaloids, solanine and chaconine, were undetectable in either the GM or unmodified lines. An analysis of the levels of biogenic amines (tyramine, tryptamine and serotonin), which are considered as potential toxins in tomato, did not show any significant differences in the respective concentrations of these compounds in raw tomatoes from either the transgenic F cultivars or the unmodified parental line. Other than reduced PG activity, the disease, pest and other agronomic characteristics of the 1401F, H282F, 11013F and 7913F cultivars were judged to be comparable to those of the unmodified hybrids. Consequently, Health Canada concluded that these cultivars did not raise concerns related to human food safety and that they were as safe and nutritious as those available from current commercial tomato cultivars.

Reed et al. (1996) conducted an assessment of the safety of delayed ripening tomatoes that had been developed by stable insertion of the gene encoding the 1-aminocyclopropane-1-carboxylic acid deaminase (ACCd) enzyme into the tomato genotype. The ACCd enzyme catalyses the metabolism of ACC, an essential precursor for ethylene biosynthesis, and removal of ACC in these tomato plants reduces ethylene production and delays fruit ripening. Two approaches were used to assess the safety of the ACCd protein for human consumption. Purified *E. coli*-produced ACCd protein, which is chemically and functionally equivalent to the ACCd protein produced in delayed ripening tomato fruit, was used. Findings showed that the ACCd protein was readily degraded under simulated mammalian digestive conditions and that the ACCd protein did not have any deleterious effects when administered to mice by acute gavage at a dosage of up to

602 mg/kg of body weight. This dosage correlates to more than a 5000-fold safety factor relative to the average daily consumption of tomatoes, assuming that all tomatoes consumed contain the ACCd enzyme. Therefore the researchers concluded that ingestion of tomato fruit expressing the ACCd enzyme does not pose any safety concerns to the human consumer.

Lin and Pan (2011) conducted a safety study on a GM tomato developed to resist cucumber mosaic virus (CMV). Both qualitative and quantitative PCR results showed no significant difference between the GM (R8) and a conventional (L4738) tomato in the rate of DNA degradation in simulated human gastric fluid. After 180 min incubation about 99.98% of the GM DNA was degraded and could not be amplified. These results indicated rapid hydrolysis of tomato DNA in the acidic conditions of the human stomach, leading to the conclusion that only a very small amount of DNA from this genetic cultivar is likely to survive passage through the human digestive system. They also suggested that any surviving GM DNA is heavily degraded and would likely lose any function if taken up by microorganisms.

Nutritional and organoleptic qualities

Substantial equivalence with conventional cultivars of tomato was one of the factors taken into account by regulatory authorities in North America in approving the cultivars described above. Analysis of nutrients in the FlavrSavr line and non-GM control lines by the FDA did not show any significant differences in the levels of macro- and micronutrients, pH, total acidity, total solids and sugars. Also, analysis of nutrients from the novel 1401F hybrid line and non-GM 1401 hybrid did not show any significant differences in the levels of macro- and micronutrients in either fresh tomatoes or processed tomato paste.

Health Canada reported that an analysis of nutrients from the novel 1345-4 line and the non-GM 91103-114 parental line did not show any significant differences in the levels of macro- and micronutrients. Ethylene in the GM tomatoes was only about 1/50th the level found in the unmodified parental line and the fruit does not fully ripen unless an external source of ethylene is applied.

There is a growing interest in producing food plants with increased amounts of flavonoids because of their potential health benefits. Tomatoes contain small amounts of flavonoids, most of which are located in the peel of the fruit. It has been shown that flavonoid accumulation in tomato flesh, and hence an overall increase in flavonoid levels in tomato fruit, can be achieved by means of simultaneous overexpression of the maize transcription factors LC and C1. Research on this topic was conducted by Gall et al. (2003). Fruit from progeny of two GM lines (2027 and 2059) was selected for a detailed analysis and individual identification of flavonoids at different stages of maturity. Nine major flavonoids were detected in the flesh of GM ripe tomatoes. The total flavonoid glycoside content of ripe GM tomatoes of line 2059 was about 10-fold higher than that of the controls, and kaempferol glycosides accounted for 60% of this. Kaempferol glycosides comprised around 5% of the flavonoid glycoside content of ripe control tomatoes (the rest was rutin and naringenin chalcone). The rutin concentration in both GM and control fruits was similar.

Summary

Several GM cultivars of fruits and vegetables have been approved for food and feed use in several countries, though not all are currently being marketed. The list of approved GM cultivars includes cantaloupe, plums and papaya but only one type of GM fruit, the Hawaiian papaya with resistance to ringspot virus infection, is currently available to consumers. This virus-resistant cultivar was approved by the regulatory authorities in North America and Japan, based on evidence of safety and nutritional equivalence with conventional cultivars of papaya. The list of approved GM cultivars of vegetables includes lucerne (alfalfa), potatoes, squash, sugar beet, sweet corn (maize) and tomatoes. The first major criticism of GM foods by a scientist was in relation to a GM potato that was undergoing tests. The criticism was later shown to be without substance. Subsequent introduction of GM potatoes has occurred successfully. An important GM crop being grown in several countries is sugar beet, as a source of sugar for human use. As with oilseeds, all of the GM DNA is removed from the sugar (or oil) during processing and remains in the residue, which is used in animal feeding. Much of the sweet corn now being grown in North America is GM since these approved cultivars provide resistance to insect attack or herbicide application. The first commercially grown GM food crop was a tomato with delayed ripening properties. Several other GM cultivars were introduced later. A consistent finding with these GM foods is that much or all of the transgenic DNA is degraded in the gastro-intestinal tract of the human or animal consumer, and that none is absorbed into the tissues. Based on the available scientific evidence it is clear that the approved cultivars of GM fruits and vegetables are as safe and nutritious as non-GM cultivars.

References

Böhme, H., Aulrich, K., Daenicke, R. and Flachowsky, G. (2001). Genetically modified feeds in animal nutrition 2nd communication: Glufosinate tolerant sugar beets (roots and silage) and maize grains for ruminants and pigs. *Archives of Animal Nutrition* 54: 197–207.

Bruening, G. and Lyons, J. M. (2000). The case of the FLAVR SAVR tomato. *California Agriculture* 54 (4): 6–7.

Canadian Food Inspection Agency (2001). DD2002-39: Determination of the Safety of Aventis CropScience Canada Inc. Glufosinate Ammonium Tolerant Sugar beet (*Beta vulgaris*) Lines Derived from Event T120-7. Ottawa, Canada.

Chen, Y., Hwang, W., Fang, T. J., Cheng, Y. and Lin, J. (2011). The impact of transgenic papaya (TPY10-4) fruit supplementation on immune responses in ovalbumin-sensitised mice. *Journal of the Science of Food and Agriculture* 91: 539–546.

Combs, D. K. and Hartnell, G. F. (2008). Alfalfa containing the glyphosate-tolerant trait has no effect on feed intake, milk composition, or milk production of dairy cattle. *Journal of Dairy Science* 91: 673–678.

El-Sanhoty, R., El-Rahman, A. A. A. and Bögl, K. W (2004). Quality and safety evaluation of genetically modified potatoes Spunta with Cry V gene: compositional analysis, determination of some toxins, antinutrients compounds and feeding study in rats. *Nahrung* 48: 13–18.

El-Sanhoty, R. M. E. S., El-Maged, A. D. A. and Ramadan, M. F. (2006). Safety assessment of genetically modified potato Spunta: degradation of DNA in gastrointestinal tract and carryover to rat organs. *Journal of Food Biochemistry* 30: 556–578.

Ewen, S. W. and Pusztai, A. (1999). Effect of diets containing genetically modified potatoes expressing Galanthus nivalis lectin on rat small intestine. *Lancet* 354: 1353–1354.

Fedoroff, N. V. (2011). Pusztai's Potatoes - Is 'Genetic Modification' the Culprit? AgBioWorld. www.agbioworld.org/biotech-info/articles/biotech-art/pusztai-potatoes.html (accessed 14 October 2013).

Fermín, G., Keith, R. C., Suzuki, J. Y., et al. (2011). Allergenicity assessment of the papaya ringspot virus coat protein expressed in transgenic rainbow papaya. *Journal of Agricultural and Food Chemistry* 59: 10006–10012.

FSANZ (2001). Food derived from insect and potato virus y-protected (New Leaf) potato lines RBMT15-101, SEMT15-02 and SEMT15-15: A safety assessment. Technical report series No. 13. Canberra, Australia, and Wellington, New Zealand.

FSANZ (2003). Food produced from Glyphosate-tolerant Sugar beet line 77: A safety assessment. Technical report series no. 24. Canberra, Australia, and Wellington, New Zealand.

FSANZ (2005). First Review Report, Application A525, Food Derived from Herbicide-tolerant Sugar beet Line H7-1. Canberra, Australia, and Wellington, New Zealand.

GAIN Report Number JA1048 (2011). Global Agriculture Information Network (GAIN), United States Foreign Agricultural Service, Washington, DC. Accessed September 21, 2013.

Gall, G. le, DuPont, M. S., Mellon, F. A., et al. (2003). Characterization and content of flavonoid glycosides in genetically modified tomato (*Lycopersicon esculentum*) fruits. *Journal of Agricultural and Food Chemistry* 51: 2438–2446.

Langkilde, S., Schrøder, M., Frank, T., et al. (2012). Compositional and toxicological analysis of a GM potato line with reduced α-solanine content - a 90-day feeding study in the Syrian Golden hamster. *Regulatory Toxicology and Pharmacology* 64: 177–185.

Lin, C. and Pan, T. (2011). Assessing the digestion of a genetically modified tomato (Solanum lycopersicum) R8 DNA in simulated gastric fluid using event-specific real-time PCR. *European Food Research and Technology* 232: 1061–1067.

Lin, H., Yen, G., Huang, T., et al. (2013). Toxicity assessment of transgenic papaya ringspot virus of 823-2210 line papaya fruits. *Journal of Agricultural and Food Chemistry* 61: 1585–1596.

McCann, M.C., J. Rogan, G. J., Fitzpatrick, S., et al. (2006). Glyphosate tolerant alfalfa is compositionally equivalent to conventional alfalfa (*Medicago sativa* L.). *Journal of Agricultural and Food Chemistry* 54: 7187–7192.

Mochida, L. H. (2007). Trade and marketing of conventional versus transgenic papayas. *Acta Horticulturae* 740: 25–30.

Oguchi, T., Onishi, M., Chikagawa, Y., et al. (2009). Investigation of residual dnas in sugar from sugar beet (*Beta vulgaris* L.). *Journal of the Food Hygienic Society of Japan* 50: 41–46.

Powell, D. A., Blaine, K., Morris, S. and Wilson, J. (2003). Agronomic and consumer considerations for Bt and conventional sweet-corn. *British Food Journal* 105: 700–713.

Powell, M., Wheatley, A., Omoruyi, F., et al. (2008). Effects of subchronic exposure to transgenic papayas (*Carica papaya* L.) on liver and kidney enzymes and lipid parameters in rats. *Journal of the Science of Food and Agriculture* 88: 2638–2647.

Quemada, H., Zarka, K., Pett, W., et al. (2010). Safety evaluations of the Cry1Ia1 protein found in the transgenic potato 'SpuntaG2'. *Journal of the American Society for Horticultural Science* 135: 325–332.

Rai, G. K., Rai, N. P., Rathaur, S., et al. (2013). Expression of rd29A::AtDREB1A/CBF3 in tomato alleviates drought-induced oxidative stress by regulating key enzymatic and non-enzymatic antioxidants. *Plant Physiology and Biochemistry* 69: 90–100.

Reed, A. J., Kretzmer, K. A., Naylor, M. W., et al. (1996). Safety assessment of 1-aminocyclopropane-1-carboxylic acid deaminase protein expressed in delayed ripening tomatoes. *Journal of Agricultural and Food Chemistry* 44: 388–394.

Rhee, G. S., Cho, D. H., Won, Y. H., et al. (2005). Multigeneration reproductive and developmental toxicity study of bar gene inserted into genetically modified potato on rats. *Journal of Toxicological and Environmental Health A* 68: 2263–2276.

Roberts, M., Minott, D., Tennant, P. and Jackson, J. (2008). Assessment of compositional changes during ripening of transgenic papaya modified for protection against papaya ringspot virus. *Journal of the Science of Food and Agriculture* 88: 1911–1920.

Rogan, G. J., Bookout, J. T., Duncan, D. R., et al. (2000). Compositional analysis of tubers from insect and virus resistant potato plants. *Journal of Agricultural and Food Chemistry* 48: 5936–5945.

Royal Society UK (1999). Review of data on possible toxicity of GM potatoes. *Annual Review of Plant Biology* 59:771–812.

Shehata, S. (2011). Market testing of labeled and unlabeled GMO papaya fruits in Honolulu chain stores. *Journal of Food Distribution Research* 42: 107–110.

Tripathi, S., Suzuki, J. Y., Carr, J. B., et al. (2011). Nutritional composition of Rainbow papaya, the first commercialized transgenic fruit crop. *Journal of Food Composition and Analysis* 24: 140–147.

UK Advisory Committee on Novel Foods and Processes (1995). Annual Report - Advisory Committee on Novel Foods and Processes 1995 pp. iv + 94 pp. Food Standards Agency, London WC2B 6NH, UK.

Wall, M. (2006). Ascorbic acid, vitamin A, and mineral composition of banana (*Musa* sp.) and papaya (Carica papaya) cultivars grown in Hawaii. *Journal of Food Composition and Analysis* 19: 434–445.

CHAPTER 7

Fish and other animals

Several GM types of animal have been developed, but to date none are being marketed as a commercial food source. Only a GM salmon seems likely to be marketed commercially as a food source in the near future. In anticipation of this event, regulatory agencies in several countries have, or are, developing comprehensive guidelines to be used in an assessment of possible risks related to food and feed safety and to the environment of using a GM animal, as well as any special animal health and welfare aspects.

Fish now provide an important human food source. According to Larsen and Roney (2013) the world reached a milestone in the evolution of the human diet in 2011. For the first time in modern history, world production of farmed fish was greater than the production of beef. The gap widened in 2012, with output from fish farming reaching a record 66 million tons compared with production of beef at 63 million tons. Larsen and Roney also estimated that 2013 may well be the first year in which people ate more fish raised on farms than caught in the wild. China may emerge as one of the leading countries in the introduction of GM food animals (Vàzquez-Salat and Salter, 2011). One important factor is that Chinese consumers appear to be more accepting of GM animals than European and American consumers (Zhang et al., 2010).

Fish

Environment Canada, the agency of the Government of Canada with responsibility for regulating environmental policies and issues, announced in 2013 that GM AquAdvantage® Salmon from AquaBounty Technologies is not harmful to the

Genetic Modification and Food Quality: A Down to Earth Analysis, First Edition.
Robert Blair and Joe M. Regenstein.

environment or to human health when produced in containment facilities. The decision was announced in a Significant New Activity Notice (*Canada Gazette*, 2013), following a risk assessment conducted by Fisheries and Oceans Canada that involved a panel of independent scientific experts in the fields of genetic modification and fish containment technology. The decision marks the first time any government has given the go-ahead to the commercial-scale production of a GM food animal.

The Atlantic salmon in question have been modified to incorporate genes from both Chinook salmon and a seal eel. These give the fish the potential to grow to market size in half the time of conventional Atlantic salmon. The hatcheries will only produce sterile, all-female eggs and the plan is to develop an environmentally sustainable alternative confinement system in place of the current net pen system in open waters used to farm salmon. The sterile, all-female population will be grown in land-based facilities with biological and physical containment systems. As a result the fish cannot escape or reproduce in the wild and pose no threat to wild salmon populations.

Countries utilising the transgenic salmon domestically or importing it would have to be satisfied that the fish met the appropriate food safety standards in that country. The United States FDA Center for Veterinary Medicine (2010) provided details of its approach to evaluating such products. First, it provides an overall summary of FDA's risk-based, scientific process for determining the safety and effectiveness of the salmon produced by Aqua Bounty Technologies Inc. The safety assessment includes the safety of the process for the recipient animal and the safety of the food from that lineage of animals. Effectiveness refers to whether the food item consistently and uniformly does what the sponsor claims it does. Second, it provides data and information that the agency will need to have to evaluate as part of the application process, as well as the agency's evaluations. In addition, the process includes an environmental assessment and the agency's related analysis. Third, the process must continue to meet the agency's commitment to transparency by providing both the Veterinary Medicine Advisory Committee (VMAC) and the public with the same data and information that the agency evaluates in the review of an application such as that for the approval related to the AquAdvantage salmon. The VMAC will be charged with providing scientific advice to the agency and the required documentation is intended to provide the scientific and regulatory information that will help in discharging those duties. Finally, in the event that the agency approves Aqua Bounty's application, the document will serve as the basis for the Freedom of Information (FOI) summary that normally accompanies new animal drug approvals. The FOI Summary will also contain additional information such as the approved label, additional information on post-market responsibilities of the company and of FDA, and other administrative information relevant to a specific proposal.

This development is likely to be the first step towards global distribution of GM salmon. In a preliminary report (2012) the FDA concluded that food from

AquAdvantage Salmon is as safe as food from conventional Atlantic salmon, and that there is a reasonable certainty of no harm from consumption of food from triploid AquAdvantage Salmon. Further, FDA has concluded that no significant food safety hazards or risks have been identified with respect to the phenotype of the AquAdvantage Salmon.

If the AquAdvantage Salmon is approved by regulatory agencies, it is likely that other types of GM and hybrid salmon as well as trout, catfish and tilapia (the other major commercial fish species) along with shrimp and other valuable seafood will be introduced into the food supply. China, for example, is expecting soon to obtain the results from their long-term research investment in producing GM animals, with their first applications almost ready to reach the market (Vàzquez-Salat and Salter, 2011).

Research findings

Safety

As pointed out in a review by Dunham (2004), concerns have been raised over the possible risks of consumption of transgenes, their resulting protein, potential production of toxins by aquatic GM organisms, changes in the nutritional composition of foods, activation of viral sequences and allergenicity of GM products. These risks have been analysed, and the greatest potential for risk and harm is allergenicity.

Cuban scientists conducted a detailed safety evaluation of transgenic tilapia (*Oreochromis hornorum* L.) for human use that had been modified to express higher levels of growth hormone (GH) (Guillén et al., 1999). They found that the tilapia GH had no biological activity when administered to non-human primates. Juvenile monkeys, *Macaca fascicularis* (macques) were injected with 1000 ng/kg of recombinant tilapia GH once per day for 30 days. This is equivalent to administering 70,000 ng (0.07 mg)/day to a 70 kg human. Blood parameters examined included hemoglobin, serum total proteins, blood glucose, packed cell column, total leukocytes and total erythrocytes. Body weight, rectal temperature, heart rate and respiratory rate were recorded daily. Head to tail length, interscapular cutaneous pleat, left-flank cutaneous pleat, cranial circumference and cranial diameter were measured. Tilapia GH did not affect animal behaviour patterns or food intake. Body weight, temperature, heart rate and respiratory rate were unaffected by tilapia GH administration. The blood profiles and somatic growth of tilapia GH-treated macques and control animals were not different. Autopsies showed that all organs, tissues and cavities were normal, and no changes relative to controls were detected for common targets of GH such as tongue, palate plate concavity, liver, muscle, heart, kidneys and other organs. Subcutaneous and abdominal fat, mesenteric fat and peritoneal fat were unchanged, normal and of usual colour. No histopathological or morphological changes were observed.

Also no effects were detected in healthy human volunteers after the consumption of GM tilapia for five days. The measurements made included blood hemoglobin, total serum proteins, glucose, creatinine, cholesterol, leukocytes and erythrocyte levels. A probable explanation for the lack of differences is that growth hormone is known to be easily degraded proteolytically in the human gut.

In his review, Dunham (2004) analysed the theoretical food safety concerns related to GM salmon and concluded that the levels of GH and IGF (insulin-like growth factor) expressed by transgenic GH salmon are not always outside the range or much greater than the upper limit of GH and IGF secretion for other fish, food animals or humans. In addition, salmon GH and IGF are not bioavailable when ingested orally (cooked or raw), and adequate cooking will denature the proteins. Even if they were totally bioavailable, the dose from one meal would only be a small fraction of total daily human production of GH and IGF. Also, growth hormone is not orally active in higher species and salmonid GH is not bioactive in humans since the primate growth hormone receptor binds only primate growth hormone and it requires both binding sites to be occupied. Studies indicate that fish GH has no biological effect on primates and short-term ingestion of GH transgenic fish has no biological effect on humans. The lack of oral activity of GH and IGF and the non-toxic nature of the residues of these compounds, even at exaggerated doses, demonstrates that salmonid GH and IGF present no human safety concern when consumed orally. Thus, viewed from a number of aspects, any increased concentrations of GH or IGF in edible salmon skeletal muscle or skin is not hazardous to human health.

Production of transgenic fish is an area of active research in China. Zhang et al. (2000) studied the effects of feeding Kunming mice on Yellow River carp (*Cyprinus carpio* L. var. *Huanghe*) containing a salmonid GH gene. No significant differences ($p > 0.01$) in growth rate, biochemical analysis of blood or tissue physiology were reported, or on reproduction and growth or ontogenesis of the progeny. In a related study, Liang et al. (2011) reported on the safety of the flesh of common carp (*Cyprinus carpio* L.) modified to contain salmonid growth hormone genes when fed to mice over two generations. The mice were fed diets containing flesh from transgenic and non-transgenic common carp at a rate of 0 (control group), 2.5, 5 and 10% of their diet for two generations. Reproductive performance was measured. No evidence of harm from the feeding of transgenic carp to male and female mice was reported.

Pigs

A transgenic line of Yorkshire pigs (Enviropigs) with the capability of digesting plant phosphorus (P) more efficiently than conventional pigs has been developed at the University of Guelph, Ontario, Canada. Environment Canada (GAIN, 2010) has determined that the transgenic pig is not deleterious to the environment under the Canadian Environmental Protection Act. Regulatory approval of these pigs as meat animals for the human market would be dependent on favourable assessments from Health Canada and the Canadian Food Inspection Agency in Canada and the corresponding agencies in other countries.

The modification is designed to address one of the environmental issues in pig operations, namely excretion of excessive amounts of P in the manure. About 50 to 75% of the P present in cereal grains is present as phytate, which is indigestible by pigs. The unabsorbed P is then excreted and can leach into freshwater ponds, streams and rivers. P is the major nutrient enabling algal growth, which is the leading cause of reduced water quality and fish mortality resulting from anoxic conditions in both marine and freshwater environments (e.g. the 'dead zone' in the Gulf of Mexico). The modified pigs produce a phytase enzyme in their salivary glands, the enzyme becomes active in the acidic environment of the stomach and releases P from the indigestible phytate making it nutritionally available to the pigs, thus lowering the need to add other sources of needed P in their diets.

Meidinger et al. (2013) conducted a detailed study on the utilisation of P from plant-based diets in pigs that had been modified to contain the phytase gene. In a first experiment, 12 boars and 12 gilts (young females) fed diets without supplemental P gained weight and exhibited feed efficiencies similar to conventional age-matched 12 boars and 12 gilts fed similar diets with sufficient supplemental P to permit proper growth. Serum concentrations of P and Ca were similar for both groups of pigs during the growing and finishing phases, indicating that the GM pigs were not limited in P intake. The finishing phase in pig production normally uses a slightly different diet designed to optimise both growth and eating quality.

In a second experiment, six GM and six conventional weanling barrows (i.e., young male pigs just after weaning) were fed one of three diets (a control; a low Ca and P diet; and a low Ca and P diet that was also low in crude protein) over all three normal pig management phases (i.e., weaning, growing and finishing). The GM pigs fed the diet without supplemental P retained 25 to 40, 77 to 91, and 27 to 56% more P during the weaning, growing and finishing phases ($p < 0.001$ in all three cases), respectively, than conventional pigs fed similar diets without supplemental P.

In a third experiment, 10 GM and 10 conventional barrows were used during the finishing stage. Retention of P in the GM pigs fed a diet without supplemental P was 34% higher ($p < 0.001$) than in conventional pigs fed the same diet ($p < 0.001$) which included 750 units of exogenous phytase/kg diet. Urinary Ca to P ratio in the GM pigs was 0.27 and in the conventional pigs was 30, indicating that the conventional pigs were excreting a very high level of calcium relative to phosphorus, suggestive of a P deficiency.

Utilisation of the major electrolyte macrominerals, K and Na, was also improved by 18 and 16% ($p < 0.05$), respectively, in the GM pigs compared with the conventional pigs fed the diet with the added phytase. However, only K showed enhanced retention. The researchers concluded that the line of GM pigs secretes sufficient phytase from the salivary glands to enable efficient digestion of plant P, Ca and the major electrolyte macrominerals. Thus, they can maintain normal growth performance when fed typical commercial pig diets without supplemental P.

Although promising results have been obtained in developing GM animals it is clear that their commercial adoption as meat animals will be dependent on the response of the regulatory landscape and its associated range of interest groups at consumer, national, regional and international levels (Vàzquez-Salat et al., 2012).

Summary

Several types of GM animals have been developed but to date none have been approved for introduction into the human food supply. It is likely that in the future salmon will be approved as the first GM food animal.

References

Canada Gazette (2013). Government Notices, Department of the Environment. Ministerial Condition No. 17269, Significant New Activity Notice (Section 110 of the Canadian Environmental Protection Act, 1999). http://gazette.gc.ca/rp-pr/p1/2013-11-23/html/notice-avis-eng.html (accessed 8 January 2014).

Dunham, R. A. (2004). A case example: safety of consumption of transgenic salmon potentially containing elevated levels of growth hormone and insulin-like growth factor. In: Dunham, R.A. (ed.) *Aquaculture and Fisheries Biotechnology: Genetic Approaches*, CABI, Wallingford: pp. 315–328.

FDA (2010). Environmental Assessment for AquAdvantage Salmon. VMAC Briefing Packet, AquAdvantage Salmon. Center for Veterinary Medicine. http://www.fda.gov/downloads/AdvisoryCommittees/CommitteesMeetingMaterials/VeterinaryMedicineAdvisoryCommittee/UCM224760.pdf (accessed February 2015).

FDA (2012). Preliminary Finding of No Significant Impact, AquAdvantage® Salmon. http://www.fda.gov/downloads/AnimalVeterinary/DevelopmentApprovalProcess/GeneticEngineering/GeneticallyEngineeredAnimals/UCM333105.pdf (accessed February 2015).

Gain (2010). Global Agricultural Information Network Report Number:CA0010. 'Enviropig' Clears Initial Regulatory Hurdle. http://canada.usembassy.gov/news-events/2010-news-and-events/march-2010.html (accessed 8 January 2014).

Guillén, I., Berlanga, J., Valenzuela, C. M., et al. (1999). Safety evaluation of transgenic tilapia with accelerated growth. *Marine Biotechnology* 1: 2–14.

Larsen, J. and Roney, J. M. (2013). Plan B Updates. Farmed Fish Production Overtakes Beef. Report 114, Earth Policy Institute, Washington, DC 20036. www.earth-policy.org/plan_b_updates/2013/update114 (accessed 8 January 2014).

Liang, L., Wang, J., Chang, Y., et al. (2011). Food safety in common carp transferred with salmon growth hormone genes - reproductive toxicity analysis. *Chinese Journal of Fisheries* 24: 1–7.

Meidinger, R. G., Ajakaiye, A., Fan, et al. (2013). Digestive utilization of phosphorus from plant based diets in the Cassie line of transgenic Yorkshire pigs that secrete phytase in the saliva. *Journal of Animal Science* 91: 1307–1320.

Vàzquez-Salat, N. and Salter, B. (2011). Policy implications of introducing genetically modified animals in the European Union. PEGASUS Deliverable 6.2 (2011) http://www.pegasus.wur.nl/UK/ (accessed 8 January 2014).

Vàzquez-Salat, N., Salter, B., Smets, G. and Houdebine, L. M. (2012). The current state of GMO governance: are we ready for GM animals? *Biotechnology Advances* 30: 1336–1343.

Zhang, X. Y., Huang, J. K., Qiu, H. G. and Huang, Z. R. (2010). A consumer segmentation study with regards to genetically modified food in urban China. *Food Policy* 35: 456–462.

Zhang F., Wang Y., Hu W., et al. (2000). Physiological and pathological analysis of mice fed with 'all-fish' gene transferred Yellow River carp. *High Technology Letter* 10: 17–19.

CHAPTER 8
Animal products

It is now common for milk, meat and eggs to be produced using animals that have been fed diets containing GM feed ingredients. All of these GM ingredients have been approved for feed use by the regulatory authorities in the various countries where they are used, following favourable assessments of their safety and efficacy in relation to their conventional counterparts. These assessments have included assessments of the quality and safety of the milk, meat and eggs produced for human consumption.

In addition to the specific assessment reviews conducted by the regulatory authorities, several non-governmental agencies have conducted independent reviews. For instance, CAST (2006) reviewed the available evidence and concluded that performance, health and nutrient use by farm animals were similar when fed either conventional or biotechnology derived (GM) crops, and/or their by-products. Furthermore, the review found that no biologically relevant differences in the composition of animal products, including meat, milk and eggs, had been reported between farm animals fed diets containing commercially available, biotechnology derived crops and those fed diets containing conventional counterparts. Also, no intact or immunologically reactive fragments of GM plant proteins or DNA had been detected in samples of meat, milk, eggs, lymphocytes, blood and organ tissue from production animals fed biotechnology derived crops modified for agronomic input traits. CAST (2006) in their review found no authenticated case of an adverse health-related incident associated with the consumption of food or feed derived from modern biotechnology, and that the regulatory processes in place to assess the safety of biotechnology derived crops were effective in safeguarding public health. The review of the available data concluded that meat, milk and eggs produced by farm animals fed biotechnology derived crops are as wholesome, safe and nutritious as similar products derived from animals fed conventional crops.

Genetic Modification and Food Quality: A Down to Earth Analysis, First Edition.
Robert Blair and Joe M. Regenstein.

Research findings

Table 8.1 summarises data on the effect of feeding GM and non-GM feed sources on the composition of milk, meat and eggs. None of the published reports indicates any change in the composition or quality of these products.

Several research studies have been conducted to determine the extent of transfer of GM plant proteins or DNA into animal products. However, before reviewing the relevant findings, it is relevant to consider three relevant issues related to this aspect.

1 To what extent does feed processing (i.e. what is done to a crop after it is harvested until it is fed to an animal) affect the levels of GM material in the diets fed to food-producing animals, poultry and fish?
2 How much of the ingested GM DNA in the feed escapes the digestive process in food-producing animals, poultry and fish and is available for absorption?
3 If the GM DNA is absorbed from the gut, is it transferred into the edible tissues of food-producing animals, poultry and fish and does it present a hazard to the human consumer?

Effects of feed processing

Much or all of the feed given to food-producing animals, poultry and fish has been processed by treatments such as ensiling, grinding, heating, mixing and steam-pelleting. Several investigations have provided data on the extent to which transgenic (GM) DNA survives these processes.

Chiter et al. (2000) showed that conservation of forages through ensiling (anaerobic storage of harvested crop materials in a closed container) and certain feed-processing methods result in considerable fragmentation of plant DNA. Even though animal feedstuffs may contain intact DNA, the quantity of this component in most food crops is less than 0.02% on a dry-matter basis (Beever and Kemp 2000).

Ensiling is commonly used in ruminant feeding. Lutz et al. (2006) showed that this relatively long-term process results in a marked reduction in the content of GM protein and protein fragments in maize silage. No marked degradation was observed during the first 43 hours, with a sharp decrease thereafter. Tests showed that 23.6% of the initial GM protein was still detectable in the silage after 61 days.

Guan et al. (2005) and Reuter et al. (2010) reported similar findings, showing that composting resulted in a substantial or complete degradation of endogenous and transgenic plant DNA.

Chiter et al. (2000) found that oil-extracted linseed, soyabeans and rapeseed, and dried sugar beet pulp, all contained completely degraded DNA, while intact high-molecular weight DNA was readily extracted from linseed leaves and dry grains, wheat leaves and grains, fresh wet sugar beet pulp, fresh ryegrass and ryegrass silage, soyabean leaves and grains, three varieties of maize grains, maize leaves and maize silage, as well as from rapeseed leaves and grains. Intact

Table 8.1 Reported effects of dietary inclusion of conventional or GM feedstuffs on the composition of meat, milk and eggs.

Animal species	Reference	GM feedstuff	Reported effects
Meat			
Beef cattle	Aulrich et al., 2001	B.t. maize	No significant differences in carcass composition.
Beef cattle	Berger et al., 2002	Herbicide-tolerant maize	No significant differences in carcass characteristics.
Beef cattle	Berger et al., 2003	B.t.-tolerant maize	No significant differences in carcass characteristics.
Beef cattle	Daenicke et al., 1999	B.t. maize silage	No significant differences in carcass characteristics.
Beef cattle	Erickson et al., 2003	B.t. maize silage	No significant differences in carcass characteristics.
Beef cattle	Simon et al., 2002.	Herbicide-tolerant maize	No significant differences in carcass characteristics or meat composition.
Sheep (lambs)	Stanford et al., 2003	Glyphosate-tolerant canola	No significant differences in carcass composition.
Dairy cattle	Hammond et al., 1996	Meal from glyphosate-tolerant soyabeans	No significant differences in milk composition.
Dairy cows	Donkin et al., 2000	Herbicide-tolerant maize, maize silage	No significant differences in milk composition.
Dairy cows	Folmer et al., 2000	B.t. maize silage, maize silage	No significant differences in milk composition.
Dairy cows	Castillo et al., 2001	Herbicide-tolerant cottonseed, B.t. cottonseed	No significant difference in milk composition.
Dairy cows	Castillo et al., 2004	Glyphosate-tolerant cottonseed, B.t. cottonseed	No significant differences in milk composition.
Dairy cows	Calsamiglia et al., 2007	B.t., herbicide-tolerant maize silage	No significant difference in milk composition.
Dairy cows	Ipharraguerre et al., 2002	Herbicide-tolerant maize, maize silage	No significant difference in milk composition.
Dairy cows	Donkin et al., 2003	B.t. maize, glyphosate-tolerant maize silage	No significant differences in milk composition.
Dairy cows	Grant et al., 2002	B.t. + glyphosate-tolerant maize grain, silage	No significant differences in milk composition.
Dairy cows	Grant et al., 2003	Gt maize -silage and grain	No significant differences in milk composition.
Dairy cows	Grant et al., 2003	Corn rootworm protected corn -silage and grain	No significant differences in milk composition.

Continued

Table 8.1 Continued

Animal species	Reference	GM feedstuff	Reported effects
Dairy cows	Singhal et al., 2006	Cry1ac+ cr2ab cottonseed.	No significant differences in milk composition.
Dairy cows	Steinke et al. 2010	B.t.-maize, maize silage	No consistent effects on milk composition or body condition.
Dairy cows	Weisbjerg et al., 2001	Glyphosate-tolerant sugar beet, sugar beet pulp, fodder beet	No significant differences in milk composition.
Pigs	Armstrong et al., 2001	Glyphosate-tolerant soyabean meal	No significant effect on pork quality.
Pigs	Bressner et al., 2003	B.t. maize	No significant differences in carcass traits.
Pigs	Cromwell et al., 2002	Herbicide-tolerant soyabean meal	Carcass traits of barrows were similar for the two types of soyabean meal. Longissimus muscle samples from barrows fed conventional soyabean meal tended to have less fat than those fed Roundup Ready soyabean meal, but water, protein and ash were similar. Sensory scores of cooked longissimus muscles were not influenced by diet.
Pigs	Cromwell et al., 2005	Glufosinate-tolerant rice	No significant differences in carcass traits.
Pigs	Fischer et al., 2002	Herbicide-tolerant maize	No significant differences in carcass measurements.
Pigs	Custodio et al., 2006	B.t. maize	No significant effect on carcass composition or meat quality.
Pigs	Fischer et al., 2003	B.t. maize	No significant differences in carcass measurements.
Pigs	Stein et al., 2009	B.t., glufosinate-resistant maize.	No significant differences in dressing peercentage, loin muscle area, backfat thickness or carcass lean content.
Pigs	Świątkiewicz et al., 2011, 2013	Glyphosate-tolerant soyabean meal and B.t. maize	No significant effects on carcass or meat quality, or on health and immunological indices.
Broilers, laying hens	Aeschbacher et al., 2005	B.t. maize	No significant effects on carcass or meat quality.
Broilers	Sidhu et al., 2000	Herbicide-tolerant maize	No significant differences in carcass composition.
Broilers	Kan et al., 2001	Meal from dehulled B.t. soyabeans	No significant differences in carcass composition.

Table 8.1 Continued

Animal species	Reference	GM feedstuff	Reported effects
Broilers	Kan and Hartnell, 2004	Glyphosate-tolerant wheat	Lower carcass yield for birds fed the conventional wheat.
Broilers	Řehout et al., 2009	B.t. maize	No significant effect of maize type on meat quality or blood biochemistry.
Broilers	Taylor et al., 2004	Glyphosate-tolerant canola	No significant differences in body composition.
Broilers	Brake and Vlachos, 1998	B.t. maize	No significant differences in body composition.
Channel catfish	Li et al., 2008.	Range of GM cottonseed meals.	Fillet composition similar.

high-molecular weight DNA was readily extracted from fresh maize leaves, dry maize grain and maize silage, but the DNA was substantially degraded in maize gluten and flaked maize. Heating maize grain to 92 °C for 5, 10 or 15 minutes greatly reduced the amount of DNA retained and heating to 94 °C for at least 5 minutes degraded it completely.

These results indicate that there is little likelihood of transmission of transgenic DNA from extracted oilseed meals and maize products such as maize gluten, which are the major feedstuffs that are traded internationally. Transmission is more likely to occur with feedstuffs not subjected to high temperatures, such as wet sugar beet pulp and silage, although some of the DNA extracted from silage is undoubtedly of microbial origin.

Other relevant data have been obtained from studies involving the processing of commercial fish feed, since these feeds are usually produced using processes such as milling and mixing, shearing, heating under high pressure and extrusion, which are likely to result in some protein damage. DNA fragments from GM feedstuffs have been reported in extruded fish diets (Sissener et al., 2011), although the fragments detected have generally been below a size likely to code for functional proteins (Jonas et al., 2001).

Guan et al. (2013) investigated the effects of dry heating, wet heating, and extrusion on the degradation of DNA in GM (B.t.) cottonseed meal. Both the sad1 DNA, ranging between 128 and 883 base pairs (bp) in size and the cry1Ab/c gene, ranging between 183 and 652 bp in size, were detectable in all dry-heated samples of cottonseed and extracted cottonseed meal. (Base pairs are defined as two nitrogenous bases paired together in double-stranded DNA by weak bonds. Specific pairing of these bases – adenine with thymine and guanine with cytosine – facilitates accurate DNA replication. When quantified – e.g. 8 bp – this refers to the physical length of a sequence of nucleotides.) During wet heating, however, the sad1 gene and the cry1Ab/c gene were thoroughly degraded at 105 and 120°C, respectively. These results indicated that feed processing markedly degrades the larger DNA fragments of sad1 and cry1Ab/c proteins, with high temperature and moisture content being the main factors involved.

These findings can be interpreted as indicating that processing feedstuffs for animal feeding has a very marked effect in degrading the amount of DNA present, particularly when the material is subjected to high temperature and moisture.

Effects of digestion in the GI tract and possible transfer of transgenic DNA into edible tissues

An important issue relating to the safety of milk, meat and eggs from animals fed GM feed ingredients is the fate of the transgenic DNA in the gastro-intestinal tract and whether it is digested differently than regular DNA from conventional food and feed. It is known that DNA (and RNA) in conventional food is broken down into mononucleotides by the nucleases deoxyribonuclease and ribonuclease (DNase and RNase) released by the pancreas.

If the transgenic DNA is digested differently and is absorbed and deposited as active molecules into the tissues of food animals, then the safety of the residual DNA to the human consumer would have to be evaluated very carefully. On the other hand, if the transgenic DNA is degraded completely in the animal's gastro-intestinal tract, then any possible harm to humans is then much less likely.

Results to date indicate that GM crops are digested by animals in the same way as conventional crops. Also, there are no reports available in the scientific peer-reviewed literature demonstrating that an active gene has transferred from GM plants into animal or human tissues (Hohlweg and Doerfler, 2001; Thomson, 2001; Van den Eede et al., 2004).

For instance, Harrison et al. (1996) measured the in vitro digestion in mice of the CP4 EPSPS protein expressed in soyabeans resistant to glyphosate and reported a rapid breakdown by proteolytic enzymes. Aumaitre et al. (2002) reported data showing that the expressed proteins of glyphosate-tolerant soyabeans do not produce any harmful effects in the digestive tract or in other tissues after ingestion by a range of farm animals.

Another important factor that is relevant to this issue is the amount of DNA from GM feed in relation to the amount of conventional DNA in the diet. For instance, it has been estimated that when cows eat a feed ration containing 40% silage and 20% grain made from GM corn varieties, approximately 0.00042 of the animal's total daily DNA intake would consist of GM DNA (Beever and Kemp, 2000).

Ruminant animals

Digestion and absorption

Lutz et al. (2005) measured the degradation of cry1Ab protein from GM maize in the bovine gastro-intestinal tract using immunoblotting assays, and found a statistically significant reduction in the amount of GM protein as the digesta passed from the rumen to the small intestine, colon and feces.

Alexander et al. (2002) used in vitro ruminal cultures to measure the persistence of recombinant and canola-specific endogenous DNA in the ruminant gut, using various canola substrates (whole seed, cracked seed, meal and diet). The canola seed was heated to an average discharge temperature of 74.4 °C and flaked to an average thickness of 0.254–0.381 mm. The flakes were then cooked to a discharge temperature of 96.1–101.7 °C, dried, and extracted in hexane heated to 54.4 °C. The meal was recovered at 93.9 °C. Plant DNA was not detected in ruminal supernatant but only in pellets containing plant debris, suggesting that the presence of plant DNA in the ruminant gut is directly related to intact plant cells. Disappearance of plant DNA fragments appears therefore to be analogous to the breakdown of plant cells in the rumen. This appears to be the limiting step for transformation of DNA by rumen bacteria because once DNA is released into the aqueous ruminal environment, it is degraded almost immediately. Plant DNA fragments reaching the intestinal tract of ruminants and being available for absorption are most likely the result of undigested plant material passing through the rumen to the intestines, where plant cells can then be degraded, releasing DNA.

Sharma et al. (2004) reported that conventional and GM (glyphosate-tolerant) canola were substantially equivalent in terms of degradation of DNA during ruminal incubation. Plant debris was evident throughout an incubation period of meals and diets for 48 hours, but the GM gene fragments were not detectable beyond 4 or 8 hours – consistent with the previous study indicating that DNA in disrupted plant cells is rapidly degraded by nucleases present in ruminal contents (Alexander et al., 2002).

In a related investigation, Alexander et al. (2004) assessed the stability of the cp4 epsps transgene from GM (glyphosate-tolerant) canola during digestion in sheep. The concentration of cp4 epsps DNA in the diet declined more rapidly in duodenal fluid at pH 5 as compared to pH 7, attributed mainly to microbial activity at pH 7 and perhaps to plant endogenous enzymes at pH 5. This study showed that digestion of plant material and release of transgenic DNA can occur in the ovine small intestine. However, free DNA is rapidly degraded at neutral pH in duodenal fluid, thus reducing the likelihood that intact transgenic DNA would be available for absorption in the distal ileum (main absorptive section of the gut).

The same research group measured the concentration of transgenic DNA in digesta, feces and blood collected from sheep fitted with ruminal and duodenal cannulae and fed forage-based or concentrate-based diets containing 15% GM (glyphosate-tolerant) canola meal (Alexander et al., 2006). Ruminal and duodenal fluids and faecal samples were collected at 3- or 4-hour intervals over four days immediately following the last feeding of the diets. Blood was collected on days 1, 5 and 9 of feeding the GM canola meal. DNA was isolated from whole ruminal fluid and duodenal fluid and from the cell-free supernatant fraction. Transgenic DNA was not detectable in feces or blood, or in microbial DNA. Diet type (forage vs. concentrate) did not affect the quantity of transgenic DNA in digesta. A greater concentration of transgenic DNA was detected in rumen fluid than in duodenal fluid. Three different fragments of transgenic DNA, in supernatant fractions from rumen fluid and duodenal fluid and ranging in size from 62 to 420 bp, were not amplifiable. These data can be interpreted as showing that transgenic DNA is completely degraded as it passes through the intestine of sheep since none is detected in the feces. Furthermore, none is absorbed into the blood.

Further work on the digestion of GM (glyphosate-tolerant) canola in sheep was conducted by Sharma et al. (2006). The animals were fed to slaughter on diets containing 6.5 or 15% GM (glyphosate-tolerant) canola. Native plant DNA (high- and low-copy-number gene fragments) and the cp4 epsps transgene that encodes 5-enolpyruvyl shikimate-3-phosphate synthase were tracked in ruminal, abomasal and large intestinal digesta and in tissue from the esophagus, rumen, abomasum, small and large intestine, liver and kidney of sheep. High-copy chloroplast-specific DNA (a 520-bp fragment) was detected in all digesta samples, the majority (89–100%) of intestinal tissues, and at least one of each visceral organ sample (liver, kidney: frequencies of 3–27%). Low-copy rubisco fragments (186- and 540-bp sequences from the small subunit) were present at slightly lower, variable frequencies in digesta (18–82%) and intestinal tissues (9–27%) and infrequently in visceral organs (1 of 88). Each of the five cp4 epsps transgene fragments (179–527 bp) surveyed was present in at least 27% of large intestinal content samples (maximum = 64%). Transgenic fragments were more common in intestinal digesta than in ruminal or abomasal content. Transgene fragments were detected in 0 (esophagus) to 3 (large intestine) gut tissues but not in visceral tissues (liver, kidney). This study confirmed that some feed-ingested DNA fragments (endogenous and transgenic) do survive to the terminal gut and that uptake into epithelial tissues does occur. However, there was no evidence to suggest that recombinant DNA is processed in the gut in a manner different from non-GM feed-ingested genetic material. Non-transgenic and transgenic DNA exhibited similar stability and persistence in the gastro-intestinal tract.

B. Transfer of transgenic DNA into edible tissues

Phipps et al. (2002) investigated the possible presence of transgenic DNA in milk from cows receiving GM (herbicide-tolerant, CP4EPSPS) soyabean meal. The Holstein/Friesian cows, of 630 kg average live weight and producing 25.3 kg milk/day, received a total mixed ration in which the forage portion comprised 55% total dry matter that contained non-GM grass and maize silage in a 1 : 3 ratio (dry-matter basis). During weeks 1–3 the dietary dry matter also contained non-GM supplements of 18.5% cracked wheat and 26.1% rapeseed meal. During weeks 4–12 GM soyabean meal replaced the rapeseed meal at 26.1% of the total diet in weeks 4–5 and 13.9% during weeks 6–12. Weekly milk samples were taken from all cows. Feed and milk samples analysed at week 3 were negative for the presence of transgenic DNA. The feed samples at weeks 4 and 12 were positive for DNA for the GM soyabean meal, but all milk samples were negative. The results showed that transgenic DNA could not be detected in milk from cows receiving up to 26.1% of their dietary dry matter as GM soyabean meal.

Jennings et al. (2003b) investigated whether transgenic or endogenous plant DNA is detectable in dairy milk or beef organs. The animals were fed a diet containing conventional or GM cottonseed or maize. Genomic DNA was extracted from milk and analysed for fragments of the cry1Ac transgene and an endogenous cotton gene, acp1, from cows fed a diet containing whole cottonseed. Additionally, milk, liver, kidney and spleen were assessed for fragments of the cry1Ab transgene and an endogenous maize gene, sh2. These results indicated that DNA, whether derived from conventional or GM cottonseed or maize, is not present at detectable levels in bovine milk or organs.

Castillo et al. (2004) conducted two experiments to evaluate dry-matter intake, milk yield and milk composition in Argentinean Holstein dairy cows fed diets containing different sources of GM whole cottonseed. All cows received the same total mixed ration but with different whole cottonseed sources. Cottonseed was included to provide 2.50 kg per cow each day (dry-matter basis) or about 100 g per kg total dietary dry matter. No sample of milk was found to contain GM or plant DNA fragments.

Pigs

A. Digestion and absorption

Beagle et al. (2006) studied the digestive fate in weanling pigs (non-ruminants) of a GM maize cultivar (gdhA) developed at Southern Illinois University, Carbondale. This cultivar has an *E. coli* gene encoding for glutamate dehydrogenase inserted into the genome to facilitate the assimilation of N. The cultivar had not been released for commercial use. The pigs were fed a commercial non-GM starter diet for two weeks, when some of the animals were slaughtered to provide baseline measures of DNA in ingesta. One pig was then randomly selected from each pen to provide control samples. The remaining animals were gradually switched to a maize-soyabean meal diet containing 58% GM maize for 6–8 days. Following slaughter, ingesta were collected from the stomach, distal ileum and large intestine. Transgenic DNA was detected in 71.43% of the stomach and 1.79% of the ileal ingesta samples, but was not detected in the large intestine. These data suggested that intestinal degradation of the gdhA transgene began in the stomach and that the transgene was not detectable in the large intestine.

A related study on the digestion of GM (glyphosate-tolerant) canola in pigs fed to slaughter was conducted by Sharma et al. (2006). The cp4 epsps transgene was tracked in the duodenum, caecum, liver, spleen and kidney. High-copy chloroplast-specific DNA (a 520-bp fragment) was detected in all digesta samples, the majority (89–100%) of intestinal tissues, and at least one of each visceral organ sample (frequencies of 3–27%). Low-copy rubisco (non-GM) fragments (186- and 540-bp sequences from the small subunit) were present at slightly lower, variable frequencies in digesta (18–82%) and intestinal tissues (17–25%) and infrequently in visceral organs (3 of 216 samples). Each of the five cp4 epsps transgene fragments (179–527 bp) surveyed was present in at least 33% of cecal content samples (maximum = 75%). Transgene fragments were detected in 0–28% of the duodenal and cecal tissues but not in the spleen. One liver and one kidney sample (different animals) were positive for a 278-bp fragment of the GM cp4 epsps protein. This study confirmed that feed-ingested DNA fragments (endogenous and GM) do survive to the terminal gut of pigs and that uptake into gut epithelial tissues does occur. Transmittance to visceral tissue was also found to occur, though at a very low frequency. The researchers concluded that there was no evidence to suggest that recombinant DNA is processed differently in the gut than endogenous feed-ingested genetic material, as concluded above with sheep.

B. Transfer of transgenic DNA into edible tissues

Beagle et al. (2006) studied the possible transfer of genes from a GM maize cultivar (gdhA) into the tissues of young pigs. As noted above, this cultivar has not yet been released for commercial use. The pigs were fed a commercial non-GM starter diet for two weeks, then gradually switched to a maize–soyabean meal diet containing 58% GM maize for approximately one week. Transgenic DNA was detected in 71.4% of the stomach and 1.79% of the ileal ingesta samples from treatment animals (as noted above) but was not detected in the large intestine, white blood cells, plasma, liver or muscle samples.

Jennings et al. (2003a) conducted a study to investigate whether GM and endogenous plant DNA and GM protein are detectable in pork loin tissue of pigs fed a diet containing GM (glyphosate-tolerant) soyabean meal. Neither the CP4 EPSPS protein nor any immunoreactive peptide fragments were detected in the loin muscle. These results showed that neither small fragments of transgenic DNA nor immunoreactive fragments of GM protein are detectable in loin muscle samples from pigs fed a diet containing glyphosate-tolerant soyabean meal.

Walsh et al. (2011) tested the presence of transgenic DNA and protein in the tissues of pigs fed a diet containing 38.9% GM or the non-GM parent-line maize for 31 days. They reported that detection of the cry1Ab gene and protein was limited to the gastro-intestinal digesta. Cry1Ab-specific IgG and IgA were not detected in the plasma or in the kidneys, liver, spleen, muscle, heart or blood.

Poultry

A. Digestion and absorption

Ash et al. (2003) collected samples of droppings from commercial flocks of laying hens fed diets containing GM (glyphosate-tolerant) soyabean meal and reported the absence of GM residues. Deaville and Maddison (2005) studied the digestion of diets containing GM maize and soyabean meal in broiler chickens. Male broiler chicks ($n = 24$) were allocated at 1 day old to each of four treatment diets designated T1–T4. T1 and T2 contained the near isogenic non-GM maize grain, whereas T3 and T4 contained GM maize grain (cry1a(b) gene). T1 and T3 also contained the near isogenic non-GM soybean meal, whereas T2 and T4 contained

GM soybean meal (cp4epsps gene). Four days prior to slaughter at 39–42 days old, 50% of the broilers on T2–T4 had the source(s) of GM ingredients replaced by their non-GM counterparts. Following slaughter, samples were obtained of breast tissue, liver, gizzard, heart, spleen, kidney and bursa. It was found that some feed-derived DNA survived complete degradation up to the large intestine. The multicopy rubisco (endogenous) gene was detected in a proportion of samples from all tissue types (23% of total across all tissues studied) and in low numbers in blood. There was no positive detection of any endogenous or GM single copy genes in either blood or tissue DNA samples. Transgenic DNA was detected in gizzard digesta but not in intestinal digesta when a comparison was made 96 hours after the last feeding of treatment diets containing a source of GM maize and/or soyabean meal.

Rossi et al. (2005) found that feed-derived DNA was progressively degraded along the digestive tract of broilers. An 1800-bp fragment of the cry1A(b) gene was detected only in the crop and gizzard of birds fed a diet containing B.t. corn. No significant difference in DNA detection was observed between birds fed B.t. and isogenic (non-GM) maize, indicating that DNA derived from GM feed undergoes the same fate as conventional feed.

Tony et al. (2003) studied the degradation of maize-DNA and its metabolic fate in broilers. The investigation involved a new cultivar of B.t. 176 maize, NX 6262- B.t. 176. Digesta from different sections of the gastro-intestinal tract (GIT) were collected. Results showed that the new cultivar was digested in a similar fashion to that of a conventional cultivar, and that transgenic DNA was degraded in different parts of the gut in the same way as the DNA of the control maize cultivar.

B. Transfer of transgenic DNA into edible tissues

Tony et al. (2003) studied the degradation of the GM protein in broilers fed a diet containing B.t. 176 maize, and the possible presence of transgenic DNA in muscle tissue and in some organs. One-day-old male broilers were fed ad libitum on either an experimental diet containing NX 6262- B.t. 176 maize or a control diet containing the conventional non-GM maize for 35 days, when they were slaughtered and samples from pectoral and thigh muscles, liver, spleen, kidney, heart muscle, bursa and thymus glands were collected. B.t. gene specific constructs from B.t. 176 maize could not be detected in blood or in any tissue samples.

Deaville and Maddison (2005) conducted a study to determine the fate of GM and endogenous plant DNA fragments in the blood and tissues of broilers. Male birds were allocated at one-day-old to diets containing GM maize (cry1a(b) gene) or near-isogenic non-GM maize until slaughter at 39–42 days of age. There was no positive detection of any endogenous or GM single copy genes in either blood or tissue DNA samples, except for a single detection of lectin (non-GM single copy gene; unsubstantiated) in the extracted DNA from one bursa tissue sample. However, the naturally occurring rubisco gene was detected in 23% of the tissues and in low numbers in blood, indicating absorption and retention of an endogenous plant gene but not of the cry1a(b) gene.

Ash et al. (2003) investigated the extent of deposition of genetically modified protein from glyphosate-tolerant soyabeans in tissues and eggs of laying hens. The GM soyabeans, soyabean meal, and complete diets were found to contain GM proteins. However, whole eggs, egg albumen, liver and droppings were all negative for residues of GM proteins.

Fish

A. Digestion and absorption

Work with fish has produced findings similar to those obtained with other animals. For instance, Sanden et al. (2004) investigated the survival of fragments of GM feedstuffs in the gut of Atlantic salmon. The fish were fed a series of test diets for six weeks, formulated to contain conventional or GM maize and conventional or GM (glyphosate-tolerant) full-fat soyabean flour. The conventional and GM soya flour replaced 17.2% of fish meal in the diets. A diet composed of fish meal as the only protein source served as a control. GM sequences (120 and 195 bp) and a lectin gene (180 bp) were detected in the diets containing GM soya, but only the smaller DNA fragment (120 bp) could be isolated from the contents of the stomach, pyloric region, mid intestine and distal intestine.

B. Transfer of transgenic DNA into edible tissues

Some of the most detailed work on the extent of uptake of transgenic DNA fragments has been done with fish.

Sanden et al. (2004) measured the presence of fragments of GM feedstuffs in the gut of Atlantic salmon and their possible transference to the tissues. The fish were fed a series of test diets for six weeks, formulated to contain conventional or GM maize and conventional or GM (glyphosate-tolerant) full-fat soyabean flour. A DNA fragment (120 bp) could be isolated from the contents of the stomach, pyloric region, mid intestine and distal intestine, but no GM or conventional soya DNA fragments could be detected in liver, muscle or brain tissues. The researchers concluded that GM sequences of glyphosate-tolerant soyabeans may survive passage through the gastro-intestinal tract but that they cannot be detected in fish tissues. A subsequent investigation (Sanden et al., 2006) found the presence of dietary DNA from GM soyabean meal in the epithelial cells in the salmon intestine, in the cellular vacuolar system and in the lamina propria (one of the layers of the intestinal epithelium) in some samples. Another investigation using tilapia that were fed diets containing GM soyabean meal found that fragments of the CaMV 35S promoter were detected in a few muscle samples after continuous feeding but none were detected two days after changing to a non-GM diet (Suharman et al., 2009). Transgenic exogenous DNA fragments were not detected in the heart, liver, stomach, intestines, brain, branchia, spleen, cholecyst or muscle of tilapia fed a non-GM diet but DNA fragments of 254 bp were detected in all organs except the cholecyst after four weeks of a GM diet (Ran et al., 2009). After a subsequent two-week fast, lower levels of transgenic DNA fragments could still be detected in all samples, except the spleen and cholecyst. It is not clear from these findings whether the authors found a differential effect of GM and non-GM DNA fragments in terms of transfer to tissues, since the main aim of the study appears to have been the detection of GM residues in fish tissues.

Other work with fish suggests that uptake of residues of GM feedstuffs does not appear to be different from the uptake of non-GM residues. In a study involving zebrafish, dietary DNA fragments from the endogenous (non-GM) rubisco gene and from GM maize (MON810) were detected in several organs, while DNA fragments from GM soyabean meal were not detected (Sissener et al., 2011).

The current findings indicate that dietary DNA fragments, both from GM and non-GM feedstuffs, can be absorbed from the fish intestine and can be detected in internal organs. However, the fragments appear to be below a size likely to code for functional proteins and are, therefore, not a safety concern. Further research needs to be conducted on this issue to confirm that conclusion.

Assessment of available findings

As indicated above, CAST (2006) reported that no intact or immunologically reactive fragments of GM plant proteins or DNA had been detected in samples of meat, milk, eggs, lymphocytes, blood, or organ tissue from production animals fed GM feed ingredients.

EFSA (2007) came to a similar conclusion, following an extensive review of the available evidence. They reported that, although biologically active genes and proteins are common constituents of foods and feed in varying amounts, after ingestion there is a rapid degradation into short DNA or peptide fragments in the gastro-intestinal tract of animals and humans. In addition they found that a large number of investigations with livestock showed that recombinant DNA fragments or proteins derived from GM plants had not been detected in tissues, fluids or edible tissues of broilers, cattle, pigs or quail.

Researchers have reached similar conclusions, as outlined above. For instance, Nemeth et al. (2004) investigated whether trace levels of feed-derived DNA fragments, endogenous and GM, are detectable in milk from dairy cows and in the muscle tissue of chickens, swine and beef steers fed diets containing GM maize. They found that transgenic DNA fragments were not detected in any tissue or milk samples. In contrast, fragments of naturally occurring chloroplast

DNA were found in 5, 15 and 53% of muscle samples from beef steers, broiler chickens and pigs, respectively, and in 86% of milk samples.

The results of investigations published following the CAST (2006) and EFSA (2007) reviews confirm the conclusions of these agencies. For instance, Sieradzki et al. (2013) studied the presence of transgenic DNA in blood, organs, tissues, eggs, excreta and intestinal bacteria in broilers, laying hens, pigs and calves fed diets containing GM (glyphosate-tolerant) soyabean meal and/or MON810 maize. None of the samples were found to contain transgenic DNA. These researchers concluded that transgenic DNA from GM maize and soyabean meal is digested in the same way as endogenous plant DNA, with no probability of its transfer to animal tissues or gut bacterial flora. In addition they concluded that a diet containing GM feedstuffs did not affect bacterial gut flora in terms of the diversity of bacterial species, the quantity of particular bacterial species, or incorporation of transgenic DNA into the bacterial genome.

Evidence relating to this issue can also be obtained from the results of researchers investigating whether milk, meat and eggs could be used as possible monitors for the inclusion of GM ingredients in the feed. The method would be useful in addressing possible labelling issues if residues of transgenic DNA were found in these foods, such as whether it was necessary to specifically label any product from animals fed GM ingredients. In addition, organic producers could use this method as a possible way of detecting fraud. However, this does not seem to be the case as these residues have not been found in milk, meat or eggs.

An important finding with all of the species investigated is that no residues of GM feedstuffs have been found in milk, meat or eggs.

Summary

Review of the available data indicates that meat, milk, and eggs produced by farm animals fed biotechnology derived crops are as wholesome, safe and nutritious as similar products derived from animals fed conventional crops. The available research findings indicate that the transgenic proteins in GM feedstuffs are substantially degraded by feed processing and by passage through the digestive tract of cattle, sheep, pigs, poultry and fish. There is no evidence to suggest that transgenic DNA is digested in the gut differently from endogenous feed-ingested genetic material. Little or no GM residues are absorbed, although some investigations have reported the presence of residues of both GM and non-GM proteins in certain organs of some species of fish. The GM residues do not appear to be absorbed preferentially, and the fragments appear to be below a size able to code for functional proteins. They are, therefore, not a safety concern although further work is recommended. No GM residues have been reported in meat, milk or eggs, suggesting that fish may differ from other species in their utilisation of plant-derived diets, possibly because of their different digestive system.

References

Aeschbacher, K., Messikommer, R., Meile, L. and Wenk, C. (2005). B.t.176 Corn in poultry nutrition: physiological characteristics and fate of recombinant plant DNA in chickens. *Poultry Science* 84: 385–394.

Alexander, T. W, Reuter, T., Okine, E., et al. (2006). Conventional and real-time polymerase chain reaction assessment of the fate of transgenic DNA in sheep fed Roundup Ready rapeseed meal. *British Journal of Nutrition* 96: 997–1005.

Alexander, T. W., Sharma, R., Deng, M. Y., et al. (2004). Use of quantitative real-time and conventional PCR to assess the stability of the cp4 epsps transgene from Roundup Ready (R) canola in the intestinal, ruminal, and fecal contents of sheep. *Journal of Biotechnology* 112: 255–266.

Alexander, T. W., Sharma, R., Okine, E. K., et al. (2002). Impact of feed processing and mixed ruminal culture on the fate of recombinant EPSP synthase and endogenous canola plant DNA. *FEMS Microbiology Letters* 214: 263–269.

Armstrong, C. L., Mikel, W. B. and Cromwell, G. L. (2001). Sensory evaluation of pork longissimus muscle from swine fed soybean meal from Roundup Ready or conventional soybean. *Journal of Animal Science* 79 Suppl 1: 374 Abstract 155.

Ash, J., Novak, C. and Scheideler, S. E. (2003). The fate of genetically modified protein from roundup ready soybeans in laying hens. *Journal of Applied Poultry Research* 12: 242–245.

Aulrich, K., Böhme, H., Daenicke, R., et al. (2001). Genetically modified feeds in animal nutrition. 1st communication: Bacillus thuringiensis (B.t.) corn in poultry, pig and ruminant nutrition. *Archives of Animal Nutrition* 54: 183–19.

Aumaitre A., Aulrich, K, Chesson, A. C., et al. (2002). New feeds from genetically modified plants: substantial equivalence, nutritional equivalence, digestibility, and safety for animals and the food chain. *Livestock Production Science* 4: 223–238.

Beagle, J. M., Apgar, G. A., Jones, K. L., et al. (2006).The digestive fate of *Escherichia coli* glutamate dehydrogenase deoxyribonucleic acid from transgenic corn in diets fed to weanling pigs. *Journal of Animal Science* 84: 597–607.

Beever, D. E. and Kemp, C. F. (2000). Safety issues associated with the DNA in animal feed derived from genetically modified crops. A review of scientific and regulatory procedures. *Nutrition Abstracts and Reviews, Series B: Livestock Feeds and Feeding* 70: 175–182.

Berger, L. L., Robbins, N. D., Sewell, J. R. and Stanisiewski, E. P. (2002). Effect of feeding diets containing corn grain with Roundup (event GA21 or NK603), control, or conventional varieties on steer feedlot performance and carcass characteristics. *Journal of Animal Science* 80 (Suppl. 1): 270 Abstract 1080.

Berger, L. L., Robbins, N. D., Sewell, J. R., et al. (2003). Effect of feeding diets containing corn grain with corn rootworm protection (event MON 863), control, or conventional varieties on steer feedlot performance and carcass characteristics. *Journal of Animal Science* 81 (Suppl.1): 214–215 Abstract M150.

Brake, J. and Vlachos, D. (1998). Evaluation of Transgenic Event 176 "B.t." Corn in Broiler Chickens. *Poultry Science* 77: 648–653.

Bressner, G. E., Hyun, Y., Stanisiewski, E. P., et al. (2003). Performance comparison of growing-finishing pigs fed diets containing Corn Root Worm Protected corn (Event MON 863) or conventional corn hybrids. *Journal of Animal Science* 81 (Suppl. 1): 207 Abstract M119.

Calsamiglia, S., Hernandez, B., Hartnell, G. F. and Phipps, R. (2007). Effects of corn silage derived from a genetically modified variety containing two transgenes on feed intake, milk production, and composition, and the absence of detectable transgenic deoxyribonucleic acid in milk in Holstein dairy cows. *Journal of Dairy Science* 90: 4718–4723.

CAST (2006). Safety of Meat, Milk, and Eggs from animals fed crops derived from modern biotechnology. Animal agriculture's future through Biotechnology, Issue Paper 34. Council for Agricultural Science and Technology, Ames, IA 50014-3447.

Castillo, A.R., Gallardo, M. R., Maciel, M., et al. (2001). Effect of feeding dairy cows with cottonseeds containing BollGard® and Roundup Ready® genes or control non-transgenic cottonseeds on feed intake, milk yield and milk composition. *Journal of Dairy Science* 84 (Suppl. 1): 413 Abstract 1713.

Castillo, A. R., Gallardo, M. R., Maciel, M., et al. (2004). Effects of feeding rations with genetically modified whole cottonseed to lactating Holstein cows. *Journal of Dairy Science* 87: 1778–1785.

Chiter, A., Forbes, J. M. and Blair, G. E. (2000). DNA stability in plant tissues: implications for the possible transfer of genes from genetically modified food. *FEBS Letters* 481: 164–168.

Cromwell, G., Henry, B., Scott, A., et al. (2005). Glufosinate herbicide-tolerant -Liberty Link-Rice vs. conventional rice in diets for growing-finishing swine. *Journal of Animal Science* 83: 1068–1074.

Cromwell, G. L., Lindemann, M., Randolph, J., et al. (2002). Soybean meal from Roundup Ready or conventional soybeans in diets for growing-finishing swine. *Journal of Animal Science* 80: 708–715.

Custodio, M. G., Powers, W. J., Huff-Lonergan, E., et al. (2006). Growth, pork quality, and excretion characteristics of pigs fed B.t. corn or non-transgenic corn. *Canadian Journal of Animal Science* 86: 461–469.

Daenicke, R., Aulrich, K., and Flachowsky, G. (1999). GMO in Animal Feedstuffs: Nutritional Properties of B.t.-Maize Unaffected. *Mais* 27 (September): 135–137.

Deaville, E. R. and Maddison, B. C. (2005). Detection of transgenic and endogenous plant DNA fragments in the blood, tissues, and digesta of broilers. *Journal of Agricultural and Food Chemistry* 53: 10268–10275.

Donkin, S., Velez, J., Stanisiewski, E. and Hartnell, G. (2000). Effect of feeding Roundup Ready® corn silage and grain on feed intake, milk production and milk composition in lactating dairy cattle. *Journal of Dairy Science* 83 (Supplement 1): 273.

Donkin, S.S., Velez, J. C., Totten, A. K., et al. (2003). Effects of feeding silage and grain from glyphosate-tolerant or insect-protected corn hybrids on feed intake, ruminal digestion, and milk composition in dairy cattle. *Journal of Dairy Science* 86: 1780–1788.

EFSA (2007). EFSA statement on the fate of recombinant DNA or proteins in meat, milk and eggs from animals fed with GM feed. Report European Food Safety Authority, 19 July 2007.

Erickson, G. E., Robbins, N. D., Simon, J. J., et al. (2003). Effect of feeding glyphosate-tolerant (Roundup-Ready events GA21 or nk603) corn compared with reference hybrids on feedlot steer performance and carcass characteristics. *Journal of Animal Science* 81: 2600–2608.

Fischer, R.L., Lewis, A. J., Miller, P. S., et al. (2002). Comparison of swine performance when fed diets containing Roundup Ready® corn (event NK603), control, or conventional corn grown during 2000 in Nebraska. *Journal of Animal Science* 80 (Suppl. 1): 224 Abstract 894.

Fischer, R. L., Miller, P. S., Hyun, Y., et al. (2003). Comparison of swine performance when fed diets containing corn root worm protected corn, parental line corn, or conventional corn grown during 2000 in Nebraska. *Journal of Animal Science* 81 (Suppl. 1): 207 Abstract M118.

Folmer, J. D., Grant, R. J., Milton, C. T. and Beck, J. F. (2000). Effect of B.t. corn silage on short term lactational performance and ruminal fermentation in dairy cows. *Journal of Dairy Science* 83: 1182 Abstract 272.

Grant, R. J., Kleinschmit, D., Sparks, A. L., et al. (2002). Influence of glyphosate-tolerant (trait NK603) corn silage and grain on feed consumption and milk production in Holstein dairy cattle. *Journal of Dairy Science* 85 (Suppl. 1): 384 Abstract 1540.

Grant, R., Fanning, K., Kleinschmit, D., et al. (2003). Influence of glyphosate-tolerant -Event NK603- and corn rootworm protected -event MON863- corn silage and grain on feed consumption and milk production in Holstein cattle. *Journal of Dairy Science* 86: 1707–1715.

Guan, J., Spencer, J. L. and Ma, B. L. (2005). The fate of the recombinant DNA in corn during composting. *Journal of Environmental Science and Health B* 40: 463–473.

Guan, Q., Wang, X., Teng, D., et al. (2013). DNA degradation of genetically modified cottonseed meal during feed processing. *Applied Biochemistry and Biotechnology* 169: 368–379.

Hammond, B., Vicini, J., Hartnell, G., et al. (1996). The feeding value of soybeans fed to rats, chickens, catfish and dairy cattle is not altered by genetic incorporation of glyphosate tolerance. *Journal of Nutrition* 126: 717–727.

Hohlweg, U. and Doerfler, W. (2001). On the fate of plant or other foreign genes upon the uptake in food or after intramuscular injection in mice. *Molecular Genetics and Genomics* 265: 225–233.

Harrison, L. A., Bailey, M. R., Naylor, M. W., et al. (1996). The expressed protein in glyphosate-tolerant soy-bean, 5-enolpyruvylshikimate-3-phosphate synthase from *Agrobacterium* sp. strain CP4, is rapidly digested in vitro and is not toxic to acutely gavaged mice. *Journal of Nutrition* 126: 728–740.

Ipharraguerre, I. R., Younker, R. S., Clark, J. H., et al. (2002). Performance of lactating dairy cows fed glyphosate-tolerant corn (event NK603). *Journal of Dairy Science* 85 (Suppl. 1): 358 Abstract 1435.

Jennings, J. C., Kolwyck, D. C., Kays, S. B., et al. (2003a). Determining whether transgenic and endogenous plant DNA and transgenic protein are detectable in muscle from swine fed Roundup Ready soybean meal. *Journal of Animal Science* 81: 1447–1455.

Jennings J. C., Whetsell, A. J., Nicholas, N. R., et al. (2003b): Determining whether transgenic or endogenous plant DNA is detectable in dairy milk or beef organs. *Bulletin 383/2003 of the International Dairy Federation*: 41–46.

Jonas, D. A., Elmadfa, I., Engel, K. H., et al. (2001). Safety considerations of DNA in food. *Annals of Nutrition and Metabolism* 45: 235–254.

Kan, C. and Hartnell, G. (2004). Evaluation of broiler performance when fed roundup-ready wheat (Event Mon71800), control, and commercial wheat varieties. *Poultry Science* 83: 1325–1334.

Kan, C. A., Versteegh, H. A. J., Uijttengoogaart, T. G., et al. (2001).Comparison of broiler performance and carcass characteristics when fed B.t., parental control or commercial varieties of dehulled soybean meal. *Poultry Science* 80 (Suppl. 1): 203 Abstract 841.

Li, M. H., Hartnell, G. F., Robinson, E. H., Kronenberg, J. M., et al. (2008). Evaluation of cottonseed meal derived from genetically modified cotton as feed ingredients for channel catfish, *Ictalurus punctatus*. *Aquaculture Nutrition* 14: 490–498.

Lutz, B., Wiedemann, S. and Albrecht, C. (2006). Degradation of transgenic Cry1Ab DNA and protein in B.t.-176 maize during the ensiling process. *Journal of Animal Physiology and Animal Nutrition* (Berl). 90: 116–123.

Lutz, B., Wiedemann, S., Einspanier, R., et al. (2005) Degradation of Cry1Ab protein from genetically modified maize in the bovine gastrointestinal tract. *Journal of Agricultural and Food Chemistry* 53: 1453–1456.

Nemeth, A., Wurz, A., Artim, L., et al. (2004). Sensitive PCR analysis of animal tissue samples for fragments of endogenous and transgenic plant DNA. *Journal of Agricultural and Food Chemistry* 52: 6129–6135.

Phipps, R. H., Beever, D. E. and Humpries, D. J. (2002). Detection of transgenic DNA in milk from cows receiving herbicide tolerant (CP4EPSPS) soybean meal. *Livestock Production Science* 74: 269–273.

Ran, T., Mei, L., Lei, W., et al. (2009). Detection of transgenic DNA in tilapias (Oreochromis niloticus, GIFT strain) fed genetically modified soybeans (Roundup Ready). *Aquaculture Research* 40: 1350–1357.

Řehout, V., Kadlec, J., Čítek, J., et al. (2009). The influence of genetically modified B.t. maize MON 810 in feed mixtures on slaughter, haematological and biochemical indices of broiler chickens. *Journal of Animal and Feed Sciences* 18: 490–498.

Reuter, T., Alexander, T. W., Xu, W., et al. (2010). Biodegradation of genetically modified seeds and plant tissues during composting. *Journal of the Science of Food and Agriculture* 90: 650–657.

Rossi F, Morlacchini, M., Fusconi, G., et al. (2005). Effect of B.t. corn on broiler growth performance and fate of feed-derived DNA in the digestive tract. *Poultry Science* 84:1022–1030.

Sanden, M., Bruce, I. J., Rahman, M. A. and Hemre, G. I. (2004) The fate of transgenic sequences present in genetically modified plant products in fish feed, investigating the survival of GM soybean DNA fragments during feeding trials in Atlantic salmon, *Salmo salar* L. *Aquaculture* 237: 391–405.

Sanden, M., Krogdahl, A., Bakke-Mckellep, A. M., et al. (2006). Growth performance and organ development in Atlantic salmon, *Salmo salar* L. parr fed genetically modified (GM) soybean and maize. *Aquaculture Nutrition* 12: 1–14.

Sharma, R., Alexander, T. W., John, J., et al. (2004). Relative stability of transgene DNA fragments from GM rapeseed in mixed ruminal cultures. *British Journal of Nutrition* 91: 673–681.

Sharma, R., Damgaard, D., Alexander, T. W., et al. (2006). Detection of transgenic and endogenous plant DNA in digesta and tissues of sheep and pigs fed Roundup Ready canola meal. *Journal of Agricultural and Food Chemistry* 54: 1699–709.

Sidhu, R. S., Hammond, B. G., Fuchs, R. L., et al. (2000). Glyphosate-tolerant corn: the composition and feeding value of grain from glyphosate-tolerant corn is equivalent to that of conventional corn (*Zea mays* L.). *Journal of Agricultural and Food Chemistry* 48: 2305–2312.

Sieradzki, Z., Mazur, M., Kwiatek, K., et al. (2013). Assessing the possibility of genetically modified DNA transfer from GM feed to broiler, laying hen, pig and calf tissues. *Polish Journal of Veterinary Science* 16: 435–441.

Simon, J. J., Van der Pol, K. J., Erickson, G. E., et al. (2002). Effect of Roundup Ready corn (event NK603) on performance in beef feedlot diets. *Journal of Animal Science* 80 (Suppl.1): 46.

Singhal, K. K., Kumar, R. S., Tyagi, A. K. and Rajput, Y. S. (2006). Evaluation of B.t. cottonseed as protein supplement in the ration of lactating dairy cows. *Indian Journal of Animal Sciences* 76: 532–537.

Sissener, N. H., Sanden, M., Krogdahl, Å., et al. (2011). Genetically modified plants as fish feed ingredients. *Canadian Journal of Fisheries and Aquatic Sciences* 68: 563–574.

Stanford, K., Aalhus, J. L., Dugan, M. E. R., et al. (2003). Effects of feeding transgenic canola on apparent digestibility, growth performance and carcass characteristics of lambs. *Canadian Journal of Animal Science* 83: 299–305.

Stein, H. H., Sauber, T. E., Rice, D. W., et al. (2009). Growth performance and carcass composition of pigs fed corn grain from DASØ15Ø7-1 (Herculex I) hybrids. *The Professional Animal Scientist* 2: 689–694.

Steinke, K., Guertler, P., Paul, V., et al. (2010). Effects of long-term feeding of genetically modified corn (event MON810) on the performance of lactating dairy cows. *Journal of Animal Physiology and Animal Nutrition* 94: 185–193.

Suharman, I., Satoh, S., Haga, Y., et al. (2009). Utilization of genetically modified soybean meal in Nile tilapia Oreochromis niloticus diets. *Fisheries Science* 75: 967–973.

Świątkiewicz, M., Bednarek, D., Markowski, J., et al. (2013). Effect of feeding genetically modified maize and soybean meal to sows on their reproductive traits, haematological indices and offspring performance. *Bulletin Veterinary Institute, Pulawy, Poland* 57: 413–418.

Świątkiewicz, M., Hanczakowska, E., Twardowska, M., et al. (2011). Effect of genetically modified feeds on fattening results and transfer of transgenic DNA to swine tissues. *Bulletin of the Veterinary Institute, Puławy, Poland* 55: 121–125.

Taylor, M. L., Stanisiewski, E. P., Riordan, S. G., et al. (2004). Comparison of broiler performance when fed diets containing roundup ready (Event RT73), nontransgenic control, or commercial canola meal. *Poultry Science* 83: 456–461.

Thomson, J. (2001). Horizontal transfer of DNA from GM crops to bacteria and to mammalian cells. *Journal of Food Science* 66: 188–193.

Tony, M.A., Butschke, A., Broll, H., et al. (2003). Safety assessment of B.t. 176 maize in broiler nutrition: degradation of maize-DNA and its metabolic fate. *Archiv fur Tierernahrung* 57: 235–252.

Walsh, M. C., Buzoianu, S. G., Gardiner, G. E., et al. (2011). Fate of transgenic DNA from orally administered B.t. MON810 maize and effects on immune response and growth in pigs. PLoS ONE 6(11): e27177.

Weisbjerg, M., Hvelplund, T., Purup, S., et al. (2001). Genetically modified beets and beet pulp as feeds for ruminants, experiments with sheep and dairy cattle at research centre Foulum, Denmark. *Proceedings: International Symposium on Genetically Modified Crops and Co-Products as Feeds for Livestock*, Nitra, Slovak Republic: pp. 37–40.

Van den Eede, G., Aarts H., Buhk, H. J., et al. (2004). The relevance of gene transfer to the safety of food and feed derived from genetically modified (GM) plants. *Food and Chemical Toxicology* 42: 1127–1156.

CHAPTER 9

Overall assessment of the safety of GM foods and feeds

The question of safety is paramount in the attitude of consumers towards GM foods. Therefore it is important to summarise the existing knowledge on this issue, based on the detailed findings outlined in previous chapters and on formal scientific reviews of the available evidence by competent authorities.

Microorganisms

Several GM microorganisms have been approved for food use in some countries, the yeast *Saccharomyces cerevisiae* being the first GM organism to be approved for use in food production. GM bacteria are used in the production of enzymes, for example milk-clotting enzymes and food/feed additives such as aspartame and L-lysine. GM yeasts are an important source of enzymes used in food production (e.g. cheese-making), and in improving the quality of beer and wine.

An important feature relating to the safety of these approved GM microorganisms is that they have been modified by utilising genetic material from within the host species and from microorganisms with GRAS (Generally Regarded as Safe) (in Europe QPS – Qualified Presumption of Safety) status.

The safety of the approved GM microorganisms in food and feed production is attested to by the fact that no authenticated reports of allergenicity or toxicity have been published in any peer-reviewed journals.

Genetic Modification and Food Quality: A Down to Earth Analysis, First Edition.
Robert Blair and Joe M. Regenstein.

Cereal crops

Several GM cultivars of cereal crops have been developed but maize (including sweet corn) is the only GM cereal being grown internationally in commercially significant quantities at this time. All of the approved cultivars have received clearance by regulatory authorities, based on safety tests and on substantial equivalence with conventional lines of maize. Globally, most of the GM maize currently is of 'stacked trait' type. The currently approved stacked-traits maize category comprises a combination of traits: a double stack with insect resistance and herbicide tolerance; a double stack with two different traits for insect resistance; and a triple stack with two different types of insect resistance, plus herbicide tolerance. A large body of data published in peer-reviewed scientific journals shows that no safety concerns have become evident when approved cultivars of GM maize are included in the diets of test animals. Reports from one laboratory of the possibility of cancer induction as a result of consuming GM maize have been discredited.

The published data suggest a lower level of mycotoxins in GM maize possessing the trait for insect resistance, which is beneficial to animals and humans consuming such maize.

Oilseed crops

Approved GM oilseed crops being grown worldwide include soyabeans, canola and cotton, which are major sources of oil and protein for human and animal feeding. Soyabeans are the most important of these crops since both the beans and extracted oil are used as human food. Only the oil from canola and cottonseed are used as human food, the fat-extracted meals being used as animal feed. Thus, the fraction consumed by humans is devoid of any GM DNA. A striking feature is the high proportion of GM cultivars of these crops being grown, in some countries approaching 100%. Several GM cultivars of these crops have been approved as food/feed sources by regulatory authorities, based on evidence of safety and nutritional composition. Results of scientific investigations confirm the approval granted to GM cultivars by regulatory authorities, and a few contrary findings published by some investigators have not been accepted as credible by the scientific community, as discussed in Chapter 5.

Fruits and vegetables

Several GM cultivars of fruit and vegetables have been approved for food and feed use in several countries, although not all of the approved ones are currently being marketed. The list of approved GM cultivars of fruits includes cantaloupe, plums and papaya but only one type of GM fruit, the Hawaiian papaya with resistance to ringspot virus infection, is currently available to consumers. This virus-resistant cultivar was approved by the regulatory authorities in North America and Japan, based on evidence of safety and nutritional equivalence with conventional cultivars of papaya.

The list of approved GM cultivars of vegetables includes lucerne (alfalfa), potatoes, squash, sugar beet, sweet corn (maize) and tomatoes. The first major criticism of GM foods by a scientist (Pusztai) was in relation to a GM potato that was undergoing testing. He reported that his experiments demonstrated that rats fed a diet of genetically modified potatoes expressing the gene coding for GNA showed stunted growth and reduced immune function. The criticism was later shown to be without substance. The subsequent introduction of GM potatoes has occurred successfully. An important GM crop being grown in several countries is sugar beet, as a source of sugar for human use. As with oilseeds, all of the transgenic DNA is removed from the sugar (or oil) during processing and remains in the residue which is used in animal feeding. Much of the sweet corn now being grown in North America is GM since these approved cultivars provide resistance to insect attack or herbicide application. The first commercially grown GM food crop was a tomato with delayed ripening properties. It was a commercial failure because it did not provide sufficient added value for consumers. Several other GM cultivars were introduced later. A consistent finding with these GM foods is that much, or all, of the transgenic DNA is degraded in the gastro-intestinal tract of the human or animal consumer, and that none is actually absorbed into the blood and tissues. Based on the available scientific evidence it is clear that the approved cultivars of GM fruit and vegetables are as safe as non-GM cultivars.

Animals and animal products

Several types of GM animals have been developed but to date none have been approved for introduction into the human food supply. It is possible that, in the future, the AquaBounty salmon will be approved as the first GM food animal. The safety of this Atlantic salmon has been approved by a few governments, but other issues have held up formal approval including political issues.

The available research investigations indicate that the transgenic proteins in GM feedstuffs are substantially degraded by feed processing and by passage through the digestive tract of cattle, sheep, pigs, poultry and fish. There is no evidence to suggest that transgenic DNA is processed in the gut differently from endogenous feed-ingested genetic material. The current findings indicate that a small proportion of dietary DNA fragments, both from GM and non-GM feedstuffs, can be absorbed by some fish from their intestines and can be detected in internal organs (Ran et al., 2009). However, the fragments appear to be below a size that is able to code for functional proteins and are, therefore, not a safety concern. Further research needs to be conducted on this issue to confirm that conclusion. The GM residues do not appear to be absorbed preferentially. No GM residues have been reported in meat, milk or eggs, suggesting that fish may differ from other species in terms of their absorption of DNA breakdown particles.

Chemical residues

The available data indicate that cereal grains contain chemical and pesticide residues, but suggest that the amounts are too small to be of concern in relation to human health. Many of the residues found are at the limit of detection. No separate data on residue levels in GM grains are available: more studies need to be carried out to provide comparable data on GM and conventionally-grown grains.

Although there is no evidence to support the claim that glyphosate-tolerant maize or soyabeans are more contaminated with glyphosate residues than conventional crops, the issue of glyphosate contamination is important to consumers. It is useful, therefore, to summarise the available evidence relating to humans and animals. Glyphosate is one of the least toxic pesticides to animals (Duke and Powles, 2008). According to these authors, glyphosate is less acutely toxic than common chemicals such as sodium chloride or aspirin, with an LD50 for rats of over 5 g/kg compared to the LD50 of sodium chloride of 3000 mg/kg(rat) or aspirin at 200 mg/kg (rat). It is not a carcinogen or a reproductive toxin, nor does it have any subacute chronic toxicity. No toxic effects were observed in a chronic feeding study with dogs fed up to 500 mg/kg body weight/day, the highest dose tested (US Environmental Protection Agency, 1992).

Williams et al. (2000) conducted a review of the data related to the safety of the herbicide Roundup and its active ingredient, glyphosate, for humans. It included assessments of aminomethylphosphonic acid (AMPA, the breakdown product of glyphospate), Roundup formulations, and the predominant surfactant polyethoxylated tallow amine (POEA) used internationally in Roundup formulations. The reports studied included those conducted for regulatory purposes as well as published research reports. These reviewers found that the oral absorption of glyphosate and AMPA is low and that both materials are eliminated essentially un-metabolised. Dermal penetration studies with Roundup showed very low absorption. The reported evidence also showed that neither glyphosate nor AMPA bioaccumulate in any animal tissue. Acute risks were assessed by comparison with the oral LD50 values with estimated maximum acute human exposure. Based on the available evidence the authors concluded that, under present and expected conditions of use, Roundup herbicide does not pose a health risk to humans.

The rat study conducted by Séralini et al. (2012) that was outlined in Chapter 4 included groups provided with drinking water containing glyphosate.

The treatments were: (i) regular water; (ii) water containing 50 ng/l of glyphosate (described as the contamination level of some regular tap waters); (iii) water containing 400 mg/kg glyphosate (described as the US MRL of glyphosate and present in some GM feed); and (iv) water containing 2.25 g/l glyphosate (described as half of the minimal agricultural working dilution). The water was changed weekly. All four groups were fed a control diet containing conventional maize. Each group consisted of 10 males and 10 females. The results (as outlined in Chapter 4) showed that total deaths (or removals due to morbidity) over the 600-day test period were, respectively, 5/20, 8/20, 9/20 and 5/20. Although these data indicated no clear evidence of a linear effect (which would be expected if glyphosate was toxic at the levels used), Séralini et al. (2012) reported pathological changes in the livers of rats given the glyphosate-supplemented water. The data to support this conclusion, as presented, do not allow an accurate assessment of liver damage and appear to again show a non-linear effect. For instance in males the incidence of non-regressive tumours was lowest in treatment 2 and highest in treatment 4. It would be reassuring to consumers if the drinking water treatments used in the Séralini et al. (2012) trial could be repeated in another independent study and an assessment made of the extent of any pathological changes in liver and other organs.

Additionally, it would be advisable for farmers to time the herbicidal treatment of growing crops with Roundup sufficiently before harvesting so that glyphosate residues in the harvested maize and soyabeans are minimised. The possibility of chemical residues in food may be of greater concern to many consumers than the possibility that the food being consumed is derived from GM crops.

The significance of possible glyphosate contamination of soyabean products is similar to that of maize crops. The currently available data suggest no proven negative effects on the health of animal or human consumers. One study, though, may merit further investigation. Vecchio et al. (2004) fed pregnant Swiss mice on a standard laboratory chow containing 14% non-GM or GM glyphosate-tolerant soyabean meal. The litters produced were fed on the parental diet (12 controls and 12 GM-fed) and were sacrificed at 2, 5 or 8 months of age. The researchers reported that Sertoli cells in the testes of the mice fed the GM diet showed enlarged vesicles of the smooth endoplasmic reticulum. Also, the nucleoli were reported to be larger and more reticulated in 2- and 5-month-old mice fed the GM diet than in the controls, but there was no difference at 8 months of age. This effect needs to be investigated further since it does not appear to have been replicated in any other independent studies. The authors speculated that the effects reported might have been due to the presence of glyphosate residues in the diet containing GM soyabean meal, although no analytical data on glyphosate residues were presented.

Arregui et al. (2003) reported that in Argentina glyphosate residues ranged from 0.1 to 1.8 mg/kg in soyabeans. AMPA residues were detected both in the leaves and in the grains, indicating some metabolism of the herbicide. As is the case with glyphosate-treated maize, consumers need to be reassured that consuming GM soyabean products or the meat of animals fed GM soyabeans is safe with respect to possible glyphosate residues.

Claims of higher contents of other pesticide residues in GM oilseeds are not supported by the scientific evidence. Based on the available scientific evidence it is clear that the approved cultivars of GM oilseeds produce food and feed that is as safe as non-GM cultivars.

Consensus

There is broad scientific consensus that genetically engineered crops currently being marketed are safe as food and feed. In 2002 the US National Research Council concluded that after 14 years of cultivation and a cumulative total of 0.81 billion hectares planted, no adverse health or environmental effects had resulted from commercialisation of genetically engineered crops (NAS-NRC, 2002). Further studies by the National Research Council and by the European Joint Research Centre have concluded that there is a comprehensive body of knowledge that adequately addresses the food safety issue of GM engineered crops (US Institute of Medicine-National Research Council, 2004; European Commission, 2008).

Other more recent reviews have reached similar conclusions. Batista and Oliveira (2009) published an interesting review under the title of 'Facts and fiction of genetically engineered food'. Their analysis found 'no unequivocal evidence supporting adverse effects of any of the currently commercialized GM food products'. However they noted a clear division between the public and the scientific community on knowledge regarding this issue.

Snell et al. (2012) reviewed data on the effects of diets containing GM maize, potato, soyabeans, rice, or triticale on animal health. They examined 12 long-term studies (of more than 90 days up to 2 years in duration) and 12 multi-generational studies (from two to five generations). They noted that many parameters had been evaluated, including biochemical analyses, histological examination of specific organs, hematology and the specific detection of transgenic DNA. Results from all of the 24 studies did not suggest any health hazards and, in general, there were no statistically significant differences for any of the parameters studied. However, some small differences were observed, although these fell within the normal variation range of the considered parameter and thus had no biological or toxicological significance. If required, a 90-day feeding study conducted with rodents, according to the OECD Test Guideline, is generally considered sufficient to evaluate the health effects of GM feed (Snell et al., 2012). The researchers concluded that the studies reviewed presented evidence to show that GM plants are nutritionally equivalent to their non-GM counterparts and can be safely used in food and feed. Snell et al. (2012)

noted also that: 'The observations of major flaws in some papers highlight the urgent need to improve the reviewing process before publication of papers addressing this subject. This would avoid spreading confusion in the general press, which may not be able to judge the real scientific quality of publications.'

ChileBio (2013) published a list of 610 scientific articles on the safety assessments of foods and feeds derived from genetically modified crops. The list contained only original papers (excluding scientific reviews) published in peer-reviewed journals and abstracted in scientific databases such as Pubmed. No health concerns were found. In an impressive review involving 1783 studies relating to the safety and environmental impacts of approved GM crops, Nicolia et al. (2014) concluded that the scientific research conducted so far has not detected any significant hazards to human or animal health. Their review covered the period from 2002 to 2012 and involved original research papers, reviews, relevant opinions and reports addressing all of the major issues that have emerged in the debate relating to GM crops. Environmental impact studies were found to be predominant in the database, comprising 68% of the 1783 studies. Nicolia et al. (2014) found little to no evidence that GM crops had a negative environmental impact. In the food and feed category (comprising 43.2% of the database), the researchers found no evidence that approved GM crops introduced any unique allergens or toxins into the food supply. The researchers also addressed the safety of transcribed RNA from transgenic DNA. Analysis of the relevant data showed that humans consume between 0.1 and 1 g of DNA per day, from both GM and non-GM ingredients. The transgenic DNA is enormously diluted by the total amount of ingested DNA (from 0.00006 to 0.00009% of total consumption of DNA) and is digested like any other DNA. This DNA is generally degraded during the food processing, and any surviving DNA is then subsequently degraded in the digestive system. It was concluded that no evidence has been obtained to date that DNA absorbed through the gut wall can be integrated into human cells and lead to a germ line transfer. Their conclusion is that transgenic DNA does not differ intrinsically or physically from any other DNA already present in foods and that the ingestion of transgenic DNA does not imply any higher risks than ingestion of any other type of DNA. One of the aims of these researchers was to make the entire database available to researchers, communicators and teachers at all levels to help create an informed, balanced public perception on the important issue of GM crop use in agriculture and food production.

Two important conclusions can be drawn from the review by Nicolia et al. (2014): there is no credible evidence that food derived from GM crops poses any unique threat to human health or to the environment; also GM foods are among the most extensively studied scientific subjects in history.

In an assessment of the safety of GM foods it is fair to point out that instances of harmful effects on humans or animals have been documented only in foods developed through conventional breeding approaches. Certain potato lines have been found to express greater resistance to diseases or pests, and have been selected as superior but not always with favourable or intended results. One such selection was the Lenape potato, which was developed using selective breeding. After a successful commercial launch, it was found to have a dangerously elevated solanine content in the tubers and had to be removed from the market. A similar high-solanine potato variety was detected and withdrawn from the market in Sweden. Another example of a toxic compound in traditional crops is psoralen in celery, a group of secondary metabolites (furanocoumarins) found in a variety of fruits and vegetables (Diawara and Kulkosky, 2003). The psoralens deter insects from feeding on the plant and also confer protection from some diseases (Beier and Oertli, 1983). Celery plants with an elevated expression of psoralens suffer less damage from disease and insect predation, making them more appealing to consumers. One selected cultivar that was produced with relatively high amounts of psoralens unfortunately caused farm workers who harvested the crop to develop a severe skin rash, an unintended consequence of this breeding strategy (US Institute of Medicine-NRC, 2004). This resulted in the new cultivar having to be being removed from the market.

Any criticism of GM foods because they have been produced using very advanced and invasive technology needs to tempered by the realisation that food derived from mutation breeding is widely used and accepted since induced mutagenesis is considered a conventional breeding technique. Mutations, defined as any change in the base sequence of DNA, can either occur spontaneously or be induced, and both methods have produced new crop varieties. The Joint FAO/IAEA Mutant Variety Database (2014) lists more than 3218 cultivars of various plant species worldwide that have been developed using induced mutagenesis agents, including ionising irradiation and ethyl methane sulfonate (which is a mutagenic, teratogenic and possibly carcinogenic organic compound that produces random mutations in genetic material by nucleotide substitution). China and Europe account for most of the registered cultivars. Crop plants account for 75% of released mutagenic species. In the United States many varieties have been developed using induced mutagenesis, such as lettuce, beans, grapefruit, rice, oats and wheat. Organic farming systems permit food from mutated varieties to be sold as organic (Wikipedia, 2013). A constant concern with the use of mutagenesis is whether genes that are not monitored but have negative consequences may unknowingly enter the food/feed supply.

All of the reports cited above conclude that the processes of genetic modification and conventional breeding are no different in terms of unintended consequences to human health and the environment and when problems are detected the affected cultivars are removed from the food/feed supply. This conclusion should provide great reassurance to consumers. Each new plant cultivar developed through genetic modification has been assessed on a case-by-case basis by regulatory agencies prior to its approval. On the other hand conventional crops are not regulated by these agencies prior to the release of new cultivars.

Factors that mitigate possible harmful effects from foods derived from GM crops

As discussed above and in more detail in earlier chapters, factors that mitigate any possible harmful effects from foods derived from GM crops and foods derived from livestock fed diets containing GM feedstuffs are outlined below.

1 All GM crops being cultivated for food or feed have been approved by regulatory authorities (e.g. FDA, Health Canada and EFSA) as being safe for human consumption before they are allowed to be marketed. Approval is only provided after a lengthy and detailed scientific assessment relating to safety, nutritional composition and effects on the environment.
2 Most of these foods/feeds are used in feeding livestock: only a small proportion is consumed directly by the human consumer. Effectively, then, farm stock can be regarded as a filter in the utilisation of foods derived from GM crops.
3 Processing of GM crops into livestock feed disrupts the cellular structure and breaks up the DNA sequences, reducing the possibility of absorption of active DNA sequences into the edible tissue from these animals.
4 Acidic and enzymatic conditions in the animal gut break down the protein structures in the cells of GM (and conventional) feeds. As a result few, if any, transgenic residues have been found in meat, milk or eggs from livestock fed diets containing GM feed.
5 The process of cooking food disrupts the cellular structures, including DNA sequences.
6 Acidic and enzymatic conditions in the human gut break down the protein and DNA sequences in the diet.
7 Extracted oils from GM crops (such as canola and maize oils) have been purified to contain zero levels of transgenic DNA.

Consequently, all genetically altered DNA in our food (if ingested) is inactivated in the human diet. It is, therefore, very unlikely that any food derived from approved GM crops or from animals or poultry fed diets containing GM feed can be regarded as posing any health risk to the human consumer.

Summary

There is broad scientific consensus that all approved foods and feeds that have been derived from GM crops being marketed currently, and from livestock, poultry and fish fed diets containing GM products, are safe to eat. This conclusion is based on the findings of independent scientists, regulatory agencies and scientific agencies in many countries. Also, there is no evidence that the use of herbicides such as glyphosate (Roundup) on GM crops result in residues that are harmful to humans. The fact that some consumers have concerns about the safety of GM foods can be attributed to the fact that they have not been made aware of the extensive database of favourable findings or choose to ignore them.

References

Arregui, M. C., Lenardon, A., Sanchez, D., et al. (2003). Monitoring glyphosate residues in transgenic glyphosate-resistant soybean. *Pest Management Science* 60: 163–166.

Batista, R. and Oliveira, M. M. (2009). Facts and fiction of genetically engineered food. *Trends in Biotechnology* 27: 277–286.

Beier, R. C. and Oertli, E. H. (1983). Psoralen and other linear furocoumarins as phytoalexins in celery. *Phytochemistry* 22: 2595–2599.

ChileBio (2013). 610 Scientific Articles Confirm Safety of Food and Feeds Derived from Genetically Modified Crops. http://www.isaaa.org/kc/cropbiotechupdate/article/default.asp?ID=11295 (accessed February 2015).

Diawara, M. M. and Kulkosky, P. (2003). Reproductive toxicity of the psoralens. *Pediatric Pathology and Molecular Medicine* 22: 247–258.

Duke, S. O. and Powles, S. B. (2008). Glyphosate: a once-in-a-century herbicide. *Pest Management Science* 64: 319–325.

European Commission (2008). Scientific and technical contribution to the development of an overall health strategy in the area of GMOs. Report EUR 23542 EN, Joint Research Centre - Institute for Health and Consumer Protection. Luxembourg: Office for Official Publications of the European Communities, 56 pp, Scientific and Technical Research series - ISSN 1018-5593.

Farm Bureau (2014). Poll identifies top consumer questions on GMOs. http://fbnews.fb.org/FBNews/Top_News/Poll_identifies_top_consumer_questions_on_GMOs.aspx (accessed 7 April 2014).

Joint FAO/IAEA Programme (2014). Mutant Variety Database. http://mvgs.iaea.org/AboutMutantVarities.aspx (accessed 21 September 2014).

Nicolia, A. Manzo, A., Veronesi F., and Rosellini, D. (2014). An overview of the last 10 years of genetically engineered crop safety research. *Critical Reviews in Biotechnology* 34: 77–88.

Ran, T., Mei, L., Wang, L., et al. (2009). Detection of transgenic DNA in tilapias (*Oreochromis niloticus*, GIFT strain) fed genetically modified soybeans (Roundup Ready). *Aquaculture Research* 40: 1350–1357.

Séralini, G-E., Clair, E., Mesnage, R., et al. (2012). Long term toxicity of a Roundup herbicide and a Roundup-tolerant genetically modified maize. *Food and Chemical Toxicology* 50: 4221–4231 (paper retracted by journal).

Snell, C., Bernheim, A., Bergé, J., et al. (2012). Assessment of the health impact of GM plant diets in long-term and multigenerational animal feeding trials: A literature review. *Food and Chemical Toxicology* 50: 1134–1148.

US Environmental Protection Agency (1992). Pesticide Tolerance for Glyphosate. Federal Register 57: 8739–8740.

NAS-NRC (2002). Environmental Effects of Transgenic Plants, the Scope and Adequacy of Regulation. National Academy Press, Washington, DC.

US Institute of Medicine-NRC (2004). Safety of Genetically Engineered Foods: Approaches to Assessing Unintended Health Effects. National Academy Press, Washington, DC.

Vecchio, L., Cisterna, B., Malatesta, M., et al. (2004). Ultrastructural analysis of testes from mice fed on genetically modified soybean. *European Journal of Histochemistry* 48: 449–453.

Wikipedia, (2013). Mutation breeding. http://en.wikipedia.org/wiki/Mutation_breeding (accessed 18 April 2014).

Williams, G. M., Kroes, R. and Munro, I. C. (2000). Safety evaluation and risk assessment of the herbicide Roundup and its active ingredient, glyphosate, for humans. *Regulatory Toxicology and Pharmacology* 31: 117–165.

CHAPTER 10

Overall assessment of the nutritional value of GM foods and feeds

A main consideration in addition to safety is whether GM foods are of the same nutritional quality as conventional foods. Accordingly it is one criterion that is evaluated by regulatory authorities before the food in question is permitted to enter the human or animal feed chain.

It is important to summarise the existing knowledge on this issue, based on the detailed findings outlined in previous chapters and on reviews of the available evidence by competent authorities.

Microorganisms

As outlined in Chapter 3, several GM microorganisms have been approved for food use in some countries, the yeast *Saccharomyces cerevisiae* being the first GM organism to be approved for use in food production. GM bacteria are used in the production of enzymes, for example milk-clotting enzymes and food/feed additives such as aspartame and L-lysine. GM yeasts are an important source of enzymes used in food production (e.g. cheese-making), and in improving the quality of beer and wine. An important feature of these approved GM microorganisms is that they have been modified by utilising genetic material from within the host species and from microorganisms with GRAS (Generally Regarded As Safe) (in Europe: QPS (Qualified Presumption of Safety)) status. No authenticated reports of a change in nutritional composition of GM microorganisms have been published in any peer-reviewed journals. This aspect

Genetic Modification and Food Quality: A Down to Earth Analysis, First Edition.
Robert Blair and Joe M. Regenstein.

does not appear to have received much detailed attention from researchers, possibly because the approved GM microorganisms are used mainly in the production of food products rather than as food items per se.

Cereal crops

As outlined in Chapter 4, several GM cultivars of cereal crops have been approved by regulatory authorities but maize (including sweet corn) is the only GM cereal being grown internationally in commercially significant quantities. All of the approved cultivars have received clearance based on evaluations of safety and on substantial equivalence with conventional cultivars.

Much of the available data on GM cereal crops has been published in reports of animal feeding studies, since most of the grain and other plant matter produced is used in the feeding of poultry and livestock, and since detailed information on nutrient composition is required before dietary mixtures can be formulated correctly for optimal production.

As outlined in Chapter 4, there is a preponderance of evidence from these studies indicating that the approved GM and related non-GM cereal crops and their products (e.g. grain and silage) are of equivalent nutritive value (Table 10.1). These studies have included maize, rice and wheat. The data presented in several of the reports are quite detailed (e.g. Aulrich et al., 2001; Brake and Vlachos, 1998; Herman et al., 2011; Obert et al., 2004; Venneria et al., 2008).

The nutritional data are not always exactly similar for GM and non-GM crops since environmental and soil conditions have an important influence on composition (Baker et al., 2006). However the data on comparability are within the acceptable biological range based on the range of values found for non-GM crops.

Confirmatory data on nutritional equivalence are shown in Table 10.2, which summarises the results of feeding diets containing GM and related non-GM food and feed products on the growth and productivity of farm stock and laboratory animals.

Several studies on laboratory animals have shown similar feed intakes and weight gains when the diet contained conventional and GM maize (e.g. Chrenková et al., 2002; Hammond et al., 2004; MacKenzie et al., 2007; He et al., 2008; Appenzeller et al., 2009; Delaney et al., 2013). Similar findings were reported with commercial livestock and poultry (e.g. Barrière et al., 2001; Reuter et al., 2002b; Donkin et al., 2003; Hyun et al., 2004; Phipps et al., 2005; Calsamiglia et al., 2007; McNaughton et al., 2007; Jacobs et al., 2008; Steinke et al., 2010).

In general, these results indicate similar growth and productivity when animals were fed GM or non-GM food and feed sources, confirming the conclusion of nutritional equivalence.

Several studies indicate better growth and productivity with diets containing insect-tolerant GM grain than non-GM grain and this has been attributed to a lower content of mycotoxins in the insect-protected grain. Mycotoxin contamination of the feed is known to have a deleterious effect on animal productivity (Flachowsky et al., 2005).

Table 10.3 summarises data on the effect of feeding GM and non-GM feed sources on the composition of milk, meat and eggs. None of the published reports indicate any change in these products.

Table 10.1 Comparative nutrient composition of GM and conventional foods/feedstuffs.

Food/Feedstuff	Reference	Reported findings
		Cereal grains
Maize		
B.t. maize	Aeschbacher et al., 2001, 2002	Nutrient composition of the two cultivars similar.
B.t. maize	Aeschbacher et al., 2005	Nutrient composition of the two cultivars similar.
B.t. maize	Aulrich et al., 2001	Nutrient composition of two grain cultivars and silage samples similar in all investigated components including proximate nutrients, amino acids, fatty acids, minerals and non-starch polysaccharides.
B.t. maize, maize silage	Barrière et al., 2001	No significant difference in the analytical composition of the two cultivars of maize.
Glyphosate-tolerant maize	Berger et al., 2002	Nutrient composition of the two cultivars similar.
Glufosinate-tolerant maize	Böhme et al., 2001	Nutrient composition of the two cultivars similar.
B.t. maize	Brake and Vlachos, 1998	No significant difference in the analytical composition of the two cultivars of maize.
B.t. maize	Brake et al., 2003	Nutrient composition of the two cultivars similar.
B.t. + Glufosinate-tolerant maize	Brake et al., 2003	Nutrient composition of the GM and isogenic cultivars similar but protein lower than in commercial maize.
VIP3 insect-resistant maize	Brake et al., 2005	Higher protein content in the control cultivar.
B.t. maize	Bressner et al., 2003	No significant difference in the analytical composition of the two cultivars of maize.
B.t. maize	Chrastinová et al., 2002	No significant difference in the nutritional composition or utilization of the two cultivars of maize, using rats.
Glyphosate-tolerant maize	Chrenková et al., 2002	Nutritional composition of the two cultivars similar.
B.t. maize, maize silage	Daenicke et al., 1999	No significant differences in nutrient composition between silage sources or between grain sources.
B.t. maize silage	Donkin et al., 2000	No significant differences in nutrient composition between silage sources.
B.t. maize silage	Donkin et al., 2003	No significant differences in nutrient composition between silage sources.

Continued

Table 10.1 Continued

Food/Feedstuff	Reference	Reported findings
B.t. maize grain, glyphosate-tolerant maize silage	Donkin et al., 2003	No significant differences in nutrient composition between GM and conventional silages or between GM and conventional grain.
Glyphosate-tolerant maize	Erickson et al., 2003	No significant differences in nutrient composition.
B.t. maize silage, maize silage	Faust and Spangler, 2000	No significant differences in nutrient composition between silage sources.
Glyphosate-tolerant maize	Fischer et al., 2002	No significant differences in nutrient composition.
B.t. maize	Fischer et al., 2003	No significant differences in nutrient composition.
B.t. maize, glyphosate-tolerant maize	Gaines et al., 2001a	No significant differences in nutrient composition between GM and conventional maize types.
B.t. maize, glyphosate-tolerant maize	Gaines et al. 2001b	No significant differences in digestible energy value between GM and conventional maize types.
B.t. maize, glyphosate-tolerant maize	Gaines et al. 2001c	No significant differences in nutritive value between GM and conventional maize types.
Glyphosate-tolerant maize-silage and grain	Grant et al., 2002	No significant differences in nutritive value between GM and conventional maize types.
B.t. + glyphosate-tolerant maize grain, silage	Grant et al., 2003	No significant differences in nutritive value between GM and conventional maize types.
Glyphosate-tolerant maize-silage and grain	Grant et al., 2003	No significant differences in nutritive value between GM and conventional maize types.
B.t. maize	Halle et al., 1998	No significant differences in nutritive value between GM and conventional maize types.
B.t., glufosinate-resistant maize.	He et al., 2008	No significant differences in nutritive value between GM and conventional maize types
B.t.-resistant, glufosinate-resistant maize.	Herman et al., 2007	No significant differences in nutritive value between GM and conventional maize forage and grain in terms of proximate constituents, fibre and minerals, amino acids, fatty acids, vitamins, secondary metabolites and anti-nutrients.
Glyphosate-tolerant maize	Hyun et al., 2004	No significant differences in nutritive value.
Glyphosate-tolerant maize, maize silage	Ipharraguerre et al., 2003	No significant differences in nutritive value between GM and conventional maize types.
B.t. maize	Masoero et al., 1999	Reduced fibre content in GM maize, levels of other nutrients not significantly different between maize types.

Table 10.1 Continued

Food/Feedstuff	Reference	Reported findings
B.t. maize	Mireles et al., 2000	No significant differences in nutritive value between maize types.
B.t. maize	Piva et al., 2001	No significant differences in nutritive value between maize types.
B.t. maize	Řehout et al., 2009	No significant differences in nutritional composition between maize types.
B.t. maize	Reuter et al., 2002a	Levels of starch, sugar, non-starch polysaccharides (NSP), amino acids, fatty acids and selected minerals not significantly different.
B.t. maize	Rossi et al., 2005	No significant differences in nutritive value between maize types, although level of phenylalanine higher in GM maize.
B.t. maize	Russel et al., 2000	No significant differences in nutritive value between maize types.
Glyphosate-tolerant maize	Sidhu et al., 2000	No significant differences in nutritive value between maize types.
Glyphosate-tolerant maize	Simon et al., 2002.	No significant differences in nutritional composition between maize types.
B.t.-maize	Stanisiewski et al., 2001	No significant differences in nutritive value.
B.t., glufosinate-resistant maize.	Stein et al., 2009	Nutritional value of the GM and non-GM near-isoline control maize similar.
B.t.-maize	Steinke et al. 2010	No significant differences in the chemical composition and estimated net energy content of the two cultivars.
Insect resistant, glyphosate-tolerant maize	Taylor et al., 2001a	No significant differences in nutritional composition between maize types.
Glyphosate-tolerant maize	Taylor et al., 2001b	No significant differences in nutritional composition between maize types.
B.t. maize	Taylor et al., 2002a	No significant differences in nutritional composition between maize types.
Insect resistant, glyphosate-tolerant maize	Taylor et al., 2002b	No significant differences in nutritional composition between maize types.
B.t. maize, glyphosate-tolerant maize	Taylor et al., 2003a	No significant differences between GM and conventional cultivars.
B.t. maize, glyphosate-tolerant maize	Taylor et al., 2003b	No significant differences between GM and conventional cultivars
B.t. maize, glyphosate-tolerant maize	Taylor et al., 2003c	No significant differences between GM and conventional cultivars
B.t.-maize	Tony et al., 2002	No significant differences in nutritive value.

Continued

Table 10.1 Continued

Food/Feedstuff	Reference	Reported findings
B.t.-maize	Venneria et al., 2008	No significant differences in fatty acid content, unsaponifiable fraction of antioxidants, total phenols, polyphenols, carotenoids, vitamin C, total antioxidant activity and mineral composition.
Rice		
Glufosinate-tolerant rice	Choi et al., 2012	Two GM cultivars compared with conventional rice. No significant differences in levels of proximates, fatty acids, amino acids, minerals, vitamins and anti-nutrients. Some small differences in a few fatty acids, minerals and trypsin inhibitor content.
Glufosinate-tolerant rice	Cromwell et al., 2005.	GM and non-GM cultivars had similar nutritional composition
Glufosinate-tolerant rice	Oberdoerfer et al., 2005	Two cultivars had similar nutritional composition. Most values for nutritional components fell within the range of values reported for commercial rice.
Wheat		
Wheat (various)	Baker and et al., 2006	The metabolome was found to be affected by environmental conditions. Differences between the control and GM lines were generally within the same range as the differences observed between the control lines grown at different sites and in different years.
Glyphosate-tolerant wheat	Kan and Hartnell, 2004a	Two cultivars had similar nutritional composition.
Glyphosate-tolerant wheat	Obert et al., 2004	Levels of proximates, amino acids, fatty acids, minerals, vitamins, secondary metabolites and anti-nutrients in two types of wheat forage similar, grain also equivalent to commercial wheat cultivars.
Glyphosate-tolerant wheat	Peterson et al., 2008	Two cultivars had similar nutritional composition.
	Oilseed crops, products	
Soyabeans		
B.t. soyabean seed, forage, fractions	Berman et al., 2009	The compositions of the two types of soyabean products were similar.
Glyphosate-tolerant soyabeans	Berman et al., 2010	The compositions of the two types of soyabeans were similar.
Glyphosate-tolerant soyabeans	Cromwell et al., 2002	The compositions of the two types of soyabeans and the processed soyabean meals were similar.
Glyphosate-tolerant soyabeans	Flachowsky et al., 2007	The compositions of the two types of soyabeans were similar.
Glyphosate-tolerant soyabeans herbicide	Herman et al., 2011	The compositions of the two types of soyabeans were similar.

Table 10.1 Continued

Food/Feedstuff	Reference	Reported findings
Glyphosate-tolerant soyabeans	McCann et al., 2005	Beans grown in different locations over three generations were compositionally equivalent.
Glyphosate-tolerant soyabeans	Padgette et al., 1996	The compositions of the two types of soyabeans and extracted meals were similar.
Glyphosate-tolerant soyabeans	Sanden et al., 2006	The compositions of the two types of soyabeans were similar.
Glyphosate-tolerant soyabeans	Sissener et al., 2009	The compositions of the two types of soyabeans were similar.
Glyphosate-tolerant soybeans	Taylor et al., 1999.	The nutritional compositions of the two types of soyabeans were similar.
Oil from glyphosate-tolerant soyabeans	Padgette et al., 1996	The compositions of the two types of oil were similar.
Meal from glyphosate-tolerant soyabeans	Chukwudebe et al., 2012	The compositions of meals from the two types of soyabeans were similar.
Meal from glyphosate-tolerant soyabeans	Hammond et al., 1996	No significant differences in nutritional composition between soya types.
Meal from dehulled B.t. soyabeans	Kan et al., 2001	No significant differences in nutritional composition between soya types.
Meal from glyphosate-tolerant soyabean meal and B.t. maize	Świątkiewicz et al., 2013a	No significant differences in nutritional composition between soyabean cultivars or between maize cultivars.
Meal from glyphosate-tolerant soybeans	Zhu et al., 2004	The nutritional compositions of the two types of soyabeans were similar, including amino acids.
Canola		
Glyphosate-tolerant canola seed	Daun, 2004	GM cultivar had a higher oil content and a lower crude protein content.
Glyphosate-tolerant canola	Maertens et al., 1996	No significant differences in nutritional composition between canola types.

Continued

Table 10.1 Continued

Food/Feedstuff	Reference	Reported findings
Glyphosate-tolerant canola	Stanford et al., 2003	No significant differences in nutritional composition between canola types.
Glyphosate-tolerant canola	Stanisiewski et al., 2002	The nutritional compositions of the two types of canola were similar.
Glyphosate-tolerant canola	Taylor et al., 2004a	The nutritional compositions of the two types of canola were similar.
Oil from glyphosate-tolerant canola	Daun, 2004	No difference in the contents of erucic acid, unsaturated fatty acids or linolenic acid between the two types.
Oil from glyphosate-tolerant canola	FSANZ, 2003	Oil derived from both cultivars was comparable in composition.
Oil from GM canola (range of cultivars)	FSANZ, 2003	Contents of 11 key fatty acid components are comparable in all of the oils tested.
Meal from glyphosate-tolerant canola	Aalhus et al., 2003	The nutritional compositions of the two types of meal were similar.
Meal from glyphosate-tolerant canola	Brown et al., 2003	The nutritional compositions of the two types of meal were similar.
Meal from glyphosate-tolerant canola	Caine et al., 2007	The nutritional compositions of the two types of meal were similar. Contents of glucosinolates also similar.
Meal from glyphosate-tolerant canola	Stanford et al., 2003	The nutritional compositions of the two types of meal were similar.
Meal from glyphosate-tolerant canola	Taylor et al., 2004a	The nutritional compositions of the two types of meal were similar.
Cottonseed		
B.t. cottonseed	Berberich et al., 1996.	The nutritional compositions of the two types of cottonseed were similar
B.t., glyphosate-tolerant cottonseed	Bertrand et al., 2005	Nutrient composition of GM and conventional cultivars similar, also digestibility in cattle.
B.t., glyphosate-tolerant cottonseed	Castillo et al., 2004	Nutrient composition of GM and conventional cultivars similar.
B.t.-cottonseed	Hamilton et al., 2004	Cottonseed and cottonseed oil from Bollgard II cotton were comparable in composition to a conventional control cotton line and commercial cultivars.
Glyphosate-tolerant cottonseed	Nida et al., 1996	Nutrient composition of GM and conventional cultivars similar.
B.t. cottonseed	Singh et al., 2003	Nutrient composition of GM and conventional cultivars similar

Table 10.1 Continued

Food/Feedstuff	Reference	Reported findings
Cry1ac+ cr2ab cottonseed	Singhal et al., 2006	Nutrient composition of GM and conventional cultivars similar.
Range of GM cottonseed meals	Li et al., (2008)	Considerable variation in proximate composition, amino acids, and gossypol concentrations. Concluded that the GM cottonseed meals were not significantly different from a conventional control and reference cottonseed meals.
Oil from B.t. cottonseed	Berberich et al., 1996.	The fatty acid compositions of the two types of cottonseed were similar.
Meal from glyphosate-tolerant cottonseed	Sissener et al., 2009	Nutrient composition of GM and conventional cultivars similar.
Meal from glyphosate-tolerant cottonseed	Sissener et al., 2011	Nutrient composition of GM and conventional cultivars similar.
B.t. cottonseed	Kumar and Singhal, 2004	Minor differences found in nutritional composition between GM and conventional cultivars.
	Fruits, vegetables	
Glyphosate-tolerant lucerne (alfalfa)	McCann et al., 2006	Forage from GM cultivar equivalent in composition to forage from conventional alfalfa.
Virus-resistant papaya	Jiao et al., 2010	Nutritional composition of two types of papayas was similar.
Virus-resistant papaya	Roberts et al., 2008	No significant difference in nutritional composition.
Virus-resistant papaya	Wall, 2006	No significant difference in nutritional composition.
Virus-resistant papaya	Tripathi et al., 2011	No significant difference in nutritional composition.
B.t. potatoes	El Sanhoty, 2004	The nutritional compositions of the two types of potatoes were similar.
Virus-resistant and insect-resistant (B.t.) potatoes	Rogan et al., 2000	Tubers produced by insect- and virus-resistant cultivars (6) were substantially equivalent to conventional potatoes in their contents of key nutrients, anti-nutrients, macronutrients and minor nutrients.
Glufosinate-tolerant potatoes	Rhee et al., 2005	The nutritional compositions of the two types of potatoes were similar.
GM potatoes (range)	Shepherd et al., 2006	Substantial equivalence of key nutrients and anti-nutrients in genetically engineered and conventional potatoes.
Reduced solanine content potatoes	Langkilde et al., 2012	The nutritional compositions of the two types of potatoes were similar.
Inulin synthesising potatoes, silage	Böhme et al., 2005	The nutritional compositions of the two types silage were similar but metabolisable energy value was lower in the GM silage.

Continued

Table 10.1 Continued

Food/Feedstuff	Reference	Reported findings
Glufosinate-tolerant sugar beet and leaf silage	Böhme et al., 2001	The nutritional compositions of the two types of sugar beet products were similar, also digestibility.
Glyphosate-tolerant sugar beet, fodder beet	Weisbjerg et al., 2001	The nutritional compositions of the two types of sugar beet and fodder beet were similar.
Tomatoes (tobacco rab gene)	Venneria et al., 2008	Total antioxidant activity and naringenin content were lower in GM tomatoes. Levels of other nutrients were similar.

Table 10.2 Reported effects of dietary inclusion of conventional or GM feedstuffs on the productivity of animals, poultry and fish.

Animal species	Reference	GM feedstuff	Reported effects
		Meat/lab animals	
Beef cattle	Aulrich et al., 2001	B.t. maize	No significant difference in nutrient digestibility.
Beef cattle	Berger et al., 2002	Herbicide-tolerant maize	No significant differences in growth performance.
Beef cattle	Berger et al., 2003	B.t. maize	No significant differences in growth performance or body score.
Beef cattle	Daenicke et al., 1999	B.t. maize silage	No significant differences in digestibility, weight gain or dry matter intake.
Beef cattle	Erickson et al., 2003	Glyphosate-tolerant maize	No significant differences in growth performance.
Beef cattle	Folmer et al., 2002	B.t. maize, maize residues, maize silage	No significant differences in growth performance.
Beef cattle	Hendrix et al., 2000	B.t. maize silage, maize residues	No significant differences in growth performance of cows, improved growth performance of steer calves.
Beef cattle	Kerley et al., 2001	B.t. maize	No significant differences in growth performance.
Beef cattle	Petty et al., 2001	Glyphosate-tolerant maize, silage	No significant differences in growth performance.
Beef cattle	Russel et al., 2000	B.t. maize, crop residues	No significant differences in growth performance.
Beef cattle	Russel et al., 2001	B.t. maize, crop residues	No significant differences in growth performance.
Beef cattle	Simon et al., 2002.	Herbicide-tolerant maize	No significant differences in growth performance.

Table 10.2 Continued

Animal species	Reference	GM feedstuff	Reported effects
Beef cattle	Van der Pol et al., 2002	Corn rootworm-protected maize	No significant differences in growth performance.
Buffaloes	Singh et al., 2003	B.t. cottonseed	No significant differences in milking performance.
Sheep	Aulrich et al., 2001	B.t. maize	No significant difference in nutrient digestibility.
Sheep	Böhme et al. 2001	Pat herbicide-tolerant sugar beet and leaf silage	No significant differences in digestibility of nutrients.
Sheep	Barrière et al., 2001	B.t. maize, maize silage	No significant differences in growth performance.
Sheep	Daenicke et al., 1999	B.t. maize silage	No significant differences in digestibility, weight gain or dry matter intake.
Sheep	Hartnell et al., 2005	Glyphosate-tolerant sugar beet, sugar beet pulp, fodder beet	Digestibilities and feeding values of GM fodder beet, sugar beet, and beet pulp similar to conventional cultivars.
Sheep	Hvelplund and Weisbjerg, 2001	Herbicide-tolerant canola	No significant differences in digestibility.
Sheep	Stanford et al., 2003	Glyphosate-tolerant sugar beet, sugar beet pulp, fodder beet	No significant differences in feed intake or growth performance (lambs).
Sheep	Weisbjerg et al., 2001	Glyphosate-tolerant sugar beet, sugar beet pulp, fodder beet	No significant differences in digestibility.
Pigs	Aalhus et al., 2003	Meal from glyphosate-tolerant canola	No significant difference in growth performance.
Pigs	Aulrich et al., 2005	Glyphosate-tolerant soybean meal	No significant difference in growth performance.
Pigs	Böhme and Alrich, 1999	Herbicide-tolerant maize	No significant effects on nutrient digestibility.
Pigs	Böhme et al. 2001	Herbicide-tolerant sugar beet	No significant difference in nutrient digestibility.
Pigs	Bressner et al., 2003	B.t. maize	No significant difference in growth performance.
Pigs	Cromwell et al., 2002	Herbicide-tolerant soyabean meal	No significant difference in growth performance.
Pigs	Cromwell et al., 2005	Glufosinate-tolerant rice	No significant differences in feed intake or growth performance.
Pigs	Custodio et al., 2004	B.t. maize	No significant difference in growth performance.

Continued

Table 10.2 Continued

Animal species	Reference	GM feedstuff	Reported effects
Pigs	Custodio et al., 2006	B.t. maize	No significant difference in growth rate but feed conversion efficiency better with control maize. P digestibility higher with control maize.
Pigs	Fischer et al., 2002	Herbicide-tolerant maize	No significant difference in growth performance.
Pigs	Fischer et al., 2003	B.t. maize	No significant differences in growth performance.
Pigs	Gaines et al., 2001b	Herbicide-tolerant maize	No significant difference in energy digestibility.
Pigs	Hyun et al., 2004	Herbicide-tolerant maize	No significant difference in growth performance.
Pigs	Hyun et al., 2005	Insect-tolerant (MON 863) maize	Equivalent growth performance and carcass quality with GM and non-transgenic maize.
Pigs	Peterson et al., 2008	Glyphosate-tolerant wheat	No significant effect of GM or near isogenic cultivar or 4 commercial types on growth performance.
Pigs	Piva et al., 2001	B.t. maize	Significantly higher weight gain with B.t. maize (lower mycotoxin contamination), no significant effect on feed : gain ratio.
Pigs	Reuter et al., 2001a	B.t. maize	No significant difference in nutrient digestibility, dry matter intake or weight gain.
Pigs	Reuter et al., 2002b	B.t. maize	Digestibility of all measured parameters similar, metabolisable energy contents similar.
Pigs	Stanisiewski et al., 2001	B.t. maize	No significant effect on growth performance.
Pigs	Stein et al., 2004	B.t. maize	No significant difference in growth performance.
Pigs	Stein et al., 2009	Double-stacked maize	No significant difference in growth performance.
Pigs	Świątkiewicz et al., 2013a, b	Glyphosate-tolerant soyabean meal and B.t. maize	No significant effects on growth or reproductive performance.
Pigs	Weber and Reichert, 2001	B.t. maize	No significant difference in growth performance.
Broilers	Aeschbacher et al., 2001	B.t. maize	No significant difference in growth performance or metabolisable energy value.

Table 10.2 Continued

Animal species	Reference	GM feedstuff	Reported effects
Broilers, laying hens	Aeschbacher et al., 2005	B.t. maize	No significant effects of diet on metabolic data or growth or laying performance.
Broilers	Aulrich et al., 2001	B.t. maize	No significant difference in digestibility.
Broilers	Brake and Vlachos, 1998	B.t. maize	No significant differences in growth performance or livability. Feed : gain ratio significantly lower with B.t. maize.
Broilers	Brake et al., 2003	B.t. maize	No significant difference in growth performance.
Broilers	Brake et al., 2003	B.t. + Glufosinate-tolerant maize	No significant difference in growth performance.
Broilers	Elangovan et al., 2003	B.t.-cottonseed meal	No significant difference in growth performance.
Broilers	Gaines et al., 2001a	B.t. maize	No significant difference in growth performance.
Broilers	Gaines et al., 2001c	Glyphosate-tolerant maize	No significant difference in growth performance.
Broilers	Halle et al., 1998	B.t. maize	No significant difference in growth performance.
Broilers	Halle et al., 1998	B.t. potatoes	No significant difference in growth performance.
Broilers	Hammond et al., 1996	Glyphosate-tolerant soyabean meal	No significant differences in growth performance or livability.
Broilers	Kan et al., 2001	Meal from dehulled B.t. soyabeans	No significant difference in growth performance.
Broilers	Kan and Hartnell, 2004a	Glyphosate-tolerant wheat	No significant difference in growth performance, but lower carcass yield with control wheat.
Broilers	Kan and Hartnell, 2004b	B.t. soyabean meal	No significant difference in growth performance.
Broilers	Mandal et al., 2004	Bollgard (BG) II, non-BG II or commercial cottonseeds	Body weight gain, feed intake, feed conversion efficiency, nutrient utilisation, certain blood constituents and carcass traits were not significantly affected by diet.
Broilers	Mireles et al., 2000	B.t. maize	No significant difference in growth performance or metabolisable energy value.

Continued

Table 10.2 Continued

Animal species	Reference	GM feedstuff	Reported effects
Broilers	Piva et al., 2001	B.t. maize	Significantly higher weight gain with B.t. maize (lower mycotoxin contamination), no significant effect on feed : gain ratio.
Broilers	Reuter et al., 2001b	B.t. maize	Improved growth rate with B.t. maize.
Broilers	Řehout et al., 2009	B.t. maize	No significant effect on growth performance.
Broilers	Sidhu et al., 2000	Glyphosate-tolerant maize	No significant effect on growth performance.
Broilers	Stanisiewski et al., 2002	Glyphosate-tolerant canola	No significant difference in growth performance.
Broilers	Tan et al., 2012	Glyphosate-tolerant maize	No treatment differences detected for any measured performance, carcass parameters or microbial population of ileum and cecum.
Broilers	Taylor et al., 2001a	Insect resistant, glyphosate-tolerant maize	No significant difference in growth performance.
Broilers	Taylor et al., 2001b	Glyphosate-tolerant maize	No significant difference in growth performance.
Broilers	Taylor et al., 2002a	B.t. maize	No significant difference in growth performance.
Broilers	Taylor et al., 2002b	Insect resistant, glyphosate-tolerant maize	No significant difference in growth performance.
Broilers	Taylor et al., 2003a	B.t. maize, glyphosate-tolerant	No significant effect on growth performance.
Broilers	Taylor et al., 2003b	B.t. maize, glyphosate-tolerant	No significant effect on growth performance.
Broilers	Taylor et al., 2003c	B.t. maize, glyphosate-tolerant	No significant effect on growth performance.
Broilers	Taylor et al., 2004a	Glyphosate-tolerant canola meal	No significant difference in growth performance.
Broilers	Taylor et al., 2004b	Glyphosate-tolerant, B.t.-tolerant maize	No significant differences in growth performance.
Broilers	Taylor et al., 2005	B.t. maize + Gt maize	No significant differences in growth performance.
Broilers	Tony et al., 2002	B.t. maize	No significant differences in growth performance.
Broilers	Yonemochi et al., 2002	B.t. maize	Growth performance in starter phase better with GM cultivar.
Laying hens	Daenicke et al., 1999	B.t. maize	No significant difference in laying performance.
Laying hens	Aeschbacher et al., 2001	B.t. maize	No significant difference in laying performance or metabolisable energy value.

Table 10.2 Continued

Animal species	Reference	GM feedstuff	Reported effects
Laying hens	Aulrich et al., 2001	B.t. maize	No significant difference in nutrient digestibility.
Quail	Halle et al., 2004	B.t. maize	No significant difference in growth or layer performance.
Rabbits	Maertens et al., 1996	Glyphosate-tolerant canola	No significant difference in digestibility between canola types.
Rabbits	Chrastinová et al., 2002	B.t. maize	No significant difference in growth performance.
Catfish	Hammond et al., 1996	Glyphosate-tolerant soyabean meal	No significant differences in feed : gain ratio or livability. Improved growth with GM soya.
Channel catfish	Li et al. 2006.	Range of GM cottonseed meals.	Rate of weight gain, efficiency of feed conversion, survival rate and fillet composition similar.
Rainbow trout	Chainark et al., 2006	Meal from glyphosate-tolerant soybeans	No significant effect on growth performance.
Rainbow trout	Brown et al., 2003	Meal from glyphosate-tolerant canola	No significant effect on growth performance.
Atlantic Salmon	Hemre et al., 2005	Glyphosate-tolerant soyabean meal	No significant effect on growth performance.
Rats	Appenzeller et al., 2009	1507x59122 maize, insect-resistant and tolerant to glufosinate.	No toxicologically significant differences observed in nutritional performance variables.
Rats	Chrenková et al., 2002	Glyphosate-tolerant maize	Similar values for protein efficiency ratio, nitrogen digestibility and net protein utilisation.
Rats	Delaney et al., (2013)	4114 maize insect-resistant and tolerant to glufosinate.	No biologically significant, treatment related differences in body weight, feed consumption, clinical pathology parameters (hematology, blood chemistry, urinalysis or organ weight) were observed.
Rats	He et al., 2008	Maize tolerant to glufosinate and insects	No significant differences in body weight and feed utilization were observed between rats consuming diets formulated with 59122 and 091 control corn. Statistical differences ($p < 0.05$) were observed in certain hematology and serum chemistry response variables between rats consuming diets formulated with 59122 or 091.

Continued

Table 10.2 Continued

Animal species	Reference	GM feedstuff	Reported effects
Rats	MacKenzie et al., 2007	B.t. maize grain with pat gene	No significant differences observed in nutritional performance variables.
Rats	Zhu et al., 2004	Meal from glyphosate-tolerant soybeans	No significant effect on growth performance.
Milk			
Dairy cows	Hammond et al., 1996	Meal from glyphosate-tolerant soyabeans	Production of fat-corrected milk greater with GM soya.
Dairy cows	Faust and Miller, 2000	B.t. maize chopped	No significant differences in dry matter intake or milk production.
Dairy cows	Rutzmoser et al., 1999	B.t. maize silage	No significant differences in dry matter intake or milk production.
Dairy cows	Donkin et al., 2000	Herbicide-tolerant maize, maize silage	No significant differences in dry matter intake or milk production.
Dairy cows	Faust and Spangler, 2000	B.t. maize silage, maize silage	No significant difference in milk production.
Dairy cows	Castillo et al., 2001	Herbicide-tolerant cottonseed, B.t. cottonseed	No significant differences in body condition or milk yield.
Dairy cows	Castillo et al., 2004	Glyphosate-tolerant cottonseed, B.t. cottonseed	No significant differences in body condition or milk yield.
Dairy cows	Calsamiglia et al., 2007	B.t., glyphosate-tolerant maize silage	No significant difference in milk production.
Dairy cows cattle	Folmer et al., 2002	B.t. maize, maize residues, maize silage	No significant differences in milk production.
Dairy cows	Folmer et al., 2002	Glyphosate-tolerant maize silage	No significant differences in milk production.
Dairy cows	Ipharraguerre et al., 2003	Herbicide-tolerant maize, maize silage	No significant difference in milk production.
Dairy cows	Grant et al., 2002	Herbicide-tolerant maize, maize silage	Intake of maize silage and milk production reduced significantly (ascribed to silage quality).
Dairy cows	Grant et al., 2003	Glyphosate-tolerant maize -silage and grain	Lower dry matter intake of cows fed the glyphosate-tolerant diet, resulting in a lower milk production.
Dairy cows	Grant et al., 2003	Corn rootworm protected corn -silage and grain	No significant difference in milk production.
Dairy cows	Donkin et al., 2003	B.t. maize, glyphosate tolerant maize silage	No significant differences in dry matter intake or milk production.
Dairy cows	Singhal et al., 2006	Cry1ac+ cr2ab cottonseed.	No significant differences in dry matter intake or milk production.

Table 10.2 Continued

Animal species	Reference	GM feedstuff	Reported effects
Dairy cows	Steinke et al. 2010	B.t.-maize, maize silage	Similar dry matter intake and milk yield.
Dairy cows	Weisbjerg et al., 2001	Glyphosate-tolerant sugar beet, sugar beet pulp, fodder beet	No significant differences in milk production.

Oilseed crops

Published data on the chemical analyses of oilseed crops are shown in Table 10.1. As outlined in Chapter 5, these data show that conventional and GM cultivars of soyabeans, canola and cottonseed, their oils and the meals obtained after oil-extraction are similar in nutritional composition, as are the contents of anti-nutritive factors.

Feeding studies have also been conducted with the different cultivars of oilseeds and their oil-extracted meals (Table 10.2). In general, these results indicate similar growth and productivity when fed GM or non-GM food and feed sources, confirming the conclusion of nutritional equivalence. These studies have included analysis of the composition and quality of meat, milk and eggs produced by livestock and poultry fed diets containing GM or non-GM oilseed products. No deleterious effects have been reported (Table 10.3), or on ileal or cecal microflora (e.g. Tan et al., 2012).

Soyabeans

Soyabeans are a major protein source for the feeding of poultry and livestock. Therefore the nutritional composition of GM cultivars has been studied by several researchers. Their findings indicate nutritional equivalence of GM and related conventional cultivars (Table 10.1). Chukwudebe et al. (2012) evaluated the nutritional and health status of rats fed diets containing GM soyabeans CV127, a near isogenic conventional soyabean cultivar (Conquista) and two other conventional cultivars (Monsoy and Coodetec) at incorporation levels of 110 and 330 g/kg. They concluded that there were no quantitative differences among these cultivars in proximate and nutrient contents including proteins, amino acids, anti-nutrients and nutritional cofactors.

The composition of the oil and meal after oil extraction has also been shown to be similar in GM and conventional cultivars.

Hammond et al. (1996) conducted feeding studies with rats, broiler chickens, catfish and dairy cows to compare glyphosate-tolerant soyabeans/meal and conventional soyabeans/meal as dietary protein sources. Both beans and processed meal were incorporated into the diets at the same concentrations as used commercially. Growth, feed conversion (rats, catfish, broilers), fillet composition (catfish), and breast muscle and fat pad weights (broilers) were compared for animals fed the conventional and GM cultivars. Milk production, milk composition, rumen fermentation and nitrogen digestibility were also compared in dairy

Table 10.3 Reported effects of dietary inclusion of conventional or GM feedstuffs on the composition of meat, milk and eggs.

Animal species	Reference	GM feedstuff	Reported effects
Meat			
Beef cattle	Aulrich et al., 2001	B.t. maize	No significant differences in carcass composition.
Beef cattle	Berger et al., 2002	Herbicide-tolerant maize	No significant differences in carcass characteristics.
Beef cattle	Berger et al., 2003	B.t.-tolerant maize	No significant differences in carcass characteristics.
Beef cattle	Daenicke et al., 1999	B.t. maize silage	No significant differences in carcass characteristics.
Beef cattle	Erickson et al., 2003	B.t. maize silage	No significant differences in carcass characteristics.
Beef cattle	Simon et al., 2002.	Herbicide-tolerant maize	No significant differences in carcass characteristics or meat composition.
Sheep (lambs)	Stanford et al., 2003	Glyphosate-tolerant canola	No significant differences in carcass composition.
Dairy cattle	Hammond et al., 1996	Meal from glyphosate-tolerant soyabeans	No significant differences in milk composition.
Dairy cows	Donkin et al., 2000	Herbicide-tolerant maize, maize silage	No significant differences in milk composition.
Dairy cows	Folmer et al., 2002	B.t. maize silage, maize silage	No significant differences in milk composition.
Dairy cows	Castillo et al., 2001	Herbicide-tolerant cottonseed, B.t. cottonseed	No significant difference in milk composition.
Dairy cows	Castillo et al., 2004	Glyphosate-tolerant cottonseed, B.t. cottonseed	No significant differences in milk composition.
Dairy cows	Calsamiglia et al., 2007	B.t., herbicide-tolerant maize silage	No significant difference in milk composition.
Dairy cows	Ipharraguerre et al., 2003	Herbicide-tolerant maize, maize silage	No significant difference in milk composition.
Dairy cows	Donkin et al., 2003	B.t. maize, glyphosate-tolerant maize silage	No significant differences in milk composition.
Dairy cows	Grant et al., 2002	B.t. + glyphosate-tolerant maize grain, silage	No significant differences in milk composition.
Dairy cows	Grant et al., 2003	Glyphosate-tolerant maize -silage and grain	No significant differences in milk composition.
Dairy cows	Grant et al., 2003	Corn rootworm protected corn -silage and grain	No significant differences in milk composition.

Table 10.3 Continued

Animal species	Reference	GM feedstuff	Reported effects
Dairy cows	Singhal et al., 2006	Cry1ac+ cr2ab cottonseed.	No significant differences in milk composition.
Dairy cows	Steinke et al. 2010	B.t. maize, maize silage	No consistent effects on milk composition or body condition.
Dairy cows	Weisbjerg et al., 2001	Glyphosate-tolerant sugar beet, sugar beet pulp, fodder beet	No significant differences in milk composition.
Pigs	Bressner et al., 2003	B.t. maize	No significant differences in carcass traits.
Pigs	Cromwell et al., 2002	Herbicide-tolerant soyabean meal	Carcass traits of barrows were similar for the two types of soyabean meal. Longissimus muscle samples from barrows fed conventional soyabean meal tended to have less fat than those fed Roundup Ready soyabean meal, but water, protein, and ash were similar. Sensory scores of cooked longissimus muscles were not influenced by diet.
Pigs	Cromwell et al., 2005	Glufosinate-tolerant rice	No significant differences in carcass traits.
Pigs	Fischer et al., 2002	Herbicide-tolerant maize	No significant differences in carcass measurements.
Pigs	Custodio et al., 2006	B.t. maize	No significant effect on carcass composition or meat quality.
Pigs	Fischer et al., 2003	B.t. maize	No significant differences in carcass measurements.
Pigs	Hyun et al., 2005	Insect-tolerant (MON 863) maize	Equivalent carcass quality with GM and non-transgenic maize.
Pigs	Stein et al., 2009	B.t., glufosinate-resistant maize.	No significant differences in dressing percentage, loin muscle area, backfat thickness or carcass lean content.
Pigs	Świątkiewicz et al., 2011, 2013a, b	Glyphosate-tolerant soyabean meal and B.t. maize	No significant effects on carcass or meat quality, or on health and immunological indices.

Continued

Table 10.3 Continued

Animal species	Reference	GM feedstuff	Reported effects
Broilers, laying hens	Aeschbacher et al., 2005	B.t. maize	No significant effects on carcass or meat quality.
Broilers	Sidhu et al., 2000	Herbicide-tolerant maize	No significant differences in carcass composition.
Broilers	Kan et al., 2001	Meal from dehulled B.t. soyabeans	No significant differences in carcass composition.
Broilers	Kan and Hartnell, 2004	Glyphosate-tolerant wheat	Lower carcass yield for birds fed the conventional wheat.
Broilers	Řehout et al., 2009	B.t. maize	No significant effect of maize type on meat quality or blood biochemistry.
Broilers	Taylor et al., 2004a	Glyphosate-tolerant canola	No significant differences in body composition.
Broilers	Brake and Vlachos, 1998	B.t. maize	No significant differences in body composition.
Channel catfish	Li et al., 2006.	Range of GM cottonseed meals.	Fillet composition similar with all types of cottonseed meal.

cows. In all studies, measured variables were similar for animals fed both types of soyabeans/meal.

Soyabean meal from glyphosate-tolerant or conventional soyabeans was compared in diets for growing-finishing pigs (Cromwell et al., 2002).The compositions of the two types of soyabeans and of the processed soyabean meal were found to be similar (Table 10.1). The rate and efficiency of weight gain, backfat thickness, longissimus area and carcass lean percentage were not significantly different for pigs fed diets containing conventional or GM soyabean meal (Table 10.2). Sensory scores of cooked longissimus muscles were not influenced by diet (Table 10.3). Similar findings have been reported by Zhu et al. (2004), Flachowsky et al. (2007), Herman et al. (2011) and Tan et al. (2012). Sissener et al. (2009, 2011) reviewed the available data on GM soyabeans as a protein source for a range of fish. The reviewed data indicated that both types of cultivar had similar nutritional quality as feed ingredients for all of the fish species studied.

The available findings indicate that GM soyabeans, soya oil and soyabean meal are essentially equivalent in composition and nutritional value to conventional soyabeans, soya oil and soyabean meal, respectively.

Canola

The database on canola seed is quite large, with several researchers reporting no significant differences in nutritional composition between GM and conventional cultivars (Table 10.1). However, Daun (2004) reported data relating to canola seed from the crop years 1996–2001, indicating that GM canola seed in some crop years had a higher oil content and a lower crude protein content than conventional non-GM canola seed.

A large study on canola oil was conducted by the Australia/New Zealand Food Authority (FSANZ, 2003). The samples included control varieties (e.g. Drakkar), different generations of male sterile lines (Ms1 and Ms8), different generations of fertility restorer lines (Rf1, Rf2 and Rf3), multiple backcrosses of Ms and Rf lines in different canola varieties and unrestored (Ms/control) and restored (Ms/Rf) hybrids. The seed samples were collected from plants grown at locations in Belgium, France, Sweden, Canada and the United Kingdom. The data showed that the contents of 11 key fatty acid components, including erucic acid (C22:1, considered an undesirable fatty acid), were comparable in all of the oils tested, both from GM and conventional (non-GM) cultivars. Variation due to environmental conditions was found to be greater than the variation between oil from GM and conventional cultivars. The agency concluded, therefore, that oil derived from the GM canola lines (T45, Topaz 19/2, Ms1, Ms8, Rf1, Rf2, Rf3 and their crosses), was equivalent to oil from conventional canola in terms of safety and nutritional properties. Daun (2004) reported that the oil from GM canola seed tended to be slightly more unsaturated than the oil from non-GM canola seed, and there was no difference between the two cultivars studied in the contents of erucic acid, linolenic acid or type of unsaturated fatty acids.

The meal from oil-extracted GM canola seed has been shown to be comparable in nutritional composition to the meal obtained from conventional canola cultivars (Table 10.1). In addition, the results of feeding tests on swine (Aalhus et al., 2003; Caine et al., 2007), trout (Brown et al., 2003), sheep (Stanford et al., 2003) and chickens (Stanisiewski et al., 2002; Taylor et al., 2004a, b, 2007) confirm that the meals obtained from GM and conventional cultivars of canola are of similar nutritive value. Sissener et al. (2011) reviewed the available data on glyphosate-tolerant canola meal as a protein source in fish feed and found that both types of cultivar had similar nutritional quality.

Cottonseed

Nida et al. (1996) reported that the composition of seed from glyphosate-tolerant cotton was equivalent to that from conventional cottonseed (Table 10.1). The nutrients measured were protein, fat, fibre, carbohydrate, calories, moisture, ash, amino acids and fatty acids. Their analysis included anti-nutrients (gossypol, cyclopropenoid fatty acids and aflatoxin). In addition, the fatty acid profile was determined.

Hamilton et al. (2004) evaluated the composition and nutrient content of GM Bollgard II (B.t.-protected) cottonseed and conventional cottonseed cultivars from a total of 14 US field sites over two years. Results showed that the cottonseed (Table 10.1) and cottonseed oil from Bollgard II cotton were comparable in their composition to a conventional control cotton line and other commercial varieties.

The compositional data were supported by nutritional studies conducted with dairy cows, catfish and quail (Table 10.2). Whole cottonseed is used commonly in some countries as a dietary component for dairy cows. For example, Castillo et al. (2004) reported that feed intake, milk yield, milk composition, body weight and body condition score did not differ when cows were fed diets based on glyphosate-tolerant, B.t. or conventional cottonseed. It was concluded that cottonseed from the GM cultivars used in this study resulted in similar performance in lactating dairy cows when compared to non-GM cultivars and reference cottonseed.

Composition of the oil from GM cultivars of cottonseed has been studied by regulatory authorities as part of the approval process. The UK Advisory Committee on Novel Foods and Processes (1998) issued a favourable opinion on the equivalence of cottonseed oil and food ingredients derived from glyphosate-tolerant cotton. The data included an extensive database of compositional analyses of cottonseed from field trials in the USA. The refined oil from RRC line 1445 and the isogenic parent line Coker 312 were found to be comparable in quality to commercially processed cottonseed oil from non-GM sources. These data indicate that the levels of the important fatty acids, as well as the toxic cyclopropenoid fatty acids, were comparable in refined cottonseed oil fractions produced from RRC line 1445 and the control cultivar. There was no detectable gossypol in the refined oils, and levels of alpha-tocopherols were also comparable in both cultivars and were within the range published for other commercial cultivars. The Committee reported also that genetic modification to provide herbicide tolerance did not alter the processing characteristics of the cottonseed or the quality of refined oil. Modification had no effect on the amino acid profile of the seeds and the Committee reported also that the values obtained for the major toxicants were statistically equivalent between the two cultivars throughout the study period.

The composition of oil and fatty acids in B.t. and conventional cotton (*G. hirsutum*) cultivars (JKCH-2245, K-5038, RCH-2, NCEN-3R, JK-Ishwar and JKCH-266) was studied by Nagappa and Khadi (2011). On average, the oil content of B.t. cultivars (174.3 g/kg) was similar to that of conventional cultivars (174.0 g/kg). The contents of unsaturated fatty acids were also similar. Based on their findings the researchers concluded that there is no significant difference in oil content or fatty acid composition of oil in B.t. cotton cultivars in comparison with conventional non-GM cultivars.

The above conclusions confirm the findings of other researchers (e.g. Berberich et al., 1996; Nida et al., 1996). Berberich et al. (1996) reported that insect-protected B.t. cultivars and a parental control contained levels of nutrients (protein, fat, carbohydrate, moisture, ash, amino acids and fatty acids) comparable to those of other commercial varieties. The levels of the anti-nutrients gossypol, cyclopropenoid fatty acids and aflatoxin in the seed from the insect-protected lines were similar to or lower than the levels present in the parental variety or reported for other commercial varieties.

Chemical analyses of various cottonseed cultivars have been conducted. Li et al. (2008) reported considerable variation in proximate composition, amino acids and gossypol concentrations. However, they found that the GM cottonseed meals were not significantly different in nutritional composition from a conventional control and reference cottonseed meals, confirming the previous findings of Berberich et al. (1996), Bertrand et al. (2005), Castillo et al. (2004), Nida et al. (1996) and Singhal et al. (2006). The fatty acid composition of the oil and meal from oil-extracted cottonseeds were found to be equivalent in GM and conventional cultivars (Berberich et al., 1996; Sissener et al., 2009, 2011).

Elangovan et al. (2003) compared the growth of broiler chickens fed diets containing 100 g/kg meal from GM (insect-resistant) cottonseed or non-GM low-gossypol cottonseed meal from day-old to six weeks of age. Feed intake, feed conversion efficiency, nitrogen retention, dressing percentage, weights of liver and heart and blood biochemical constituents did not differ significantly between the dietary treatments.

Li et al. (2008) conducted three experiments with channel catfish to assess the relative nutritional quality of cottonseed meals from several GM cotton crops (Roundup Ready Flex, Bollgard Roundup Ready, Bollgard II Roundup Ready, and Bollgard II Roundup Ready Flex). Although there was considerable variation (Table 10.1) in proximate composition, amino acids and gossypol concentrations, the researchers concluded that the GM cottonseed meals were not significantly different from a conventional control and reference cottonseed meals. Growth rate of juvenile catfish, efficiency of feed conversion, survival rate and fillet composition were found to be similar. Based on the findings, the researchers concluded that the GM cottonseed meals were nutritionally equivalent to conventional cottonseed meals. Sissener et al. (2009, 2011) reviewed the available data on glyphosate-tolerant cottonseed meal as a protein source for a range of fish and concluded that both types of cultivar had similar quality as feed ingredients for fish.

These data indicate that seed, oil and meal from GM cotton cultivars are as nutritious as those from the parental and other conventional non-GM cotton cultivars that are available commercially.

Fruits and vegetables

Several GM cultivars of fruits and vegetables have been approved for food and feed use in several countries, though not all are being marketed currently (Chapter 6). The list of approved GM cultivars includes cantaloupe, plums and papaya but only one type of GM fruit, the Hawaiian papaya with resistance to ringspot virus infection, is currently available to consumers. This virus-resistant cultivar was approved by the regulatory authorities in North America and Japan, based on evidence of safety and nutritional equivalence with conventional cultivars of papaya. The list of approved GM cultivars of vegetables includes lucerne (alfalfa), potatoes, squash, sugar beet, sweet corn (maize) and tomatoes.

Lucerne (alfalfa)

McCann et al. (2005) showed that that the forage from glyphosate-tolerant lucerne J101 × J163 was equivalent in composition to forage from conventional alfalfa. While some differences were found, the data overall were within the range observed for this crop and the differences were therefore not considered to be of biological significance.

Papaya

The nutritional composition of GM Rainbow papaya at three ripening stages was reported by Tripathi et al. (2011). Overall, the conventional and GM fruits were similar in nutritional composition when compared at three ripeness stages. These data confirm earlier findings (Table 10.1) on the composition of GM papaya (Jiao et al., 2010; Roberts et al., 2008; Wall, 2006).

Jiao et al. (2010) found that that both the nutrient composition and content of toxicants in two cultivars of virus-resistant GM papaya grown in China were very similar to those in conventional papaya at two ripening stages, although there was some seasonal variation. The data published by Roberts et al. (2008) relate to fruit ripened for six days. No significant differences in nutritional composition were reported. The data published by Wall (2006) relate to fruit typically harvested at the colour break at approximately the 1/8th ripe stage for the commercial market. Although considerable variation was recorded in this study, no significant differences in nutritional composition due to genetic type were observed. The mean vitamin C content was 512 mg/kg, with no difference among cultivars. The mean content of provitamin A carotenoids was 2323 mg β-carotene and 5943 mg β b-cryptoxanthin per kg, and vitamin A ranged from 187 to 740 mg RAE (retinol activity equivalents) per kg. Lycopene was not detected in the yellow-fleshed cultivars – Kapoho, Laie Gold and Rainbow – but the red-fleshed Sunrise and SunUp fruit contained 13,500 to 36,740 mg lycopene per kg.

Some other fruits have been approved in several countries but are not available in commercial quantity to consumers. Others are at various stages of being introduced to the commercial market.

Potatoes

Rogan et al. (2000) presented detailed information on the composition of potatoes from insect and virus resistant potato cultivars (Table 10.1). The potatoes were transgenic cultivars of Russet Burbank and Shepody potatoes and had been approved by Canadian and US regulatory agencies for commerce in North America. Their results showed that tubers produced by insect- and virus-protected cultivars were substantially equivalent to conventional potatoes in their contents of key nutrients, anti-nutrients, macronutrients and minor nutrients.

El-Sanhoty et al. (2004) compared the composition of GM Spunta potatoes with the Cry V gene (G2 and G3) and a Spunta control line and found no significant differences in proximate chemical composition. The levels of glycoalkaloids were also similar in the GM and conventional cultivars. No significant differences were found in protease inhibitor activity or total phenol content.

Shepherd et al. (2006) assessed the potential for unintended effects in genetically modified potatoes and concluded that GM potatoes were substantially equivalent to conventional potatoes in their content of key nutrients and anti-nutrients. The range of GM cultivars analysed included those modified in primary carbohydrate metabolism, polyamine biosynthesis and glycoprotein processing. Although some differences were found between GM and non-GM cultivars, they were smaller than the differences found between two non-GM cultivars (Record and Desiree).

Langkilde et al. (2012) studied the relative composition of a GM potato cultivar SGT 9-2, which has a reduced α-solanine content. Although the data indicated some differences in composition between the GM cultivar and its conventional parent, the recorded values were within the range reported by other researchers and were not considered substantive.

Squash

Approval of two cultivars (CZW3 SEM-OCZW3-2 1997 and Virus Resistance ZW20 SEM-ZW20-7 1995) by the FDA was based in part on an assessment of evidence related to the nutritional analysis that had been submitted by the applicant company. It was concluded that the GM cultivars were substantially equivalent to conventional squash in terms of nutritional composition.

Health Canada, in its assessment of a virus-resistant squash (http://www.hc-sc.gc.ca/fn-an/gmf-agm/appro/ofb-098-106-b-eng.php), determined the contents of nutrients (protein, fat, ash, total dietary fibre, carbohydrates, calories, fructose, glucose, sucrose, lactose, maltose, vitamin C, β-carotene, vitamin A, calcium, iron and sodium) and reported only small differences that were within the range of variability normally reported for squash. The assessment, which included a taste test, found that the GM squash and the parental yellow crookneck squash contained very low levels of cucurbitacins and were non-bitter.

Sugar beet

The FDA in 2004, as part of an assessment of whether glyphosate tolerant sugar beet cultivar H7-1 is as nutritious as conventional sugar beet varieties currently being produced, compared the composition of H7-1 sugar beet and corresponding non-GM related control lines and eight commercial sugar beet varieties. Samples were collected at five locations in Europe (France, Germany, Italy, Spain and the United Kingdom). The H7-1 plants were treated with glyphosate herbicide during the growing season. Tops (leaves) and brei (pulp from processed roots) were analysed chemically for contents of ash, crude fibre, fat, crude protein, dry matter, carbohydrates (by calculation), amino acids, acid detergent fibre, neutral detergent fibre and (for pulp only) percent sucrose (by polarisation), invert sugar, sodium, potassium and α-amino nitrogen. No statistically significant differences were found in the mean levels of nutrients between pulp from GM sugar beet and non-GM controls. Statistically significant differences were found between cultivar H7-1 and non-GM controls in the contents of 2 out of 18 amino acids analysed, with alanine higher and glutamic acid lower in cultivar H7-1. However, these differences were not deemed to be biologically significant since the mean values of these two amino acids were within the ranges reported for both the non-GM control and conventional varieties. It was therefore concluded that the pulp from GM sugar beet H7-1 is compositionally equivalent to that of the non-GM control and other commercial sugar beet varieties. A similar conclusion was reached for the tops. Statistical analysis of the chemical data for sugar beet tissue showed that the saponin levels in H7-1 sugar beet were not significantly different from the levels in non-GM controls and were comparable to the levels in the tissues of the commercial reference varieties.

Canadian regulators in 2001 approved the introduction of glyphosate-tolerant sugar beet as a source of sugar for human consumption and the products derived from cultivar T120-7 as animal feed (http://www.hc-sc.gc.ca/fn-an/gmf-agm/appro/august100_final-eng.php). Beet pulp and molasses are the major feed by-products of sugar beets used in livestock feeding. Approval was based on proximate, amino acid, mineral and sugar analyses of leaf and root samples from each of 32 cultivars of conventional sugar beets and two hybrids from T120-7 grown in Europe. These data showed that for the leaves, there was no difference in nutrient composition between the control and hybrid lines. For the root samples, there were no differences in proximate analysis between the control and the T120-7 hybrids. Of a total of 63 variables measured, 3 of them (Ca, Mn, Arg) were observed to be different in the T120-7 hybrids compared with the control lines. Another set of data related to sugar beets grown in the USA. Sugar beet roots and tops from T120-7 and a control non-GM variety from seven sites in three locations were analysed. The data indicated that there were no overall differences between the T120-7 lines and the control. Further analysis of processed fractions, that is pulp, molasses and refined sugar, were also conducted, and confirmed that there were no nutritionally significant differences between the T120-7 lines and the control line.

Böhme et al. (2001) studied the chemical composition and digestion of GM (glufosinate-tolerant) sugar beets in pigs and cattle. No significant differences were found in the nutrient composition of beets or beet tops with the conventional, GM or non-GM parent line sugar beets. Digestibility of GM and non-GM sugar beets differed slightly in pigs but the differences were not considered to be meaningful biologically. Digestibility of sugar beet roots and sugar beet top silage for cattle was also found to be within the biological range reported by other researchers. Weisbjerg et al. (2001) reported similar findings with sugar beet and with fodder beet.

Based on the above findings it can be concluded that the approved cultivars of GM sugar beet and fodder beet are similar to related conventional non-GM beets in chemical composition and nutritive value for farm animals.

Sweet corn

The assessment of GM sweet corn by regulatory authorities includes compositional and nutritional analyses. Approval of the GM cultivars signifies that the regulatory authorities are satisfied that the GM cultivars are substantially equivalent in composition and nutrient content to the related non-GM cultivars. No independent publications relating to composition and nutrient content appear to be available in peer-reviewed scientific journals.

Tomatoes

Analysis of nutrients in the FlavrSavr™ line and non-GM control lines by the FDA did not show any significant differences in the levels of macro- and micronutrients, pH, total acidity, total solids and sugars. Also, analysis of nutrients from the novel 1401F hybrid line and non-GM 1401 hybrid did not show any significant differences in the levels of macro- and micronutrients in either fresh tomatoes or processed tomato paste.

Health Canada reported that an analysis of nutrients from the novel 1345-4 line and the non-GM 91103-114 parental line did not show any significant differences in the contents of macro- and micronutrients.

Venneria et al. (2008) investigated the nutritional composition of tomatoes modified by insertion of a tobacco rab gene, designed to delay ripening in the fruit. No significant differences in fatty acid content, unsaponifiable fraction of antioxidants, total phenols, polyphenols, carotenoids, vitamin C, total antioxidant activity and mineral composition between the GM and related non-GM cultivars were reported (Table 10.1).

There is a growing interest in producing food plants with increased amounts of flavonoids because of their potential health benefits. Tomatoes from two GM lines (2027 and 2059) were subjected to detailed analysis and identification of flavonoids, at different stages of maturity (Gall et al., 2003). Nine major flavonoids were detected in the flesh of GM ripe tomatoes. The total flavonoid glycoside content of ripe GM tomatoes of line 2059 was about 10-fold higher than that of the controls, with kaempferol glycosides accounting for about 60%. Kaempferol glycosides comprised around 5% of the flavonoid glycoside content of ripe control tomatoes. The rutin concentration in both GM and control fruits was similar.

Animal products

Research has been conducted to investigate whether the inclusion of approved GM feed ingredients in the diets of livestock, poultry and fish affects the nutritional quality of meat, milk and eggs (Table 10.3). To date no adverse effects have been reported, either in peer-reviewed journals or in the published assessments of regulatory agencies.

The conclusion that can be drawn from the data in Table 10.1, 10.2 and 10.3 above is that the nutritional composition and nutritive value of approved GM food and feed sources are essentially equivalent to those of their related, non-GM cultivars.

Reviews

Several reviews have been published on the relative composition of GM crops and conclude that their nutritional value is equivalent to that of conventional cultivars. The reviews include assessments of substantial equivalence of anti-nutrients and inherent plant toxins in GM food products (e.g. Novak and Haslberger, 2000). The possibility of unintended consequences of genetic modification has also been assessed (e.g. Herman and Price, 2013).

These conclusions confirm the assessments of substantial nutritional equivalence by the regulatory authorities in several countries in approving these products for food and feed use.

A review by Aumaitre et al. (2002) found that the nutritional composition of GM crops has consistently been shown to fall within the range of established values. They also found that equivalence in digestible energy and crude protein contents between isogenic and transformed plants expressing a wide range of modifications also had been clearly demonstrated in different species. In none of these experiments was animal performance, whether measured as growth rate, feed efficiency or carcass merit in beef cattle, egg mass in laying hens, milk production, milk composition and quality in dairy cows or digestibility in rabbits, affected by feeding transformed plants compared to animals fed control or related non-GM plants.

Flachowsky et al. (2002, 2005, 2012) reviewed the published evidence and concluded that GM feeds were substantially equivalent to their related non-GM counterparts. Their study covered maize, sugar beets and fodder beets, wheat, canola, soyabeans, cottonseed and potatoes and included over 100 animal feeding studies involving dairy cattle, beef cattle, sheep, other ruminants, pigs, broilers, laying hens, quail, fish, rabbits and laboratory animals. They concluded that there is good agreement from these studies that feed from GM crops did not significantly influence feed digestibility, animal health, biologically relevant effects on animal performance, composition of animal products, or result in unintended effects with the exception that lower mycotoxin concentrations have repeatedly been found in *B. thuringiensis* toxin-expressing GM crops when compared to animals fed isogenic non-GM cultivars.

Harrigan et al. (2010) reported that compositional equivalence of GM crops and their conventional non-GM varieties has been demonstrated in potato, cotton, soyabean, corn, rice, wheat and lucerne. Reviews by the Society of Toxicology (2003), Catchpole (2005), Council for Agricultural Science and Technology (CAST, 2006), Faust and Glenn (2002), Flachowsky and Aulrich (2001), Quarantelli et al. (2007), Shepherd et al., (2006), Shewry et al. (2006), Lemaux (2008) and Van Eenennaam (2013) agree with the conclusion that GM and non-GM crops are substantially equivalent in nutritional composition.

Harrigan et al. (2010) conducted a large study to determine the variation in composition between GM and related non-GM crops in relation to the variation found between conventional, non-GM crops. Their analysis evaluated compositional data from seven GM maize and soyabean cultivars grown in nine different countries (France, Italy, Spain, Germany, Romania, the United States, Argentina, Brazil and Chile) over 11 growing seasons. The GM products evaluated represented a range of traits conferring insect protection, herbicide resistance or drought tolerance. The compositional data used were generated under Organization for Economic Co-operation and Development (OECD) guidelines as part of the comparative safety assessment process used to support regulatory approvals and commercialisation. Components analysed included proximate constituents, macro- and micronutrients, toxicants and anti-nutrients as well as other crop-specific secondary metabolites. Their analysis indicated that compositional differences between GM cultivars varieties and their conventional non-GM cultivars were within the natural variability of the conventional crop and that the composition of GM and conventional crops was similar. Overall, for maize, a total of 2350 (number of sites × number of compositional components) statistical comparisons between the GM varieties and their corresponding conventional controls were conducted. Results showed that 91.5% were not significantly different ($p > 0.05$). Less than 1% of all comparisons where a significant difference ($p < 0.05$) was observed had a relative magnitude difference greater than 20%. For soyabeans, involving a total of 1840 statistical comparisons between the GM products and their corresponding conventional controls, 88.5% were not significantly different ($p > 0.05$). As with maize, the statistically significant differences between the GM and conventional soyabean components generally represented small differences in relative magnitude.

Ridley et al. (2011) conducted an assessment of the nutritional composition of the forage and grain of GM maize with combinations of traits in comparison with near-isogenic, conventional control cultivars. In total, 241 statistical comparisons between the multi-trait transgenic crop and its corresponding conventional controls were conducted. Of these comparisons 192 (79.7%) were not statistically significantly different ($p > 0.05$), and all 49 of the differences were within the 99% tolerance interval for commercial hybrids grown in the same field or related field trials.

As reported by Herman and Price (2013), over the previous 20 years the US FDA has found that all of the 148 GM cultivars evaluated were substantially equivalent to their conventional counterparts. Japanese regulators came to the same conclusion with respect to 189 submissions (the latter including combined-trait products). Over 80 peer-reviewed publications also concluded this same compositional equivalence of GM crops. These studies have included maize, soyabeans, cotton, canola, wheat, potato, alfalfa, rice, papaya, tomato, cabbage, pepper, raspberry and a mushroom, with traits of herbicide tolerance, insect resistance, virus resistance, drought tolerance, cold tolerance, nutrient enhancement and expression of protease inhibitors. In addition, numerous studies had found that variation resulting from traditional breeding and environmental factors dwarfed any changes observed in the composition due to introducing a trait through transgenesis.

Summary

Evidence from a large number of peer-reviewed scientific investigations and from reviews of the available evidence by many different competent authorities indicates that the nutritional composition and nutritive value of approved GM food and feed sources are equivalent to those of their related, non-GM cultivars. Some variation in composition between GM and related non-GM crops has been reported but the differences are generally small and within the range of variation found for conventional crops. Environmental and soil conditions have been shown to have a greater influence on composition than genetic type. GM feed sources have been shown to be substantially equivalent to their non-GM counterparts when included in the diets of a number of different laboratory or farm animals.

References

Aalhus, J. L., Dugan, M. E. R., Lien, K. A., et al. (2003). Effects of feeding glyphosate-tolerant canola meal on swine growth, carcass composition and meat quality. *Journal of Animal Science* 81: 3267 (Abstract).

Aeschbacher, K., Meile, L., Messikommer, R. and Wenk, C. (2001). Genetically modified maize in diets for chickens and laying hens: influence on performance and product quality. *Proceedings: International Symposium on Genetically Modified Crops and Co-products as Feeds for Livestock*, September, Nitra, Slovak Republic: pp. 41–42.

Aeschbacher, K., Meile, L., Messikommer, R. and Wenk, C. (2002). Influence of genetically modified maize on performance and product quality of chickens. *Proceedings Society of Nutritional Physiology* 11: 196.

Aeschbacher, K., Messikommer, R., Meile, L. and Wenk, C. (2005). B.t.176 corn in poultry nutrition: Physiological characteristics and fate of recombinant plant DNA in chickens. *Poultry Science* 84(3): 385–394.

Appenzeller, L. M., Linda Malley, L., MacKenzie, S. A., et al. (2009).Subchronic feeding study with genetically modified stacked trait lepidopteran andcoleopteran resistant maize grain in Sprague-Dawley rats. *Food and Chemical Toxicology* 47: 1512–1520.

Aulrich, K., Bohme, H., Daenicke, R., et al. (2001). Genetically modified feeds in animal nutrition. 1st communication: *Bacillus thuringiensis* (B.t.) corn in poultry, pig and ruminant nutrition. *Archives of Animal Nutrition* 54: 183–195.

Aumaitre, A., Aulrich, K., Chesson, A., et al. (2002). New feeds from genetically modified plants: substantial equivalence, nutritional equivalence, digestibility, and safety for animals and the food chain. *Livestock Production Science* 74: 223–238.

Baker, J. M., Hawkins, N. D., Ward, J. L., et al. (2006). A metabolomic study of substantial equivalence of field-grown genetically modified wheat. *Plant Biotechnology Journal* 4: 381–392.

Barrière, Y., Verite, R., Brunschwig, P., et al. (2001). Feeding value of silage maize estimated with sheep and dairy cows is not altered by genetic incorporation of B.t. 176 resistance to Ostrinia nubilalis. *Journal of Dairy Science* 84: 1863–1871.

Berberich, S. A., Ream, J. E., Jackson, T. L., et al. (1996). The composition of insect-protected cottonseed is equivalent to that of conventional cottonseed. *Journal of Agricultural and Food Chemistry* 44: 365–371.

Berger, L. L., Robbins, N. D., Sewell, J. R. and Stanisiewski, E. P. (2002). Effect of feeding diets containing corn grain with Roundup (event GA21 or NK603), control, or conventional varieties on steer feedlot performance and carcass characteristics. *Journal of Animal Science* 80 (Suppl. 1): 270 Abstract 1080.

Berger, L. L., Robbins, N. D., Sewell, J. R., et al. (2003). Effect of feeding diets containing corn grain with corn rootworm protection (event MON 863), control, or conventional varieties on steer feedlot performance and carcass characteristics. *Journal of Animal Science* 81 (Suppl.1): 214–215 Abstract M150.

Berman, K. H., Harrigan, G. G., Riordan, S. G., et al. (2009). Compositions of seed, forage, and processed fractions from insect protected soybean MON 87701 are equivalent to those of conventional soybean. *Journal of Agricultural and Food Chemistry* 57: 11360–11369.

Berman, K. H., Harrigan, G. G., Riordan, S. G., et al. (2010). Compositions of forage and seed from second-generation glyphosate-tolerant soybean MON 89788 and insect-protected soybean MON 87701 from Brazil are equivalent to those of conventional soybean (*Glycine max*). *Journal of Agricultural and Food Chemistry* 58: 6270–6276.

Bertrand, J. A., Sudduth, T. Q., Condon, A., et al. (2005) Nutrient content of whole cottonseed. *Journal of Dairy Science* 88: 1470–1477.

Böhme, H. and Aulrich, K. (1999). Inhaltsstoffe und Verdaulichkeit von transgenen Zuckerrben bzw. Krnermais im Vergleich zu den isogenen Sorten beim Schwein. (Ingredients and digestibility of transgenic sugar beets and corn in comparision to the isogenic varieties in the case of pigs). *VDLUFA Conference Proceedings 1999*, 111th VDLUFA Conference, 13–17 September 1999, Halle/Saale, Germany: pp. 289–292.

Böhme, H., Aulrich, K., Daenicke R. and Flachowsky, G. 2001. Genetically modified feeds in animal nutrition 2nd communication: glufosinate tolerant sugar beets (roots and silage) and maize grains for ruminants and pigs. *Archives of Animal Nutrition* 54: 197–207.

Böhme, H., Hommel, B. and Flachowsky, G. (2005). Nutritional assessment of silage from transgenic inulin synthesizing potatoes for pigs. *Journal of Animal and Feed Science* 14 (Suppl. 1): 333–336.

Brake, J., Faust, M. and Stein, J. (2003). Evaluation of transgenic event B.t.11 hybrid corn in broiler chickens. *Poultry Science* 82: 551–559.

Brake, J., Faust, M. and Stein, J. (2005). Evaluation of transgenic hybrid corn (VIP3a) in Broiler chickens. *Poultry Science* 84: 503–512.

Brake, J. and Vlachos, D. (1998). Evaluation of transgenic event 176 "B.t." corn in broiler chickens. *Poultry Science* 77: 648–653.

Bressner, G. E., Hyun, Y., Stanisiewski, E. P., et al. (2003). Performance comparison of growing-finishing pigs fed diets containing Corn Root Worm Protected corn (Event MON 863) or conventional corn hybrids. *Journal of Animal Science* 81 (Suppl. 1): 207 Abstract M119.

Brown, P. B., Wilson, K. A., Jonker, Y. and Nickson, T. E. (2003). Glyphosate tolerant canola meal is equivalent to the parental line in diets fed to rainbow trout. *Journal of Agricultural and Food Chemistry* 51: 4268–4272.

Caine, W. R., Aalhus, J. L., Dugan, M. E. R., et al. (2007). Growth performance, carcass characteristics and pork quality of pigs fed diets containing meal from conventional or glyphosate-tolerant canola. *Canadian Journal of Animal Science* 87: 517–526.

Calsamiglia, S. S., Hernandez, B., Hartnell, G. F. and R. Phipps, R. (2007). Effects of corn silage derived from a genetically modified variety containing two transgenes on feed intake, milk production, and composition, and the absence of detectable transgenic deoxyribonucleic acid in milk in Holstein dairy cows. *Journal of Dairy Science* 90: 4718–4723.

CAST (2006). Safety of Meat, Milk, and Eggs from Animals Fed Crops Derived from Modern Biotechnology. Issue Paper 34. Council for Agricultural Science and Technology, Ames, Iowa.

Castillo, A.R., Gallardo, M. R., Maciel, M., et al. (2001). Effect of feeding dairy cows with either BollGard®, BollGard II®, Roundup Ready® or control cottonseeds on feed intake, milk yield and milk composition. *Journal of Dairy Science* 84 (Suppl. 1): 413 Abstract 1712.

Castillo, A., Gallardo, M., Maciel, M., et al. (2004). Effects of feeding rations with genetically modified whole cottonseed to lactating Holstein cows. *Journal of Dairy Science* 87: 1778–1785.

Catchpole, G. S. (2005). Hierarchical metabolomics demonstrates substantial compositional similarity between genetically modified and conventional potato crops. *Proceedings of the National Academy of Sciences* 102: 14458.

Chainark, P., Satoh, S., Hino, T., et al. (2006). Availability of genetically modified soybean meal in rainbow trout Oncorhynchus mykiss diets. *Fisheries Science* 72: 1072–1078.

Choi, H., Moon, J.-K., Park, B.-S., et al. (2012). Comparative nutritional analysis for genetically modified rice, Iksan483 and Milyang204, and nontransgenic counterparts. *Journal of the Korean Society for Applied Biological Chemistry* 55: 19–26.

Chrastinová, L., Sommer, A., Rafay, J., et al. (2002). Genetically modified maize in diets for rabbits – influence on performance and product quality. *Proceedings Society of Nutrition Physiology* 11: 195.

Chrenková, M., Sommer, A., Čerešňáková, Z., et al. (2002). Nutritional evaluation of genetically modified maize corn performed on rats. *Archives of Animal Nutrition* (*Archiv fur Tiernahrung*) 56: 229–235.

Chukwudebe, A., Privalle, L., Reed, A., et al. (2012). Health and nutritional status of Wistar rats following subchronic exposure to CV127 soybeans. *Food and Chemical Toxicology* 50: 956–971.

Cromwell, G. L., Lindemann, M., Randolph, J., et al. (2002). Soybean meal from Roundup Ready or conventional soybeans in diets for growing-finishing swine. *Journal of Animal Science* 80: 708–715

Cromwell, G., Henry, B., Scott, A., et al. (2005). Glufosinate herbicide-tolerant -Liberty Link- rice vs. conventional rice in diets for growing-finishing swine. *Journal of Animal Science* 83: 1068–1074.

Custodio, M. G., Powers, W. J., Huff-Lonergan, E., et al. (2004). Growth and carcass characteristics of pigs fed biotechnologically derived and non-biotechnologically derived corn and harvested at different weights. *Journal of Animal Science* 82 (Suppl. 1): 456 Abstract.

Custodio, M., Powers, W., Huff-Lonergan, E., et al. (2006). Growth, pork quality, and excretion characteristics of pigs fed B.t. corn or non-transgenic corn. *Canadian Journal of Animal Science* 86: 461–469.

Daenicke, R., Aulrich, K. and Flachowsky, G. (1999). GMO in animal feedstuffs: nutritional properties of B.t.-maize unaffected. *Mais* 27 (September): 135–137.

Daun, J. K. (2004). Quality of genetically modified (GM) and conventional varieties of canola (spring oilseed rape) grown in western Canada, 1996–2001. *Journal of Agricultural Science* 142: 273–280.

Delaney, B., Karaman, S., Roper, J., et al. (2013). Thirteen week rodent feeding study with grain from molecular stacked trait lepidopteran and coleopteran protected (DP-ØØ4114-3) maize. *Food and Chemical Toxicology* 53: 417–427.

Donkin, S., Velez, J., Stanisiewski, E. and Hartnell, G. (2000). Effect of feeding Roundup Ready® corn silage and grain on feed intake, milk production and milk composition in lactating dairy cattle. *Journal of Dairy Science* 83 (Supplement 1): 273.

Donkin, S., Velez, J., Stanisiewski, E. and Hartnell, G. (2003). Effects of feeding silage and grain from glyphosate-tolerant or insect-protected corn hybrids on feed intake, ruminal digestion, and milk production in dairy cattle. *Journal of Dairy Science* 86: 1780–1788.

Eenennaam, A. L. van. (2013). GMOs in animal agriculture: time to consider both costs and benefits in regulatory evaluations. *Journal of Animal Science and Biotechnology* 4: 37.

El-Sanhoty R., El-Rahman, A. A. and Bögl, K. W. (2004). Quality and safety evaluation of genetically modified potatoes spunta with Cry V gene: compositional analysis, determination of some toxins, antinutrients compounds and feeding study in rats. *Nahrung* 48: 13–18.

Elangovan, A., Mandal, A. and Johri, T. (2003). Comparative performance of broilers fed diets containing processed meals of B.t., parental non-B.t. line or commercial cotton seeds. *Asian-Australian Journal of Animal Sciences* 16: 57–62.

Erickson, G. E., Robbins, N. D., Simon, J. J., et al. (2003). Effect of feeding glyphosate-tolerant (roundup-ready events GA21 or nk603) corn compared with reference hybrids on feedlot steer performance and carcass characteristics. *Journal of Animal Science* 81: 2600–2608.

Faust, M. A. and Glenn, B. P. (2002). Animal feeds from crops derived through biotechnology: Farm animal performance and safety. In J. A. Thomas and R. L. Fuchs (eds) *Biotechnology and Safety Assessment*, third edition. Academic Press, San Diego, California: pp. 144–191.

Faust, M. A. and Miller, L. (2000). Study finds no B.t. in milk. *Iowa State University Integrated Crop Management Newsletter*/IC-478, Special Livestock Edition, Ames, IA.

Faust, M. A. and Spangler, S. M. (2000). Nutritive value of silages from MON810 B.t. and non-B.t. near-isogenic corn hybrids. *Journal of Dairy Science* 83: 1184 (Abstract).

Fischer, R. L., Lewis, A. J., Miller, P. S., et al. (2002). Comparison of swine performance when fed diets containing Roundup Ready corn (event NK603), control or conventional corn grown during 2000 in Nebraska. *Journal of Animal Science* 80 (Suppl.1): 224.

Fischer, R. L., Miller, P. S., Hyun, Y., et al. (2003). Comparison of swine performance when fed diets containing corn root worm protected corn, parental line corn, or conventional corn grown during 2000 in Nebraska. *Journal of Animal Science* 81 (Suppl. 1): 207 Abstract M118.

Flachowsky, G. and Aulrich, K. (2001). Nutritional assessment of GMO in animal nutrition. *Journal of Animal Feed Science* 10 (Suppl. 1): 181–194.

Flachowsky, G., Aulrich, K., Berk, A., et al. (2002). Nutritional assessment of feeds from genetically modified (GM) crops. *Proceedings of Society of Nutrition Physiology* 11: 183–186.

Flachowsky G., Aulrich K., Böhme H. and Halle I. (2007): Studies on feeds from genetically modified plants (GMP) Contributions to nutritional and safety assessment. *Animal Feed Science and Technology* 133: 2–30.

Flachowsky, G., Chesson A. and Aulrich K. (2005). Animal nutrition with feeds from genetically modified plants. *Archives of Animal Nutrition* 59: 1–40.

Flachowsky, G., Schafft, H. and Meyer, U. (2012). Animal feeding studies for nutritional and safety assessments of feeds from genetically modified plants: a review. *Journal für Verbraucherschutz und Lebensmittelsicherheit (Journal of Consumer Protection and Food Safety)* 7: 179–194.

Folmer, J., Grant, R., Milton, C. and Beck, J. (2002). Utilization of B.t. corn residues by grazing beef steers and B.t. corn silage and grain by growing beef cattle and lactating dairy cows. *Journal of Animal Science* 80 (5): 1352–1361.

FSANZ (2003). Oil Derived From Glufosinate-Ammonium Tolerant And Pollination Controlled Canola, A Safety Assessment. Technical Report Series No. 16. Food Standards Australia New Zealand, June 2003.

Gaines, A. M., Allee, G. L. and Ratliff, B. W. (2001a). Nutritional evaluation of B.t. (MON810) and Roundup Ready® corn compared with commercial hybrids in broilers. *Poultry Science* 80 (Suppl. 1): 51 Abstract. 214.

Gaines, A. M., Allee, G. L. and Ratliff, B. W. (2001b). Swine digestible energy evaluations of B.t. (MON810) and Roundup Ready corn compared to commercial varieties. *Journal of Animal Science* 79 (Suppl.1)/*Journal of Dairy Science* 84 (Suppl. 10): 109.

Gaines, A. M., Allee, G. L. and Ratliff, B. W. (2001c). Nutritional evaluation of B.t. (MON810) and Roundup Ready corn compared to commercial hybrids in broilers. *Journal of Animal Science* 79 (Suppl.1)/Journal of Dairy Science 84 (Suppl. 10): 51.

Gall, G. le, DuPont, M. S., Mellon, F. A., et al. (2003). Characterization and content of flavonoid glycosides in genetically modified tomato (*Lycopersicon esculentum*) fruits. *Journal of Agricultural and Food Chemistry* 51: 2438–2446.

Grant, R. J., Kleinschmit, D., Sparks, A. L., et al. (2002). Influence of glyphosate-tolerant (trait NK603) corn silage and grain on feed consumption and milk production in Holstein dairy cattle. *Journal of Dairy Science* 85 (Suppl. 1): 384 Abstract 1540.

Grant, R., Fanning, K., Kleinschmit, D., et al. (2003). Influence of glyphosate-tolerant-event NK603- and corn rootworm protected-event MON863- corn silage and grain on feed consumption and milk production in Holstein cattle. *Journal of Dairy Science* 86: 1707–1715.

Halle, I., Aulrich, K. and Flachowsky, G. (1998). Einsatz von Maiskörnen der Sorte Cesar und des gentechnisch veränderten B.t.-Hybriden in der Broiler mast. *Proceedings 5. Tagung, Schweine- und Geflhgelernährung*, December 1998, Wittenberg, Deutschland: pp. 265–267.

Halle, I., Aulrich, K. and Flachowsky, G. (2004). Four Generations of feeding of GMO-corn to breeder quail. (Fütterung von gentechnisch verändertem Mais an Zuchtwachtein über vier Generationen). *Proceedings of Society of Nutritional Physiology* 13: 124.

Hamilton, K. A., Pyla, P. D., Breeze, M., et al. (2004). Bollgard II cotton: Compositional analysis and feeding studies of cottonseed from insect-protected cotton (*Gossypium hirsutum* L.) producing the Cry1Ac and Cry2Ab2 proteins. *Journal of Agricultural and Food Chemistry* 52: 6969–6976.

Hammond, B., Dudek, R., Lemen, J. and Nemeth, M (2004). Results of a 13 week safety assurance study with rats fed grain from glyphosate tolerant corn. *Food and Chemical Toxicology* 42: 1003–1014.

Hammond, B., Vicini, J., Hartnell, G., et al. (1996). The feeding value of soybeans fed to rats, chickens, catfish and dairy cattle is not altered by genetic incorporation of glyphosate tolerance. *Journal of Nutrition* 126: 717–727.

Harrigan, G. G., Lundry, D., Drury, S., et al. (2010). Natural variation in crop composition and the impact of transgenesis. *Nature Biotechnology* 402 volume 28: 402–404.

Hartnell, G., Hvelplund, T. and Weisbjerg, M. (2005). Nutrient digestibility in sheep fed diets containing Roundup Ready® or conventional fodder beet, sugar beet, and beet pulp. *Journal of Animal Science* 83: 400–407.

He, X. Y., Huang, K. L., Li, X., et al. (2008). Comparison of grain from corn rootworm resistant transgenic DAS-59122-7 maize with non-transgenic maize grain in a 90-day feeding study in Sprague-Dawley rats. *Food and Chemical Toxicology* 46: 1994–2002.

Hemre, G., Sanden, M., Bakke-McKeller, A., et al. (2005). Growth, feed utilization and health of atlantic salmon *Salmo salar L.* fed genetically modified compared to non-modified commercial hybrid soybeans. *Aquaculture Nutrition* 11: 157–167.

Hendrix, K. S., Petty, A. T. and Lotgreen, D. L. (2000). Feeding value of whole plant silage and crop residues from B.t. or normal corns. *Journal of Animal Science* 78 (Suppl.1): 273.

Herman, R. A., Phillips, A. M., Lepping, M. D. and Sabbatini, J. (2011). Compositional safety of DAS-68416 (AAD-12) herbicide-tolerant soybean. *Journal of Nutrition and Food Sciences* DOI: 10.4172/215-9600.1000103.

Herman, R. A. and Price, W. D. (2013). Unintended compositional changes in genetically modified (GM) crops: 20 years of research. *Journal of Agricultural and Food Chemistry* 61: 11695–11701.

Hvelplund, T. and Weisbjerg, M. R. (2001). Comparison of nutrient digestibility between Roundup Ready beets and pulp derived from Roundup Ready beets and conventional beets and pulps. *Journal of Animal Science* 79 (Suppl.1): 417.

Hyun, Y., Bressner, G., Ellis, M., et al. (2004). Performance of growing-finishing pigs fed diets containing Roundup Ready ® corn - event NK603, a nontransgenic genetically similar corn, or conventional corn lines. *Journal of Animal Science* 82: 571–580.

Hyun, Y., Bressner, G., Fischer, R., et al. (2005). Performance of growing-finishing pigs fed diets containing YieldGard rootworm corn (MON 863), a nontransgenic genetically similar corn, or conventional corn hybrids. *Journal of Animal Science* 83: 1581–1590.

Ipharraguerre, I., Younker, R., Clark, J., et al. (2003). Performance of lactating dairy cows fed corn as whole plant silage and grain produced from a glyphosate-tolerant hybrid (event NK603). *Journal of Dairy Science* 86: 1734–1741.

Jacobs, C. M., P. L. Utterback, C. M. Parsons, D. et al. (2008). Performance of laying hens fed diets containing DAS-59122-7 maize grain compared with diets containing nontransgenic maize grain. *Poultry Science* 87: 475–479.

Jiao, Z. Deng, J., Li, G., et al. (2010). Study on the compositional differences between transgenic and non-transgenic papaya (*Carica papaya L.*). *Journal of Food Composition and Analysis* 23: 640–647.

Kan, C. and Hartnell, G. (2004a). Evaluation of broiler performance when fed Roundup-Ready wheat (event Mon71800), control, and commercial wheat varieties. *Poultry Science* 83: 1325–1334.

Kan, C. and Hartnell, G. (2004b). Evaluation of broiler performance when fed insect-protected, control, or commercial varieties of dehulled soybean meal. *Poultry Science* 83: 2029–2038.

Kan, C. A., Versteegh, H. A. J., Uijttengoogaart, T. J., et al. (2001). Comparison of broiler performance and carcass characteristics when fed B.t., parental control or commercial varieties of dehulled soybean meal. *Poultry Science* 80 (Suppl. 1): 203 Abstract 841.

Kerley, M. S., Felton, E. E. D., Lehmkuhler, J. W. and Shillito, R. (2001). B.t. corn that is genetically modified to prevent insect damage is equal to conventional corn in feeding value for beef cattle. *Journal of Animal Science* 79 (Suppl. 2): 98 Abstract 301.Kumar, R. and Singhal, K. (2004). Chemical composition and nutritional evaluation of transgenic cotton for ruminants. *Indian Journal of Animal Sciences* 74: 868–871.

Langkilde, S., Schrøder, M., Frank., T., et al. (2012). Compositional and toxicological analysis of a GM potato line with reduced α-solanine content – a 90-day feeding study in the Syrian Golden hamster. *Regulatory Toxicology and Pharmacology* 64: 177–185.

Lemaux, P. (2008). Do genetically engineered foods have changes in nutritional content? In Review: Genetically engineered plants and foods: a scientist's analysis of the issues (Part I). *Annual Review of Plant Biology* 59: 771–812.

Li, M. H., Hartnell, G. F., Robinson, E. H., et al. (2008). Evaluation of cottonseed meal derived from genetically modified cotton as feed ingredients for channel catfish, *Ictalurus punctatus*. *Aquaculture Nutrition* 14: 490–498.

Maertens, L., Luzi, F. and Huybechts, I. (1996). Digestibility of non-transgenic and transgenic oilseed rape in rabbits. *Proceedings of 6th World Rabbit Congress*, Toulouse, Volume I: 231–235.

Mandal, A. B., Elangovan, A. V., Shrivastav, A., et al. (2004). Comparison of broiler chicken performance when fed diets containing meals of Bollgard® II hybrid cotton containing Cry-x Gene(Cry1Ac and Cry2Ab Gene), parental line or commercial cotton. *British Poultry Science* 45: 657–663.

Masoero, F., Moschini, M., Rossi, F., et al. (1999). Nutritive value, mycotoxin contamination and in vitro rumen fermentation of normal and genetically modified corn (Cry1A(b)) grown in Northern Italy. *Maydica* 44: 205–209.

MacKenzie, S. A., Lamb, I., Schmidt, J., et al. (2007). Thirteen week feeding study with transgenic maize grain containing event DAS-Ø15Ø7-1 in Sprague–Dawley rats. *Food and Chemical Toxicology* 45: 551–562.

McCann, M. C., Liu, K., Trujillo, W. A. and Dobert, R. C. (2005) Glyphosate-tolerant soybeans remain compositionally equivalent to conventional soybeans (*Glycine max* L.) during three years of field testing. *Journal of Agricultural and Food Chemistry* 53: 5331–5335.

McNaughton, J., Roberts, M., Rice, D., et al. (2007). Feeding performance in broiler chickens fed diets containing DAS-59122-7 maize grain compared to diets containing non-transgenic maize grain. *Animal Feed Science and Technology* 132: 227–239.

Mireles, A., Kim, S., Thompson, R. and Amudsen, B. (2000). GMO (B.t.) corn is similar in composition and nutrient availability to broilers as non-GMO corn. *Poultry Science* 79 (Suppl.1): 65-66.

Nagappa, H. and Khadi, B. M. (2011). Studies on composition of oil and fatty acid in B.t. and non B.t. cotton (G. hirsutum). *Proceedings World Cotton Research Conference 5*, Mumbai, India, 7–11 November 2011: pp. 568–571.

Nida, D. L., Patzer, S., Harvey, P., et al. (1996). Glyphosate-tolerant cotton: The composition of the cottonseed is equivalent to that of conventional cottonseed. *Journal of Agricultural and Food Chemistry* 44: 1967–1974.

Novak, W. K. and Haslberger, A. G. (2000). Substantial equivalence of anti-nutrients and inherent plant toxins in genetically modified novel foods. *Food and Chemical Toxicology* 38: 473–483.

Oberdoerfer, R. B., Shillito, R. D., De Beuckeleer, M. and Mitten D. H. (2005). Rice (Oryza sativa L.) containing the bar gene is compositionally equivalent to the nontransgenic counterpart. *Journal of Agricultural and Food Chemistry* 53: 1457–1465.

Obert, J. C., Ridley, W. P., Schneider, R. W., et al. (2004). The composition of grain and forage from glyphosate tolerant wheat MON 71800 is equivalent to that of conventional wheat (*Triticum aestivum* L.). *Journal of Agricultural and Food Chemistry* 52: 1375–1384.

Padgette, S. R., Taylor, N. B., Nida, D. L., et al. (1996). The composition of glyphosate-tolerant soyabean seeds is equivalent to that of conventional soyabeans. *Journal of Nutrition* 126: 702–716.

Peterson, B. A., Hyun, Y., Stanisiewski, E. P., et al. (2008). Performance of growing-finishing pigs fed diets containing Roundup Ready wheat (MON 71800), a non-transgenic genetically similar wheat, or conventional wheat varieties. *Animal* 2: 1602–1609.

Petty, A., Hendrix, K. S., Stanisiewski, E. P. and Hartnell, G. F. (2001). Performance of beef cattle fed Roundup Ready corn harvested as whole plant silage or grain. *Journal of Animal Science* 79 (Suppl. 2): 102 Abstract 321.

Phipps, R., Jones, A. K., Tingey, A. and Abeyasekera, S. (2005). Effect of corn silage from an herbicide-tolerant genetically modified variety on milk production and absence of transgenic DNA in milk. *Journal of Dairy Science* 88: 2870–2878.

Piva, G., Morlacchini, M., Pietri, A., et al. (2001). Performance of broilers and piglets fed B.t. corn. *Proceedings: International Symposium on Genetically Modified Crops and Co-products as Feeds for Livestock*, Nitra, Slovak Republic, September 2001: pp. 27–30.

Quarantelli, A., Veroli, E., Romanelli, S., et al. (2007). GMO utilization in animal nutrition. (2007). *Annali della Facoltà di Medicina Veterinaria*, Università di Parma 27: pp. 211–228.

Řehout, V., Kadlec, J., Čítek, J., et al. (2009). The influence of genetically modified B.t. maize MON 810 in feed mixtures on slaughter, haematological and biochemical indices of broiler chickens. *Journal of Animal and Feed Sciences* 18: 490–498.

Reuter, T., Aulrich, K., Berk, A. and Flachowsky, G. (2001a). Nutritional evaluation of B.t.-Corn in pigs. *Journal of Animal Science* 79: 1073.

Reuter, T., Aulrich, K., Berk. A. and Flachowsky, G. (2001b). Influence of B.t.-corn on growing and slaughtering results of pigs. *Proceedings of the Society of Nutrition Physiology* 10: 150.

Reuter, T., Aulrich, K., Berk, A. and Flachowsky, G. (2002a). Investigations on genetically modified maize (B.t.-maize) in pig nutrition: chemical composition and nutritional evaluation. *Archives of Animal Nutrition* 56: 23–31.

Reuter, T., Aulrich, K. and Berk, A. (2002b). Investigations on genetically modified maize - B.t. maize in pig nutrition - fattening performance and slaughtering results. *Archives of Animal Nutrition* 56: 319–326.

Rhee, G., Cho, D., Won, Y., et al. (2005). Multigeneration reproductive and developmental toxicity study of bar gene inserted into genetically modified potato on rats. *Journal of Toxicology and Environmental Health, Part A* 68: 2263–2276.

Ridley, W. P., Harrigan, G. G., Breeze, M. L., et al. (2011). Evaluation of compositional equivalence for multitrait biotechnology crops. *Journal of Agricultural and Food Chemistry* 59: 5865–5876.

Roberts, M., Minott, D. A., Tennant, P. F. and Jackson, J. C. (2008). Assessment of compositional changes during ripening of transgenic papaya modified for protection against papaya ringspot virus. *Journal of the Science of Food and Agriculture* 88: 1911–1920.

Rogan, G. J., Bookout, J. T., Duncan, D. R., et al. (2000). Compositional analysis of tubers from insect and virus resistant potato plants. *Journal of Agricultural and Food Chemistry* 48: 5936–5945.

Rossi, F., Morlacchini, M., Fusconi, G., et al. (2005). Effect of B.t. corn on broiler growth performance and fate of feed-derived DNA in the digestive tract. *Poultry Science* 84: 1022– 1030.

Russell, J. R., Farnham, D., Berryman, R. K., et al. (2000). Nutritive Value of the Crop Residues From B.t.-Corn Hybrids and Their Effects on Performance of Grazing Beef Cows. 2000 Beef Research Report -Iowa State University: pp. 56–61.

Russell, J. R., Hersom, M. J., Haan, M. M., et al. (2001). Effects of grazing crop residue from B.t. -corn hybrids on pregnant beef cows. Abstracts: *American Society of Animal Science Midwestern Section*: March 19–21, 2001 Des Moines, IA. (Abstract 300): 74–75.

Rutzmoser, K., Meyer, J. and Obermaier, A. (1999). Verfütterung von Silomais der Sorten "Pactol" und "Pactol CB" (gentechnisch veränderte B.t.-Hybride) an Milchkühe. *Bodenkultur und Pflanzenbau, Schriftenreihe der Bayer. Landesanstalt für Bodenkultur und Pflanzenbau* 3: 25–34.

Sanden, M., Krogdahl, A., Bakke-Mckellep, A. M., et al. (2006). Growth performance and organ development in Atlantic salmon, *Salmo salar* L. parr fed genetically modified (GM) soybean and maize. *Aquaculture Nutrition* 12: 1–14.

Shepherd, L. V. T., McNicol, J. W., Razzo, R., et al. (2006). Assessing potential for unintended effects in genetically modified potatoes perturbed in metabolic and developmental processes. Targeted analysis of key nutrients and anti-nutrients. *Transgenic Research* 15: 409–425.

Shewry, P. R., Baudo, M., Lovegrove, A., et al. (2006). Are GM and conventionally bred cereals really different? *Trends in Food Science Technology* 18: 201–209.

Sidhu, R., Hammond, B., Fuchs, R., et al. (2000). Glyphosate-tolerant corn: the composition and feeding value of grain from glyphosate-tolerant corn is equivalent to that of conventional corn (*Zea Mays* L.). *Journal of Agricultural and Food Chemistry* 48: 2305–2312.

Simon, J. J., Van der Pol, K. J., Erickson, G. E., et al. (2002). Effect of Roundup Ready corn (event NK603) on performance in beef feedlot diets. *Journal of Animal Science* 80 (Suppl.1): 46.

Singh, M., Tiwari, D., Kumar A. and Kumar, M. (2003). Effect of feeding transgenic cottonseed vis-à-vis non-transgenic cottonseed on haematobiochemical constituents in lactating Murrah buffaloes. *Asian-Australian Journal of Animal Science* 16: 1732–1737.

Singhal, K., Kumar, S., Tyagi, A. and Rajput, Y. (2006). Evaluation of B.t. cottonseed as a protein supplement in the ration of lactating dairy cows. *Indian Journal of Animal Sciences* 76: 532–537. Sissener, N. H., Sanden, M., Bakke, A. M., et al. (2009). A long term trial with Atlantic salmon (*Salmo salar* L.) fed genetically modified soy; focusing general health and performance before, during and after the parr-smolt transformation. *Aquaculture* 294: 108–117.

Sissener, N. H., Sanden, M., Krogdahl, Å., et al. (2011). Genetically modified plants as fish feed ingredients. *Canadian Journal of Fisheries and Aquatic Sciences* 68: 563–574.

Society of Toxicology (2003). The safety of genetically modified foods produced through biotechnology. *Toxicological Sciences* 71: 2–8.

Stanford, K., Aalhus, J. L., Dugan, M. E. R., et al. (2003). Effects of feeding transgenic canola on apparent digestibility, growth performance and carcass characteristics of lambs. *Canadian Journal of Animal Science* 83: 299–305.

Stanisiewski, E. P., Hartnell, G. F. and Cook, D. R. (2001). Comparison of swine performance when fed diets containing Roundup Ready® corn (GA21), parental line or conventional corn. *Journal of Animal Science* 79 (Suppl. 1): 319–320.

Stanisiewski, E. P., Taylor, M. L., Hartnell, G. F. et al. (2002). Broiler performance when fed Roundup Ready® (event RT73) or conventional canola meal. *Poultry Science* 81 (Suppl. 1): 95 abstract 408.

Stein, H. H., Rice, D. W., Smith, B. L., et al. (2009). Evaluation of corn grain with the genetically modified input trait DAS-59122-7 fed to growing-finishing pigs. *Journal of Animal Science* 87: 1254–1260.

Stein, H. H., Sauber, T., Rice, D., et al. (2004). Comparison of corn grain from biotech and non-biotech counterparts for grow-finish pig performance. *Journal of Animal Science* 82 (Suppl. 1): 328–329 Abstract W65.

Steinke, K., Guertler, P., Paul, V., et al. (2010). Effects of long-term feeding of genetically modified corn (event MON810) on the performance of lactating dairy cows. *Journal of Animal Physiology and Animal Nutrition* 94: 185–193.

Świątkiewicz, M., Bednarek, D., Twardowska, M., et al. (2013a). Genetically modified HT soybean meal and B.t. maize in pig feeding. *Wiadomości Zootechniczne* 51: 31–47.

Świątkiewicz, M., Bednarek, D., Markowski, J., et al. (2013b). Effect of feeding genetically modified maize and soybean meal to sows on their reproductive traits, haematological indices and offspring performance. *Bulletin of the Veterinary Institute in Puławy, Poland* 57: 413–418.

Świątkiewicz, M., Hanczakowska, E., Twardowska, M., et al. (2011). Effect of genetically modified feeds on fattening results and transfer of transgenic DNA to swine tissues. *Bulletin of the Veterinary Institute, Puławy, Poland* 55: 121–125.

Tan, J., Liu, S., Sun, Z., Zhang, H., et al. (2012). Comparison of broiler performance, carcass yields and intestinal microflora when fed diets containing transgenic (Mon-40-3-2) and conventional soyabean meal. *African Journal of Biotechnology* 11: 12371–12378. Taylor, N. B., Fuchs, R. L., MacDonald, J., et al. (1999) Compositional analysis of glyphosate-tolerant soybeans treated with glyphosate. *Journal of Agricultural and Food Chemistry* 47: 4469–4473.

Taylor, M. L., George, B., Hyun, Y., et al. (2004a). Comparison of broiler performance when fed diets containing insect-protected (MON 88017 or MON 88017 x MON 810), control, or commercial corn. *Poultry Science* 83 (Suppl. 1): 322 Abstract W39.

Taylor, M. L., Hartnell, G. F., Nemeth, M. A., et al. (2001a). Comparison of broiler performance when fed diets containing YieldGard® and Roundup Ready® corn, parental lines, or commercial corn. *Poultry Science* 80 (Suppl. 1): 319 Abstract 1321.

Taylor, M. L., Hartnell, G. F., Nemeth, M. A., et al. (2001b). Comparison of broiler performance when fed diets containing Roundup Ready® corn event NK603, parental line, or commercial corn. *Poultry Science* 80 (Suppl. 1): 320 Abstract 1323.

Taylor, M.L., Hartnell, G. F., Astwood, J. D., et al. (2002a). Comparison of broiler performance when diets containing MON863 (Corn Rootworm Protection), non-transgenic control, or commercial corn. *Poultry Science* 81 (Suppl. 1): 95 Abstract 406.

Taylor, M. L., Hartnell, G. F., Astwood, J. D., et al. (2002b). Comparison of broiler performance when fed diets containing YieldGard® (MON810) x Roundup Ready® (NK603), nontransgenic control, or commercial corn. *Poultry Science* 81 (Suppl. 1): 95 Abstract 405.

Taylor, M., Hartnell, G., Nemeth, M., et al. (2005). Comparison of broiler performance when fed diets containing corn grain with insect-protected (Corn Rootworm and European Corn Borer) and herbicide-tolerant (glyphosate) traits, control corn, or commercial reference corn - revisited. *Poultry Science* 84: 1893–1899.

Taylor, M., Hartnell, G., Nemeth, M., et al. (2007) Comparison of broiler performance when fed diets containing grain from second-generation insect-protected and glyphosate-tolerant, conventional control or commercial reference corn. *Poultry Science* 86: 1972–1979.

Taylor, M., Hartnell, G., Riordan, S., et al. (2003a). Comparison of broiler performance when fed diets containing grain from Roundup Ready® - NK603, YieldGard® X Roundup Ready® - MON810 X NK603, non-transgenic control, or commercial corn. *Poultry Science* 82: 443–453.Taylor, M., Hartnell, G., Riordan, S., et al. (2003b). Comparison of broiler performance when fed diets containing grain from Yieldgard® MON810, Yieldgard® X Roundup Ready® - GA21,nontransgenic control, or commercial corn. *Poultry Science* 82: 823–830.Taylor, M., Hyun, Y., Hartnell, G., et al. (2003c). Comparison of broiler performance when fed diets containing grain from YieldGard® Rootworm - MON863, YieldGard® Plus - MON810 x MON863, nontransgenic control, or commercial reference corn hybrids. *Poultry Science* 82: 1948–1956.

Taylor, M., Stanisiewski, E., Riordan, S., et al. (2004b). Comparison of broiler performance when fed diets containing Roundup Ready® -Event RT73-, nontransgenic control, or commercial canola meal. *Poultry Science* 83: 456–461.

Tony, M., Broll, H., Zagon, J., et al. (2002). Detection and impact of B.t. 176 maize on broiler health and performance. *Proceedings of Society of Nutritional Physiology* 11: 197.

Tripathi, S., Suzuki, J. Y., Carr, J. B., et al. (2011). Nutritional composition of Rainbow papaya, the first commercialized transgenic fruit crop. *Journal of Food Composition and Analysis* 24: 140–147.

UK Advisory Committee on Novel Foods and Processes (1998). Annual Report 1998. MAFF Publications. ADMAIL 6000, London SW1A 2XX.

Vander Pol, K. J., Erickson, G. E., Macken, C. N., et al. (2002). Effect of corn root worm protected corn (event MON863) on performance in beef feedlot diets. *Journal of Animal Science* 80 (Suppl. 1): 46 Abstract 180.

Venneria. E., Fanasca, S., Monastra, G., et al. (2008). Assessment of the nutritional values of genetically modified wheat, corn, and tomato crops. *Journal of Agricultural and Food Chemistry* 56: 9206–9214.

Wall, M. M. (2006). Ascorbic acid, vitamin A, and mineral composition of banana (Musa sp.) and papaya (Carica papaya) cultivars grown in Hawaii. *Journal of Food Composition and Analysis* 19: 434–445.

Weber, T. E. and Richert, B. T. (2001). Grower-finisher growth performance and carcass characteristics including attempts to detect transgenic plant DNA and protein in muscle from pigs fed genetically modified "B.t." corn. *Journal of Animal Science* 79 (Suppl. 1): 67 Abstract 162.

Weisbjerg, M., Hvelplund, T., Purup, S., et al. (2001). Genetically modified beets and beet pulp as feeds for ruminants, experiments with sheep and dairy cattle at Research Centre Foulum, Denmark. *Proceedings: International Symposium on Genetically Modified Crops and Co-Products as Feeds for Livestock*, Nitra, Slovak Republic: pp. 37–40.

Yonemochi, C., Fujisaki, H., Harada, C., et al. (2002). Evaluation of transgenic event CBH 351 (Starlink) corn in broiler chicks. *Animal Science Journal* 73: 221–228.

Zhu, Y., Li, D., Wang, F., et al. (2004). Nutritional assessment and fate of DNA of soybean meal from Roundup Ready or conventional soybeans using rats. *Archives of Animal Nutrition* 58: 295–310.

CHAPTER 11
Addressing consumer issues

It is quite clear from reports in the popular media and on the Internet that there is a major gap between public opinion about the safety and quality of genetically modified (GM) foods and the scientific facts. Consequently, it is logical to conclude that the public needs to be better informed about the whole issue.

A good example of how being better informed can result in a change of opinion is that of Mark Lynas, who went from being an anti-GM crusader to an advocate for GM foods. As he said at the 2013 Oxford Farming Conference:

> For the record, here and upfront, I apologise for having spent several years ripping up GM crops. I am also sorry that I helped to start the anti-GM movement back in the mid-1990s, and that I thereby assisted in demonising an important technological option which can be used to benefit the environment ... So I did some reading. And I discovered that one by one my cherished beliefs about GM turned out to be little more than green urban myths.

It should be pointed out, though, that this man had the expertise and opportunity to read and understand the scientific literature. The average consumer could not be expected to have these advantages, and would require the facts to be available in unbiased and non-technical language. But, why is there such a gap? As the journal *Nature Biotechnology* (Anon., 2013) asked in an editorial: 'Three decades after transgenes were first introduced into plants, why do so many consumers remain so negative about genetically modified (GM) food?'

> It does not matter that no adverse health effects have been recorded from eating them. Nor does it matter that august agencies, such as the World Health Organization, the US National Academy of Sciences, the European Commission or the American Medical

Genetic Modification and Food Quality: A Down to Earth Analysis, First Edition.
Robert Blair and Joe M. Regenstein.

> Association, have come out with ringing endorsements of their safety. The fact is, negative attitudes remain entrenched and widespread. And changing them will require a concerted and long-term effort to develop GM foods that clearly provide convincing benefits to consumers - something that seed companies have conspicuously failed to do over the past decade.

In its analysis, the editorial outlined various factors that have contributed to the situation. One is that reports claiming negative effects on health are seized upon by anti-GM activists 'no matter how flimsy the evidence or flawed the study design' and 'all too often, an uncritical and sensationalist media leaps upon negative findings, continuing the cycle of scares, urban myths and downright mistruths about GM food, all of which serve to stoke consumer paranoia'.

The editorial finds also that the public now mistrusts most mainstream sources of data on GM foods, including large corporations, regulators, governments and even scientists. In addition European regulators have deepened the existing prejudices 'by creating a regulatory system that singles out GM products as sufficiently threatening to merit special attention'. The editorial did not elaborate on this point but it may refer to the EU labelling requirement. For instance, canola oil made from GM canola is required to be labelled as 'oil from GM canola (rapeseed)'. On the other hand, cheese made using rennin derived from a GM microorganism is not required to be labelled as GM. This distinction is hard to justify on scientific grounds since no GM material is present either in the cheese or the oil, but it does re-enforce to the consumer the practice of using GM cultivars of canola in the production of canola oil and ignores the fact that the cheese industry in Europe (and elsewhere) uses GM microorganisms.

Surveys

International

Colson and Rousu (2013) reviewed the international evidence related to consumer perceptions and willingness to pay for GM foods. They found that researchers over the previous 15 years had delivered well over 100 estimates of the willingness of consumers to pay for GM foods, using surveys and experimental methods. They concluded that three general findings had been consistently supported by the literature.

1 Consumers are willing to pay a premium for non-GM food products or would pay less if the product is genetically modified. One meta-analysis of 25 studies found a premium of almost 29% for non-GM food products. Another meta-analysis found a median premium of 18% and a 45% mean premium for non-GM food products. The finding of preferences in opposition to GM foods is largely consistent across countries, elicitation mechanisms and food types.
2 US consumers are more accepting of GM foods than European consumers. This is found both in individual studies and in major reviews of the topic. They

interpreted this finding as being consistent with public policy. While European countries have mandated the labelling of GM food products, public support in the US has been insufficient in shifting US national policy from voluntary to mandatory labelling of GMO products.

3 The magnitude of consumer discount for GM foods depends upon the type of genetic modification, the type of food product, and how the genetic modification alters the final product. Some research showed that consumers are less accepting of genetic modification involving animals than plants. Other research found that consumers are more accepting of genetically modified foods when the product has a direct tangible benefit to consumers, such as improved nutritional quality or reduced pesticide use.

Europe

The most recent Eurobarometer Report (2010) found that Europeans continued to be largely unaware of the use of biotechnology in food production, and that the European public was still wary of genetically engineered food. A survey showed that only 17% of Europeans had heard of the science and the level of acceptance was found to be correspondingly low. Other opinions of survey respondents were that decisions about biotechnology should be left to scientific experts but that strong regulation by government was necessary. They did not see any benefits of genetically modified food, considered genetically modified foods to be probably unsafe or even harmful and were not in favour of development of genetically modified food. In addition they did not see the benefits of horizontal gene transfer, had strong reservations about safety, and felt that special labelling of GM food products was necessary. The results suggested that five criteria based on public opinion were necessary to ensure the success of an emerging technology: sustainability, benefits, proper regulation, security and fair distribution of risks and benefits.

A recent survey of 2301 consumers in the UK confirmed the misgivings that many consumers have about GM foods (YouGov, 2014). As the report concluded, while the new GM cultivar of the Desiree potato that is resistant to blight might have prevented the Irish potato famine of 1845 had it been available then, proponents have yet to convince a sceptical public of the benefits of the technology. Only 6% of the public reported their views towards GM foods becoming 'more positive' over the previous 12 months, almost the same as the 5% who said their views had become more negative. And as for those whose views haven't changed at all, they remained decisively negative: 41% negative to 17% positive. A large proportion (31%) responded 'don't know' when asked how their views on GM had evolved. Additionally, British adults by a margin of almost two to one said the government should not be promoting the adoption of GM technology. Currently GM foods in the UK are regulated by the EU, being used mainly in animal feed.

An outcome of the public mistrust of GM foods is an effect on the plans of British farmers to plant GM crops. According to a 2013 survey of 625 independent farmers conducted by the *Farmers Weekly* journal, a majority (61%) of farmers in the UK would grow genetically modified (GM) crops if it were legal to do so. The research showed that farmers in the UK are aware of consumer attitudes towards GM crops. The explanation for those farmers not willing to grow GM crops related to the perceived public resistance to the technology.

Willingness-to-pay surveys are probably more valuable than regular consumer surveys in that they are closer to a real-life situation. Several researchers have investigated this approach and suggest that the findings are more meaningful. For instance, Noussair et al. (2001) studied French consumer behaviour related to GM foods and found that actual consumer behaviour indicated a lower level of hostility toward GM foods than that suggested by the survey. Experimental studies have also demonstrated that consumers – including those in Europe – are more likely to select GM foods over organic foods in the event that such foods offer a 'price advantage coupled with a consumer benefit' (Knight et al., 2007).

Fear of GM foods has been shown to be a factor in their acceptance by consumers, one study (Laros and Steenkamp, 2004) finding that Dutch consumers feel significantly more fearful of GM foods than of other new food types. No strong correlations were found between consumer socio-demographic makeup and fear, indicating that fear of this technologically new type of food is an emotion that cuts across society. The researchers found that the fear of GM foods was correlated positively with concern for the environment and negatively by their faith in technology in food production.

Ceccoli and Hixon (2012) found that support for GM foods by European consumers was associated positively with scientific literacy and an understanding of the technology. They found that media exposure also mattered, frequent reading about science and (rather surprisingly) the Internet as a source of information about science being positively associated with support for GM foods. The influence of television was not statistically related to support for GM foods, and the influence of newspapers and magazines had a negative effect.

Another finding was that support for GM foods was related to agricultural policy in the EU. Farming in several European countries is based mainly on a large number of small privately owned farms that have traditionally relied on organic production, assisted by subsidies. These farmers could not be expected to adopt biotechnology with the same enthusiasm as larger, corporate farms.

Sleenhoff and Osseweijer (2013) pointed out one of the limitations of survey responses, namely the discrepancy between intentions and actual buying behaviour. With a mandatory labelling scheme for GM food in Europe since 2004 it is now possible to measure consumer choice in practice. These researchers studied the actual purchase behaviour of European consumers of GM food products and compared this with their stated intentions to buy GM foods. They found that despite a majority of consumers voicing a negative attitude toward GM food, over 50% of respondents did not actively avoid the purchase of GM foods and 6% actually purchased one of the few available GM labelled food products. The results of this study imply that a stated negative attitude of consumers in response to questionnaires about their buying intentions is not a reliable guide for what consumers actually buy in supermarkets.

One of the outcomes of the negative attitude towards GM foods in Europe is that the European Union reached a political agreement in June 2014 allowing its member states to restrict or ban GM crops in their territory. Previously the states had to receive the support of the European Food Safety Authority (EFSA) before they could object to a farmer planting a GM crop, after submitting adequate scientific evidence for the objection. In a related development, an EU court in 2013 issued an injunction blocking the cultivation of GM potatoes, one of the only two GM crops allowed for cultivation in the Union.

North America

Information on US consumer attitude toward GM food commodities was provided in a survey conducted by Hallman et al. (2003). Results showed that most Americans were not fully aware of GM foods at that time because only about 12% of the respondents knew a lot about food biotechnology, as opposed to 45% knowing something and 43% knowing nothing about the subject. The survey also found that about 68% of the respondents had never been engaged in any discussion about GM foods, 26% of the respondents did not know whether or not they had ever eaten GM food, and almost 50% were unaware that food products made with GM-derived ingredients are currently on supermarket shelves.

A 2006 Pew Research Initiative survey (Pew, 2007) on US consumer awareness and understanding of biotech foods reported a 10% increase in net perception of biotech food product safety, compared with earlier studies. The study also found that 45% of Americans felt more comfortable about the safety of biotech food products, compared with only 29% who thought they were unsafe. Of interest was the finding that Americans were willing to change their opinions once provided with additional information about biotech food products.

The introduction of country-of-origin labelling (COOL) has allowed this information to be tested as a factor in the choice of GM foods by consumers. Xie et al. (2013) found that consumers preferred GM foods that were produced domestically over GM foods imported from foreign countries. They found also that individuals with knowledge of the benefits of GM benefits to consumers, producers and the environment were willing to pay more than individuals without that knowledge. However, the effect varied by type of knowledge, location and country of origin of the products. Other research confirms that consumers prefer foods produced in their own country or region (e.g. Loureiro and Umberger, 2005; Chambers et al., 2007).

A recent Canadian survey conducted by Ipsos-Reid on behalf of BioAccess Commercialization Centre and its partners examined Canadian consumer perceptions regarding healthy food, ingredients and natural health products (Ag-West Bio, 2014). The survey found that 43% of those surveyed were more likely to buy a product that is 'GMO-free', indicating a desire for GM foods to be labelled. Another finding was that the growing and divisive debate in the media was leaving consumers either confused or forming opinions on the issue without a good understanding of genetic modification.

In 2014 the British Columbia Growers' Association published the results of a Canadian national survey on GM food products (BC Fruit Growers' Association, 2014). The survey was commissioned after a BC biotech company applied to have the 'Arctic apple' approved for release. The Arctic apple is genetically modified to prevent browning when cut. In response to the question 'Are you in favour of the approval of this [genetically engineered] apple by Health Canada?', 69% of respondents were not in favour. Of the 1501 respondents in the national survey, 76% said that the Canadian government has not provided adequate information about genetically modified food. A further 9% said they had not heard of genetically modified food prior to the survey and 91% of respondents agreed that mandatory labelling of genetically modified food should be required by the Canadian government. If the food were clearly labelled as being genetically modified, 14% of respondents would purchase the food without hesitation, 45% said it would depend on the type of food being purchased, and 35% said that nothing could convince them to purchase genetically modified food.

Examples of consumer attitudes can be found in reports from other countries.

Africa

A survey was conducted with 179 respondents (55 farmers and 124 consumers) in both rural and urban areas of Kenya and found that consumers who were familiar with government policy and had a basic knowledge of GM crops and who shared that information were more likely to approve of the technology (Kagai, 2011). Farmers were concerned with the environmental risks associated with GM technology and its possible effect on marketing crops both locally and abroad. Consumers expressed concerns about possible health risks, the ability of the government to protect them, and the acceptance of GM products in the local market. Disapproval of GM products by both farmers and consumers was influenced by the perception of high risks and low benefits.

Australia/New Zealand

Results of a recent food and agribusiness survey were reported by Hall (2014), with 53% of respondents indicating that they had become more receptive to GM foods over the past two years and 71% indicating a likelihood of increasing their purchase of GM foods over the next two years. Respondents believed that a lack of information about GMOs is at the heart of consumers' concerns. From an environmental perspective, respondents stated that the public needed to be better educated about the use of genetically modified farming and agriculture, particularly as it can lead to hardier and more weather resistant crops that rely less on chemicals to withstand the effects of pests and disease.

A 2002 survey of 139 undergraduate students from a university in New Zealand indicated a negative attitude to GM foods, even when the foods had added benefits (Fortin and Renton, 2003).

China

A report by Hu and Chen (2004) outlined the intentions of Beijing consumers to purchase vegetable oil that had been made from genetically modified oilseeds (referred to as GM vegetable oil). The researchers found that the intentions to purchase were low, indicating a considerable scepticism toward the GM vegetable oil. A potential reason for this scepticism was that Chinese consumers are not well informed about GM technology. The results indicated that over 67% of respondents had concerns about GM foods, and about 20% believed that GM food was harmful to their health. China requires that food derived from GM crops be labelled. For products made from GM crops but containing no GM ingredients, the label reads: 'This product is made from GM ... while no longer containing GM ingredients'. The latter label applies to GM vegetable oil.

Addressing the issues effectively

The informed consumer

Consumers obviously need to be better informed on the whole issue of the safety of GM foods, preferably before the issue of labelling is addressed. Evidence of the situation in China, as outlined above, indicates that a GM label on foods is likely to deter consumers from purchasing these foods because they lack balanced information about them. The situation is probably similar in all countries, including the EU and other countries that have mandatory labelling, the evidence suggesting that most consumers are poorly or misinformed about the true nature of GM foods.

Educating consumers will not be an easy task, however, given the current climate of confusion and distrust, exemplified in the survey evidence above. Who, for instance, is the consumer likely to believe? It will be interesting to follow developments in China, where the government has launched a media campaign in support of GM crops, as it battles a wave of negative publicity over a technology it hopes will play a major role in boosting its food security. The agriculture ministry in September, 2014, announced it would try to educate the public on GMO via TV, newspapers and the Internet. China has been a long-time proponent of GMOs, which it sees as broadly safe and as potentially key in helping feed the world's largest population (Patton, 2014).

In planning to provide consumers with balanced information about GMO foods it would be desirable for all sectors of the food production industry to be consulted. For instance, the USDA (United States Department of Agriculture) organic regulations 7CFR205.105(e) prohibits the use of excluded methods relating to GM organisms in organic production and handling. However, 'mutagenesis (treatment of plants with radiation or chemicals to induce random mutation is considered part of traditional breeding programs' and is permitted (USDA, 2013). According to this provision a distinction is made in that alteration of the genetic makeup of the plant by recombinant DNA technology is prohibited in the organic regulations but alteration by mutagenesis is allowed. It is debatable whether this distinction is valid scientifically. The effect is the potential for consumers to be misled when purchasing organic produce because it is 'GMO-free'.

The results of a survey undertaken by CSIRO (Commonwealth Scientific and Industrial Research Organisation, Australia's national science agency) provide some interesting ideas on how the educational process might be designed (Cormick, 2013).

The approach in their survey was to go beyond the science of food biotechnology because of the knowledge that our values, or world views, filter our receptiveness to messages. As explained in the introduction to the report:

> When information is complex people tend to make emotionally-based judgements, driven by values rather than by the information presented to them. Messages that don't align with people's values or world views tend to be rejected or dismissed. Our attitudes to genetic modification are based on how we feel about risk, technology and the pace of change.

The study found that four value-based segments of the population are better predictors of people's attitudes to GM foods than age, gender, or other standard demographics. Understanding these segments helps to understand what drives support or opposition to things like GM foods.

- **Segment 1 (20%)** – the concerned and disengaged: Segment 1 was the least enthusiastic about the benefits of science and technology. They had the highest agreement that 'the pace of technological change is too fast to keep up with' and were the most likely to agree that 'science and technology creates more problems than it solves', that 'scientific advances tend to benefit the rich more than the poor' and that 'we rely too much on science and not enough on faith'.
- **Segment 2 (23%)** – the risk averse: This segment tended to be less positive towards the benefits of science and technology generally, and biotechnology specifically. They were also more concerned with related risks. But in contrast to Segment 1, they had relatively high awareness of the term 'biotechnology' and various biotechnological applications. They were least likely to agree that 'human activities have a significant impact on the planet' and least likely to agree that 'not vaccinating children puts others at risk'.
- **Segment 3 (28%)** – the cautiously keen: Segment 3 was defined by a relatively high interest in science and agreement that 'the benefits of science are greater than any harmful effects'. In relation to GM, this segment was the second most positive. While awareness of biotechnology was relatively high for Segment 3 they, like Segment 1, had relatively low levels of self-reported knowledge. They also had the highest agreement that 'children should be protected from all risks'.

- **Segment 4 (23%)** – the science fans: This group was the most positive towards science and technology. They expressed greater agreement that 'science is such a big part of our lives that we should all take an interest', that 'new technologies excite me more than they concern me' and that 'the benefits of science are greater than any harmful effects'. Equally, there was disagreement that 'science and technology creates more problems than it solves' and that 'we depend too much on science and not enough on faith'.

When asked if they supported modifying the genes of plants to produce food by introducing the genes of a plant of a different species, the results by segments were: Segment 1, 29% were in favour; Segment 2, 25%; Segment 3, 37% and Segment 4, 59%.

When asked if they supported modifying the genes of plants to produce food by introducing the genes of a plant of the same species, the results were: Segment 1, 49%; Segment 2, 44%; Segment 3, 61% and Segment 4, 81%.

The report concluded that broad attitudes towards science, technology and nature can influence consumer attitudes towards GM foods, and that pro-science and technology values are a strong predictor of support for GM foods. Tendencies towards conspiracy theories predict an anti-GM position. Attempts to better the understanding about GM foods should go beyond discussing the quality of the science and address how the science might align with the different values held by people. The message should also be pitched in language understood by the average consumer.

Additional research to investigate the reactions of consumers offered foods labelled GM and with an explanation for the genetic modification would be helpful in achieving the goal of a better-informed consumer. Some researchers have adopted this approach, with interesting results. One such study was conducted to test Hawaiian consumer response to GM papayas (Shehata, 2011). These papayas were developed to resist a virus disease that had almost wiped out the crop in Hawaii. At each of six grocery stores in Oahu, papayas labelled 'Hawaiian Grown GMO Papayas' or 'Hawaiian Grown Papayas – Genetically Modified Organism' were displayed next to non-labelled GM papayas in the produce department, and the sales were monitored. The results showed that labelling papaya fruits increased their sale. This was attributed possibly to a consumer perception that a labelled product implies a better quality item, a desire by the consumer to sample transgenic fruits, a desire to support the Hawaiian economy, or to a general lack of knowledge regarding GMOs. Given the importance of papaya to the economy of this region and the publicity related to the introduction of transgenic cultivars there, it is difficult to believe that Hawaiian consumers are unaware of GMO foods. It is interesting to note that Hawaii in 2013 banned the cultivation of GM crops on the island of Oahu, but the GM papaya industry was declared exempt!

Another study that examined the buying behaviour of better-informed consumers was conducted on sweet corn by Canadian researchers (Powell et al.,

2003). In this trial, GM (B.t.) sweet corn and B.t. potatoes were grown side-by-side with conventional varieties in the 2000 growing season at a farm and market in Hillsburgh, Ontario. The sweet corn harvested throughout the trial was separated and labelled, and direct consumer evaluation of purchasing preferences was conducted. Overall, the B.t. sweet corn outsold the conventional sweet corn by a margin of 680 dozen (or 8160 cobs) to 452.5 dozen (or 5430 cobs). A limited number of intercept interviews was conducted after consumers made their purchasing decision. The majority of consumers interviewed said they were more concerned about pesticides than genetic engineering. However, taste and quality also had a strong influence on purchasing decisions.

Labelling

An important concern expressed by consumers in many countries is the 'right to know', namely whether or not food items are derived from GM sources. Currently the US and Canada have adopted a voluntary labelling policy in their national regulations. Manufacturers are not required to label foods containing GM ingredients. Foods may be labelled 'GM-free' if less than 5% of the total ingredients by weight are GM. In contrast, Europe, Australia/New Zealand, South Korea, Japan and China have a mandatory GM food labelling policy. Products that contain a percentage of GM ingredients that is greater than the threshold prescribed by national regulators must have a GM label. Products produced with GM technology (cheese produced with GM enzymes, for example) do not have to be labelled. No country has a requirement that products such as meat, milk and eggs from animals fed on GM animal feed are required to be labelled.

As noted above, it is preferable that consumers are fully informed about GM foods and that agreement is reached within all sectors of the food industry as to which foods need and need not be labelled, before such a label requirement is introduced.

Summary

There is a major gap between public opinion about the safety and quality of genetically modified (GM) foods and the scientific facts, indicating that on the whole consumers are poorly informed (or misinformed) of the real facts relating to GM foods. Consequently, it can be concluded that consumers need to be better informed about the whole issue. Many consumers would prefer to have information about GM status on the food label, but given the current climate it is likely that a mandatory labelling requirement for GM foods is likely to deter consumers from purchasing these foods, either from a lack of knowledge or from prejudice. The whole issue of what foods should be excluded and included in any mandatory label of GM foods needs to precede any discussion on labelling.

Some countries require that products of GM crops (such as vegetable oil and starch) require a label indicating that they have been derived from GM plants, but do not require a similar label for foods (such as cheese) that have been produced using enzymes derived from GM organisms. That distinction would be hard to understand by the average consumer. Another issue that needs to be clarified is the practice of regarding plants produced by mutagenesis (treatment with radiation or chemicals to induce random mutation) as having been derived by traditional breeding programmes. The result is that alteration of the genetic makeup of plants by recombinant DNA technology is prohibited in the organic regulations but alteration of the genetic makeup by mutagenesis is allowed. The effect of this practice is the potential for consumers seeking to purchase organic produce because it is GMO-free to be misled.

References

Ag-West Bio (2014). Perceptions of GMO foods: Surveying the Canadian consumer. www.agwest.sk.ca/blog/posts/perceptions-of-gmo-foods-surveying-the-canadian-consumer.htm (accessed 10 August 2014).

Anon. (2013). Editorial. Contrary to popular belief: Three decades after transgenes were first introduced into plants, why do so many consumers remain so negative about genetically modified (GM) food? *Nature Biotechnology* 31: 767.

BC Fruit Growers' Association (2014). News Release: Survey finds risk of consumer backlash on gmos. The British Columbia Fruit Growers' Association, 1473 Water Street, Kelowna, BC V1Y 1J6, Canada.

Ceccoli, S. and Hixon, W. (2004). Explaining European attitudes toward genetically modified foods. Paper presented at the annual meeting of the Midwest Political Science Association, Palmer House Hilton, Chicago, Illinois, Apr 15, 2004. http://www.allacademic.com/meta/p82626_index.html (accessed August 9, 2014).

Chambers, S., Lobb, A., Bulter, L., et al. (2007). Local, national and imported foods: a qualitative study. *Appetite* 49: 208–213.

Colson, G. and Rousu, M.C. (2013). GM crops and food. *Biotechnology in Agriculture and the Food Chain* 4: 158–165.

Cormick, C. (2013). The Conversation. How values affect our attitudes to genetically modified food. http://theconversation.com/how-values-affect-our-attitudes-to-genetically-modified-food-18763 (accessed 2 August 2014).

European Commission (2010). Europeans and Biotechnology in 2010. Winds of Change? Special Eurobarometer report on biotechnology. Brussels, European Commission.

Farmers Weekly (2013). Survey results: What farmers really think about GM. http://www.fwi.co.uk/articles/12/06/2013/139481/survey-results-what-farmers-really-think-about-gm.htm (accessed 10 August 2014).

Fortin, D. R. and Renton, M. S. (2003). Consumer acceptance of genetically modified foods in New Zealand. *British Food Journal* 105: 42–58.

Hall, G. (2014). Australia: Food and agribusiness survey: Genetically modified crops. Norton Rose Fulbright, Australia. http://www.mondaq.com/australia/x/330492/agriculture+land+law/Food+and+agribusiness+survey+Genetically+modified+crops (accessed 10 August 2014).

Hallman, W. K., Hebden, W. C., Aquino, H. L., et al. (2003). Public Perceptions of Genetically Modified Foods: A National Study of American Knowledge and Opinion. Food Policy Institute

Report Number RR-1003-004. Food Policy Institute, Rutgers University, New Brunswick, NJ. http://www.foodpolicy.rutgers.edu/docs/pubs/2003_Public_Perceptions_of_Genetically_Modified_Foods.pdf (accessed 8 August 2014).

Hu, W. and Chen, K. (2004). Can Chinese consumers be persuaded? The case of genetically modified vegetable oil. *AgBioForum* 7: 124–132.

Kagai, K. K. (2011) Assessment of public perception, awareness and knowledge on genetically engineered food crops and their products in Trans-Nzoia County, Kenya. *Journal of Developments in Sustainable Agriculture* 6: 164–180.

Knight, J., Mather, D., Holdsworth, D. and Ermen, D. (2007). Acceptance of GM food – An experiment in six countries. *Nature Biotechnology* 25: 507–508.

Laros, F. J. M. and Steenkamp, J. E. M. (2004). Importance of fear in the case of genetically modified food. *Psychology & Marketing*. Special Issue: Fear Appeals in Social Marketing Campaigns 21: 889–908.

Loureiro, M. L. and Umberger, W. J. (2005). Assessing consumer preferences for country-of-origin labeling. *Journal of Agricultural and Applied Economics* 37: 49–63.

Lynas, M. (2013). Lecture to Oxford Farming Conference, 3 January 2013. http://www.marklynas.org/2013/01/lecture-to-oxford-farming-conference-3-january-2013 (accessed 2 December 2013).

Noussair, C., Robin, S. and Ruffieux, B. (2001). Consumer behaviour with regard to GMOs in the food supply: an experimental study. Comportement des consommateurs face aux aliments "avec OGM" et "sans OGM": une étude expérimentale. *Économie Rurale* 266: 30–44.

Patton, D. (2014). China Launches Media Campaign to Back Genetically Modified Crops. *Scientific American*, September 30, 2014.

PEW (2007). The Organic & Non-GMO Report. http://pewagbiotech.org/research/2006update/.#sthash.w3pkFtYZ.dpuf (accessed 9 August 2014).

Powell, D. A., Blaine, K., Morris, S. and Wilson, J. (2003) Agronomic and consumer considerations for B.t. and conventional sweet-corn. *British Food Journal* 105: 700–713.

Shehata, S. (2011). Market testing of labeled and unlabeled GMO papaya fruits in Honolulu chain stores. *Journal of Food Distribution Research* 42: 107–110.

Sleenhoff, S. and Osseweijer, P. (2013). Consumer choice. Linking consumer intentions to actual purchase of *GM labeled food products. GM Crops and Food: Biotechnology in Agriculture and the Food Chain* 4: 166–171.

USDA (2013). Cell Fusion Techniques used in Seed Production. Policy Memo 13-1. United States Department of Agriculture, National Organic Program. Washington, DC.

Xie, J., Kim, H. and House, L. (2013). Valuing information on GM foods in the presence of country-of-origin labels. *International Journal of Food System Dynamics* 4: 170–183.

YouGov, 2014. Many in Britain still sceptical of GM foods. http://yougov.co.uk/news/2014/02/21/many-britain-remain-sceptical-gm-foods/ (accessed 5 August 2014).

CHAPTER 12

Overall conclusions

> My conclusion here today is very clear: the GM debate is over. You are more likely to get hit by an asteroid than to get hurt by GM food.
>
> *Mark Lynas, British writer and environmentalist. Oxford Farming Conference, Oxford University, January 3, 2013*

Biotechnology has played, and continues to play, a very important role in the production of new cultivars of crops.

The International Service for the Acquisition of Agri-Biotech Applications (ISAAA) in 2014 released a report indicating that over 18 million farmers in 27 countries planted biotech crops in 2013, reflecting a 5 million (3%) increase in global biotech crop hectarage. According to this agency the global biotech crop hectarage has increased from 1.7 million hectares in 1996 to over 175 million hectares in 2013. The USA continues to lead global biotech crop plantings, with 70.1 million hectares or 40% of total global hectares in 2013. According to the report, over 90% (16.5 million) of the farmers planting biotech crops are small and resource-poor. Of the countries planting biotech crops, 8 are industrial countries and 19 are developing countries. For a second year in a row developing countries planted more hectares of biotech crops than industrialised countries.

The importance of biotechnology in the production of new crops for food and feed was recognised in the award of the 2013 World Food Prize to a team of scientists who had made outstanding contributions to this endeavour: Drs Marc Van Montagu (Belgium), Mary-Dell Chilton (USA) and Robert T. Fraley (USA). According to the citation, these distinguished scientists '... share the 2013 World Food Prize for their independent, individual breakthrough achievements in founding, developing, and applying modern agricultural biotechnology.

Genetic Modification and Food Quality: A Down to Earth Analysis, First Edition.
Robert Blair and Joe M. Regenstein.

Their research is making it possible for farmers to grow crops with: improved yields; resistance to insects and disease; and the ability to tolerate extreme variations in climate.'

A necessary component of this development in crop production is the need to ensure that the GM cultivars are as safe and nutritious for food and feed use as their conventional counterparts. This assessment task is assigned to regulatory authorities in those countries planning the cultivation and/or the importation of GM food products. The evidence used in the assessments by the regulatory authorities indicates quite clearly that the approved GM foods and feeds are substantially equivalent to their non-GM counterparts in terms of their safety and nutritional composition.

There is broad scientific consensus that all approved foods and feeds that have been derived from GM crops being marketed currently are safe to eat. This conclusion is based on the findings of independent scientists, regulatory agencies and scientific agencies in many countries. In its assessment, the American Association for the Advancement of Science (AAAS, 2012) found that the World Health Organization, the American Medical Association, the US National Academy of Sciences, the British Royal Society, and every other respected organisation that has examined the evidence has come to the same conclusion: consuming foods containing ingredients derived from GM crops is no riskier than consuming the same foods containing ingredients from crop plants modified by conventional plant improvement techniques. A consistent finding with these foods is that much, or all, of the transgenic DNA is degraded in the gastrointestinal tract of the human or animal consumer, and that none is absorbed into the tissues. The fact that some consumers have concerns about the safety of GM foods can be attributed largely to the fact that they have not been made aware of the extensive database of favourable findings. In some cases consumers prefer not to know about this extensive database or negate its relevance.

Evidence from a large number of peer-reviewed scientific investigations and from reviews of the available evidence by competent authorities indicates that the nutritional composition and nutritive value of approved GM food and feed sources are equivalent to those of their related, non-GM cultivars. Some variation in composition between GM and related non-GM crops has been reported but the differences are generally small and within the range of variation found for conventional crops grown at different times in different places using different cultivars. Environmental and soil conditions have been shown to have a greater influence on the composition of the edible product than its genetic type. GM feed sources have been shown to be substantially equivalent to their non-transgenic counterparts when included in the diets of laboratory or farm animals. The details are set out in the relevant chapters of this book.

The available data indicate that cereal grains contain chemical and pesticide residues, but that the amounts are too small to be of concern in relation to human health. Many of the residues found are at the limit of detection. No separate data on residue levels in GM grains are available: more studies need to be carried out to provide comparable data on commercially available GM and conventional cultivars of grains. The published data indicate a lower level of mycotoxin contamination in GM maize possessing the trait for insect resistance. No studies appear to have been conducted on the organoleptic qualities of GM maize (or other grains), relative to conventionally grown maize.

The World Health Organization (2002) in a document prepared in response to questions and concerns by a number of WHO member state governments with regard to the nature and safety of genetically modified food stated that: 'No effects on human health have been shown as a result of the consumption of such foods by the general population in the countries where they have been approved.'

The European Commission funded research that included 130 research projects involving 500 independent research groups over 25 years. They concluded that: 'There is, as of today, no scientific evidence associating GMOs with higher risks for the environment or for food and feed safety than conventional plants and organisms' (European Commission, 2010).

In a recent review, Van Eenennaam and Young (2014) assessed the scientific literature on performance and health of animals consuming feed containing GM ingredients and composition of products derived from them, since on a global basis food-producing animals consume 70–90% of the biomass of GM crops. Data on livestock productivity and health were collated from publicly available sources from 1983, before the introduction of GM crops in 1996, and subsequently through 2011, a period with high levels of predominately GM animal feed. These field datasets, representing over 100 billion animals following the introduction of GM crops, did not show any unfavourable trends of concern in livestock health and productivity. Also, the researchers found that no study has shown any differences in the nutritional profile of animal products derived from GM-fed animals. Because DNA and protein are normal components of the diet that are digested, there are no detectable or reliably quantifiable traces of GM components in milk, meat or eggs following consumption of GM feed.

It is true that several reports claim to show evidence of safety concerns related to some GM foods. All have the same deficiencies: faulty design and experimental techniques, questionable statistics and more importantly no confirmation of the findings by independent scientists in other laboratories. It would be prudent, however, not to dismiss these claims completely. The authors, and other interested scientists, should attempt to replicate their original findings on a consistent basis. Confirmation of any safety concern would then have to be followed by re-assessment of the product by the regulatory authorities and possibly by revocation of approval of the GM product in question.

The evidence on nutritional equivalence is such that this provision may eventually be modified as a requirement for regulatory approval. Several scientists have made this suggestion, for example Herman and Price (2013). These researchers assessed the available evidence and concluded that compositional equivalence studies uniquely required for GM crops may no longer be justified on the basis of scientific uncertainty. The assessment by these authors, one a biotechnologist and the other a former high-ranking member of the FDA involved in GM approval applications, indicated overwhelming evidence that genetic modification is less disruptive of crop composition than traditional breeding. Modifying the requirement for nutritional equivalence would allow the regulatory efforts, and the substantial expenses involved, to be concentrated on the more important criterion of safety. Nutritional equivalence can be predicted fairly accurately by chemical analysis which is relatively inexpensive, but the associated animal tests required to investigate the comparative utilisation of the nutrients are extremely expensive. Modifying the requirement for nutritional equivalence would have the safeguard that any new GM cultivar that was later shown to be nutritionally deficient in relation to its non-GM counterpart would probably fail in the marketplace.

One region that has not adopted biotechnology to the same extent as other regions is Europe. According to recent assessments by leading European scientists this can be attributed to its current policy which discriminates against the cultivation of GM crops (e.g. Masip et al., 2013). At present the EU has a moratorium on the introduction of new GM crops, based on a stated uncertainty of the technology and the public attitude to GM foods. This policy has been interpreted as a trade barrier to the importation of cereal grains and grain products from countries using GM technology. As a result the USA filed a World Trade Organization (WTO) case against the EU's failure to implement a timely, science-based approval system for food and feed products enhanced through biotechnology. Several other countries, including Argentina, Australia, Canada, Chile, Colombia, El Salvador, Honduras, Mexico, New Zealand, Peru and Uruguay joined the USA in the complaint because they also wanted to ensure that science-based determinations are applied to regulatory decisions. In 2006 the WTO ruled that the EU had failed to meet its WTO obligations of implementing a timely, science-based system for the approval of biotech-enhanced agricultural products. As Van Eenennaam and Young (2014) pointed out, there is a pressing need for international harmonisation of both the regulatory frameworks for GE crops and governance of advanced breeding techniques to prevent widespread disruptions in the international trade of livestock feedstuffs in the future.

The assessment by Masip et al. (2013) highlighted the fact that some of the regulations are arbitrary, excessive and are politically motivated rather than reflecting scientific reality. They stated that suppression of GM agriculture in the EU is widely recognised as ideological rather than scientific, driven to a large extent by the organic food industry in an effort to protect organic food premiums at the expense of overall competitiveness. They concluded that differences in the EU regulations relating to foods grown in EU countries and identical imported products show that the EU is undermining its own competitiveness in the agricultural sector, damaging both the EU and its humanitarian activities in the developing world. They recommended the adoption of rational, science-based principles for the harmonisation of agricultural policies, to prevent economic decline and lower standards of living across the continent.

These comments were echoed in an important assessment of EU policy by EASAC, the European Academies Science Advisory Council (2013). This body was formed by the national science academies of the EU member states to enable them to collaborate in giving advice to European policy-makers. It provides a means for the collective voice of European science to be heard. Their review of the available data found no validated scientific evidence that the cultivation of GM crops has caused any environmental harm. They also pointed out several other authoritative reports confirming the safety of approved GM food products. For instance, a recent comprehensive assessment by the Swiss National Science Foundation (2012) of over 2000 studies found that no health or environmental risks related to GM technology had been identified. In addition, a recent statement by the American Association for the Advancement of Science (AAAS, 2012) reaffirmed that GM crops are the most extensively tested crops ever added to the food supply chain, with the evidence showing that crop improvement by biotechnology is safe and that GM and their non-GM counterparts are nutritionally equivalent.

According to the EASAC assessment there has been little incentive to exploit science and technology for agricultural innovation in the EU, in comparison with developments in other parts of the world. This is in spite of the fact that the conclusion emerging from the aggregate evidence collected in other countries is that GM technology has no greater adverse impact than any other technology used in plant breeding.

The EASAC report also emphasised that any regulatory system had to be evidence-based. The report found 'an inconsistent and inefficient connection' between the recommendations of EFSA (the European Food Safety Authority) and political action for final expeditious approval of GM cultivars. In particular, some member state politicians ignore the decisions of the advisory committees and the European Commission. The regulatory system in the EU focuses on the technology used in the development of GM crops, whereas countries like Canada regulate new crop cultivars on the basis of the traits expressed and not on the basis of the method used to introduce the traits, whether done using conventional breeding, mutagenesis or GM technology. The US policy is similar to that in Canada in that it is focused on the product of genetic modification techniques, not the process itself; that regulation is based on verifiable scientific risks, and that GM products are on a continuum with existing products and, therefore, existing statutes are sufficient to assess and review the products. This approach acknowledges the fact that it is the product, and not the process, that warrants regulation because it is the presence of novel traits in a new crop that potentially pose an environmental or health risk, and not how the traits were specifically introduced. Regulations for biotechnology derived crops should therefore be focused on those cultivars that possess traits sufficiently different from the same or similar species as to require an assessment of risk.

Another issue raised by this body was that GM labelling and segregation requirements add to the costs of imports to the EU and hence increase food prices, and that the EU's demand to source non-GM food or feed imports may be progressively limited by declining availability and/or increased costs of conventional crops in major commodity exporting nations. Labelling of GM products is a controversial topic and EASAC found it scientifically indefensible for products that are substantially the same as those of non-GM origin. The study received backing from the national academies of all EU member states, plus Norway and Switzerland.

The European Union's chief scientific advisor Anne Glover fully supported the report published by EASAC (http://www.euractiv.com/science-policymaking/chief-eu-scientist-backs-damning-news-530693):

> There is no evidence that GM technologies are any riskier than conventional breeding technologies and this has been confirmed by thousands of research projects. In my view, consumers can believe in the overwhelming amount of evidence demonstrating that GM technology is not any riskier than conventional plant breeding technology. The EASAC Report is a major contribution to this debate as it reflects the view of Europe's most eminent scientists.

This background helps to explain the restrictive approach taken on GM products in the EU. It is not based on the scientific evidence relating to safety and nutritional composition but on political considerations. This policy is demonstrated in trade issues and in the labelling approach adopted in the EU, which appears designed to dissuade consumers from purchasing GM food products.

Many consumers in several countries would like to have information about the GM status on the food label, but given the current climate it is likely that a mandatory labelling requirement for GM foods will have the effect of deterring consumers from purchasing these foods, either from a lack of knowledge or from prejudice. The whole issue of what foods should be excluded and included in any mandatory label of GM foods needs to precede any discussion on labelling. Some countries require that products of GM crops (such as vegetable oil and starch) have a label indicating that they have been derived from GM plants, but do not require a similar label for foods (such as cheese) that have been produced using enzymes derived from GM organisms. That distinction would be hard to understand by the average consumer.

It seems unlikely that the federal government in the USA will introduce mandatory labelling of GM food products in the foreseeable future, although this may be done only to stop efforts to do this state-by-state and the resulting chaos that would ensue. Currently, both the USA and Canada have a voluntary labelling policy. The American Association for the Advancement of Science (AAAS, 2012) stated that:

> it is the long-standing policy of the Food and Drug Administration (FDA) that special labeling of a food is required if the absence of the information provided poses a special health or environmental risk. The FDA does not require labeling of a food based on the

> specific genetic modification procedure used in the development of its input crops. Legally mandating such a label can only serve to mislead and falsely alarm consumers.

The US Council for Agricultural Science and Technology (CAST, 2014) recently commissioned a study on the topic. According to this assessment, people have the right to know what is in their food. The mandated calorie and nutritional content panel on packaged foods is an example of labels that inform consumers about food composition. However, the right to know what is in food is different from the right to know how it was produced. Furthermore, this uniquely singles out GM technology – not other production methods and processes – for such a right-to-know claim. There is no science-based reason to single out GM foods and feeds for mandatory process-based labelling. Also, mandatory labelling based on process abandons the traditional US practice of providing for consumer food preferences through voluntary product differentiation and labelling: market-driven voluntary labelling measures are currently providing consumers with non-GM choices; and mandatory labelling may violate the First Amendment to the US Constitution and WTO requirements for international trade.

Another issue that needs to be clarified before discussions on mandatory labelling is the practice in some countries of regarding new plant cultivars produced by mutagenesis (treatment with radiation or chemicals to induce random mutation) as having been derived by traditional breeding programmes. The result is that alteration of the genetic makeup of plants by recombinant DNA technology is prohibited in the organic regulations but alteration of the genetic makeup by mutagenesis is allowed, at least in the USA. The effect of this practice is the potential for consumers seeking to purchase organic produce because it is GMO-free to be misled.

As discussed in the preceding chapter, there is a major gap between public opinion about the safety and quality of genetically modified (GM) foods and the scientific facts, indicating that on the whole consumers are poorly informed (or misinformed) about the real facts relating to GM foods. That gap probably applies also to many politicians, who are in a position to frame legislation relating to GM food products. In planning to provide consumers, and those involved in developing legislation, with balanced and accurate information about GM foods, it would be desirable for all sectors of the food production industry to contribute.

The gap between public opinion and the scientific facts probably extends to other areas of modern biotechnology. For instance, do diabetics who have to take insulin injections do so knowing that the insulin has been derived from GM organisms? And what are the views of these consumers in relation to GM products that enter their bodies in a different way, that is, food?

It would be helpful in discussions of that nature if the Churches could set out their views on the role of biotechnology in modern living and, in particular, their views on GM foods. Omobowale et al. (2009) reviewed the teachings of

Judaism, Islam and Christianity and were unable to identify any established tenets that could be interpreted as precluding the adoption of GM foods. They concluded that these religions, while assisting adherents in forming opinions, can only be one of the many factors expected to influence consumers' decisions on GM food technology. The Churches appear to be relatively silent on these issues, although there are indications that the Vatican may yet make an announcement to its followers. A panel of scientists involving 7 members of the Pontifical Academy of Sciences and 33 outside experts made a passionate endorsement in 2009 of GMOs for global food security and development. Their report (Meldolesi, 2011) stated that 'there is a moral imperative' to make the benefits of genetic engineering technology 'available on a larger scale to poor and vulnerable populations who want them', urging opponents to consider the harm that withholding this technology will inflict on those who need it most. 'Technology, in this sense, is a response to God's command to till and to keep the land (cf. Gen 2:15) that he has entrusted to humanity, and it must serve to reinforce the covenant between human beings and the environment, a covenant that should mirror God's creative love.' Both Jews and Muslims through their formal processes of food certification for compliance with their food laws (kosher and halal, respectively) have accepted microbial and plant-based GM products.

Summary

Biotechnology continues to play a very important role in the production of new cultivars of crops, and this was recognised in the award of the 2013 World Food Prize to a team of scientists who had made outstanding contributions to this science. Over 18 million farmers in 27 countries planted GM crops in 2013, the global biotech crop hectarage increasing from 1.7 million hectares in 1996 to over 175 million hectares in 2013. Over 90% (16.5 million) of farmers planting biotech crops are small and resource-poor. A necessary component of this development in crop production is the need to ensure that the GM cultivars are as safe and nutritious for food and feed use as their conventional counterparts. There is broad scientific consensus that all approved foods and feeds that have been derived from GM crops being marketed currently are safe to eat. Many consumers in several countries would like to have information about the GM status on the food label, but given the current climate it is likely that a mandatory labelling requirement for GM foods will have the effect of deterring consumers from purchasing these foods, either from a lack of knowledge or from prejudice. The whole issue of what foods should be excluded and included in any mandatory label of GM foods needs to precede any discussion on labelling. Ethical concerns about GM foods might best be addressed by religious leaders.

References

AAAS (2012). Statement by the American Association for The Advancement of Science Board of Directors on Labeling of Genetically Modified Foods. http://www. aaas.org/news/releases/2012/media/AAAS_GM_statement.pdf (accessed February 2015).

CAST (2014). The Potential Impacts of Mandatory Labeling for Genetically Engineered Food in the United States. Issue Paper Number 54, April 2014. Council for Agricultural Science and Technology, Ames, Iowa.

European Commission (2010). Commission publishes compendium of results of EU-funded research on genetically modified crops. European Commission - IP/10/1688 09/12/2010, Brussels, Belgium http://ec.europa.eu/research/quality-of-life/gmo/.

Herman, R., A. and Price, W. D. (2013). Unintended compositional changes in genetically modified (GM) Crops: 20 years of research. *Journal of Agricultural and Food Chemistry* 61: 11695–11701.

ISAAA (2014) Brief 46-2013. Global Status of Commercialized Biotech/GM Crops: 2013. Ithaca, New York 14853. http://www.isaaa.org/resources/publications/briefs/46/executivesummary/ (accessed 2 October 2014).

Masip, G., Sabalza, M., Perez-Massot, E., et al. (2013). Paradoxical EU agricultural policies on genetically engineered crops. *Trends in Plant Science* 18: 312–324.

Meldolesi, A. (2011). Vatican panel backs GMOs. *Nature Biotechnology* 29: 11.

Omobowale, E. B., Singer, P. A. and Daar, A. S. (2009). The three main monotheistic religions and gm food technology: an overview of perspectives. *BMC International Health and Human Rights* 9: 18–35.

Swiss Academies (2013). *Gentechnisch veranderte Nutzpflanzen – und ihre Bedeutung fur eine nachhaltige Landwirtschaft in der Schweiz.* Swiss Academies of Arts and Sciences. http://www.akademien-schweiz.ch/index/Publikationen/Berichte.html (accessed February 2015).

Van Eenennaam, A. L. and Young, A. E. (2014). Prevalence and impacts of genetically engineered feedstuffs on livestock populations. *Journal of Animal Science* 92: 4255–4278.

WHO (2002). 20 Questions on Genetically Modified (GM) Foods. World Health Organization, the United Nations' Health Agency. Geneva, Switzerland. http://www.who.int/foodsafety/publications/biotech/20questions/en/ (accessed 3 September 2014).

Index

Genetic Modification and Food Quality: A Down to Earth Analysis, First Edition.
Robert Blair and Joe M. Regenstein.

Printed and bound by CPI Group (UK) Ltd, Croydon, CR0 4YY
16/09/2026
14963477-0001